Body/Meaning/Healing

CONTEMPORARY ANTHROPOLOGY OF RELIGION
Robert Hefner, Series Editor
Boston University

Published by Palgrave Macmillan

*Body/Meaning/Healing*
   By Thomas J. Csordas

# Body/Meaning/Healing

Thomas J. Csordas

BODY/MEANING/HEALING
© Thomas J. Csordas, 2002

First published 2002 by
PALGRAVE MACMILLAN
Houndmills, Basingstoke, Hampshire RG21 6XS and
175 Fifth Avenue, New York, N.Y. 10010
Companies and representatives throughout the world

PALGRAVE MACMILLAN is the global academic imprint of the Palgrave
Macmillan division of St. Martin's Press, LLC and of Palgrave Macmillan Ltd.
Macmillan® is a registered trademark in the United States, United Kingdom
and other countries. Palgrave is a registered trademark in the European
Union and other countries.

ISBN 0–312–29391–7 hardback
ISBN 0–312–29392–5 paperback

Library of Congress Cataloging-in-Publication Data

Csordas, Thomas J.
    Body/meaning/healing/by Thomas J. Csordas
        p. cm.
    Includes bibliographical references and index.
    ISBN 0–312–29392–5 (alk. paper)—ISBN 0–312–29391–7
        1. Spiritual healing. I. Title.

BL65.M4 C75 2002
291.3'1—dc21

2001046164

A catalogue record for this book is available from the British Library.

Design by Newgen Imaging Systems (P) Ltd., Chennai, India.

First edition: June, 2002
10 9 8 7 6 5 4 3 2 1

291. 31 CSO
Transferred to digital printing 2005

080055656

*For Graham, Vanessa, and JJ,*
*who bring beauty to my life*

# CONTENTS

# CREDITS

Chapter One is a synthesis of "The Rhetoric of Transformation in Ritual Healing," originally in *Culture, Medicine, and Psychiatry* (1983) 7:333–375 and "Elements of Charismatic Persuasion and Healing," *Medical Anthropology Quarterly* (1988) 2:121–142.

Chapter Two was originally published as "Embodiment as a Paradigm for Anthropology, the 1988 Stirling Award Essay," *Ethos* (1990) 18:5–47.

Chapter Three was originally published as "A Handmaid's Tale: The Rhetoric of Personhood in American and Japanese Healing of Abortions," in Carolyn Sargent and Caroline Brettell, eds., *Gender and Health: An International Perspective*. Englewood Cliffs: Prentice-Hall (1996), pp. 227–241. Reprinted by permission of Pearson Education, Inc., Upper Saddle River, N.J.

Chapter Four was originally published as "The Affliction of Martin: Religious, Clinical, and Phenomenological Meaning in a Case of Demonic Oppression," in Atwood Gaines, ed., *Ethnopsychiatry*, pp. 125–170. Reprinted by permission from *Ethnopsychiatry: The Cultural Constructions of Professional and Folk Psychiatries* by Atwood D. Gaines (Ed.), the State University of New York Press. Copyright 1992 State University of New York. All rights reserved.

Chapter Five was originally published as "Ritual Healing and the Politics of Identity in Contemporary Navajo Society," *American Ethnologist* (1999) 26:3–23.

Chapter Six was prepared for this volume.

Chapter Seven was originally published as "The Sore that Does Not Heal: Cause and Concept in the Navajo Experience of Cancer," *Journal of Anthropological Research* (1989) 45:360–386.

Chapter Eight was originally published as "Words from the Holy People: A Case Study in Navajo Cultural Phenomenology," in Thomas J. Csordas, ed., *Embodiment and Experience: The Existential Ground of Culture and Self.* London: Cambridge University Press (1994), pp. 269–289.

Chapter Nine was originally published as "Somatic Modes of Attention," *Cultural Anthropology* (1993) 8:135–156.

Chapter Ten was originally published as "Computerized Cadavers: Shades of Representation and Being in Virtual Reality," in Paul Brodwin, ed., *Biotechnology and Culture: Bodies, Anxieties, Ethics.* Bloomington: Indiana University Press (2000), pp. 173–192.

# Introduction

As a child in the 1950s I sat mesmerized in front of the TV screen watching the early generation of television faith healers like Oral Roberts and Kathryn Kuhlmann ply their prayers across the airways. Something of the lilt and cadence of their language, something of their invocation of divine power and compassion, something of the pain and exultation of their ardent audiences, something of the difference of all this from anything I knew in my own Roman Catholic environment rendered me spellbound. The question of what was really happening in the supplicants' experience of their illness-wracked and suffering bodies left me with an enduring puzzlement. Decades later, on deciding to become an anthropologist, I made it my business to be professionally fascinated with the world's multiplicity of answers to the question "what does it mean to be human?" Once again I found myself attracted to the problem of how religion attempts to provide meaning through healing. The early wonderment was renewed with the observation that so many of the world's peoples tried to answer the question of what it means to be human by invoking powers or entities that were by definition wholly other than human.

My studies led to a world of healers including shamans, medicine men, curanderos, spirit mediums, faith healers, lamas, road men, pirs, physicians. With some irony, when in 1973 I was ready to initiate my own first-hand study of such a phenomenon, faith healing like that I had watched as a child had recently begun to be practiced in a movement called the Charismatic Renewal, in which the participants were Catholics whose cultural background was quite similar to my own. I decided to begin my inquiry with this kind of healing, one that was in some sense familiar to me, and from which I could later move on to less familiar forms of healing.

This book traces the trajectory of that inquiry. It is about the meaning of being human, the meaning of our existence as bodily beings, the way that meaning is sometimes created in the experience of the sacred, and the meaning of the transformations that can take place in such experience of the sacred. Based on this core concern, the chapters that follow develop two closely intertwined agendas, one more empirical and one more theoretical. The empirical agenda is to present an analysis of religious healing, with the goal of articulating an understanding of therapeutic process generalizable across healing traditions and faithful to the experience of both patients and healers. It is an agenda committed not only to the idea that such forms of healing are analogous to those of formal medicine and psychotherapy, but that there is also something explicitly religious about religious healing, something that has a profound capacity to effect the transformation of culture and self. The related theoretical agenda is to elaborate the notion of embodiment as a stance from which to understand the nature of human experience in culture. It emerged from a moment in anthropological theory when "experience" was suspect on the grounds that it was either undefinable or inaccessible. Resisting this tendency, I have come to understand experience as the meaningfulness of meaning, immediate both in the sense of its concreteness, its subjunctive openness, its breakthrough to the sensory, emotional, intersubjective reality of the present moment; and in the sense in which it is the unmediated, unpremeditated, spontaneous or unrehearsed upwelling of raw existence. The anthropological challenge is, accordingly, not to *capture* experience, but to *give access* to experience as the meaningfulness of meaning.

My empirical agenda for the study of healing evolved from three observations about the way this topic was discussed in the scholarly literature. First, it was commonly asserted that there was an analogy between various forms of religious healing and psychotherapy. Could that analogy be made specific through detailed study? At an existential level, what features do religious and medical phenomena share, and what distinguishes them from one another? Second, much of the anthropological literature focused on what healers said and did, assuming that this had an effect on their patients without really trying to understand the experience of those patients. Could that experience be documented in such a way as to account for any effects of the healing rituals performed? Third, much of the literature was preoccupied with proposing psychological mechanisms by which healing had its effect, such as catharsis, suggestion, placebo effect, trance, or hypnosis. So what was explicitly religious about religious healing? I took up these questions from the standpoint that religious healing could be seen as a kind of highly persuasive cultural performance. When I began these studies in the 1970s the most influential definition of culture was that it is a system of symbols—indeed, it was common among anthropologists at the time to use the term "symbolic healing" in reference to religious healing. Unfortunately, on the one had this term sometimes had the connotation that what was symbolic was therefore not real, and on the other hand it suggested that if healing was primarily about the manipulation of symbols then there was little need to inquire about the actual experience of participants. Fortunately, there was a link between symbol and experience in the notion that there is a rhetoric in performance by means of which symbols shaped meaning for participants. Rhetoric in this sense is the power of persuasion immanent in symbolic action and ritual performance.

The theoretical agenda that subsequently evolved was the product of a sense of uneasiness that persisted throughout my studies of therapeutic process in religious healing. Although I remained convinced that these studies showed progress toward capturing the specificity of human experience in healing, including experience of the sacred, I was wary that the approach did not go far enough toward addressing the broader question of what it means to be human. My uneasiness took shape in the context of several fundamental shifts in anthropological theorizing and thinking about the nature of ethnography that were occurring during the 1980s. One was a critique of culture conceptualized in terms of coherence, pattern, and holism from the standpoint of a vital awareness of margins, border, and boundaries in cultural life. Another was a move from the interpretation of *meaning* to the critique of *power* as the central figure animating social life. A third was a rethinking of culture and self that moved from an emphasis on symbolic action to an emphasis on embodiment and bodily experience; it is the latter shift that is most visible in this book. Throughout these changes, adjectives like "shared" and "meaningful" in descriptions of culture were replaced by ones like "shifting" and "fragmented." Cultural phenomena were no longer interpreted but interrogated, no longer understood in terms of community but in terms of contestation. The newly discovered immediacy of embodiment offered anthropological theory a specific appreciation of the perpetual dialectic between representation and being-in-the-world, indeterminacy and objectification, continuity and transformation, subjectivity and intersubjectivity.

My departure toward embodiment began with two realizations. One was that the starting point of my analyses had been language—symbols, rhetoric, performance, persuasion, narrative, and ritual utterance were the substance of healing rituals and of patients' reflective accounts of those rituals. All of these were primarily forms of *representation,* and stopped just short of capturing the existential richness of *being-in-the-world.* Understanding healing in terms of representation is not adequate because, even though concepts such as performance and persuasion have substantial experiential force, ultimately representation appeals to the model of a text. No matter how successful literary scholars might be in animating texts, in bringing them to life, textual(ist) interpretations remain inflections of experience, slightly to the side of immediacy. The missing ingredient is supplied by the notion of being-in-the-world, from phenomenological philosophy, insofar as it speaks of immediacy, indeterminacy, sensibility—all that has to do with the vividness and urgency of experience. My attempt to place these ideas in dialogue rests on the proposition that if studies of representation are carried out from the standpoint of textuality, then complementary studies of being-in-the-world can be carried out from the standpoint of embodiment.

I also realized that the object of healing is not elimination of a thing (an illness, a problem, a symptom, a disorder) but transformation of a person, a self that is a bodily being. Recognizing that our bodily being is a product of culture no less than of biology has the potential to transform our understanding of both body and culture. On the one hand, if the body can be shown to be the existential ground of culture and self rather than simply their biological substrate, the way would be clear for understanding the body as not only essentially biological, but as equally religious, linguistic, historical, cognitive, emotional, and artistic. On the other hand, if even language can be shown to be a surging forth of embodiment and not just the representative function of a Cartesian cogito, the way would be clear for defining culture not only in terms of symbols, schemas, traits, rules, customs, texts, or communication, but equally in terms of sense, movement, intersubjectivity, spatiality, passion, desire, habit, evocation, and intuition.

Putting these two realizations together led me to a conceptualization of self grounded in embodiment. The argument is that by collapsing the distinction between body and mind, subject and object, the somewhat mysterious and organic endogenous processes that are rhetorically controlled in ritual healing (Chapter One) become comprehensible as self processes grounded in embodiment (Chapter Two). Language itself becomes comprehensible as such a self process when it is understood not as representation but as enactment of a mode of being-in-the-world. Whether these moves are successful or not remains for the reader to judge. In any case, the net result is that throughout these essays the status of ritual healing oscillates between being more a case example in terms of which to address the problem of human meaning through elaboration of the notion of embodiment and being the primary focus in its own right, sometimes a means to an end rather than an end in itself. Accordingly, a good deal of space in the essays included here is dedicated to elaborating these realizations into what I refer to as cultural phenomenology, a task that, though challenging, bears promise to inform cultural analysis well beyond the domain of ritual healing.

Preparing a sequence of chapters written over the course of nearly two decades is itself a rhetorical act. I have eschewed a simple chronology and have chosen not to separate the essays into those more concerned with the rhetoric of therapeutic process and those concerned with embodiment, in recognition that these two themes are related not only as a progression, but continue to react dialectically upon one another throughout the development of my work. I have chosen to begin with chapters grounded in my ethnographic studies of Catholic Charismatic healing. I then move on to my work on Navajo healing, where again the themes of therapeutic process and embodiment are closely intertwined. Finally, I include two chapters on modulations of embodiment, one that proposes a construct intended to further comparative studies of bodily experience and culture, another that takes a new empirical turn in posing questions of representation and being-in-the-world.

Chapter One formulates a rhetorical model or framework for understanding therapeutic process as transformative experience, based on data from Charismatic healers and retrospective accounts of healing experiences.

In the first part of this chapter I suggest that the rhetoric of healing practice evokes a variety of *endogenous processes* that are the locus of whatever
efficacy religious healing might have. I formulate this rhetoric as having
three components including predisposition, empowerment, and transformation and suggest that together these amount to a model of therapeutic
process that could be applied to any form of healing to determine its efficacy: If all three components were convincingly enacted or fulfilled, healing could be said to have taken place. In the second part of the chapter I
refine this understanding of therapeutic process based on experiential
accounts of patients as they progressed through a series of healing sessions.
Tracing what the patients—or supplicants, since the healing process is
predicated on divine supplication through prayer—identified as the most
significant events in these sessions and how they integrated the results into
their lives, I conclude that the idea of a miraculous cure necessarily fails to
capture the kind of effects I observed, and that what occurred was actually
a process of *incremental change*. Healing is much more like planting a seed,
or like nudging a rolling ball to slightly change its trajectory so that it
ends up in a different place, than it is like lightning striking or mountains
moving.

Chapter Two is a general attempt to formulate the notion of embodiment, drawing on several strands of theoretical work to lay the conceptual
groundwork for understanding embodiment as an approach to the analysis
of culture and self. I argue that it is essential to collapse methodological
dualities, but unlike much contemporary scholarly literature that highlights
the Cartesian dichotomy of body and mind, I focus on the relations
between subject and object, and between structure and practice. Uneasy
with an abstract account that is not played out in empirical analysis, I adopt
the hermeneutic strategy of developing these ideas in specific discussions of
Charismatic ritual healing and ritual language. I then return to a more general discussion of the implication of this kind of analysis for the study of
religious experience, and of the relations between mind and body, thought
and emotion, self and other, and subjectivity and objectivity in the human
sciences. Perhaps ironically, I conclude with the observation that beginning
with embodiment leads not to the irreducible objective reality of a biological body, nor to the indeterminacy of endlessly iterated subjectivity, but to
a necessarily indeterminate objective reality.

In Chapter Three I apply the framework for understanding therapeutic
process introduced in Chapter One, now enhanced by a fuller theoretical
understanding of bodily experience, to a specific case. Again working with
Catholic Charismatics, I examine a kind of healing prayer done for women
who have undergone abortions. Here the vivid experience of embodied
imagery is played out in a situation where cultural meanings surrounding
life and the body are vigorously contested. Personal effects of ritual healing
stand out in relation to their broader cultural significance, since the controversial nature of abortion in contemporary society allows a perspective on
how gender politics as well as personal suffering is played out in the arena

of ritual healing. I draw out the importance of cultural themes in bodily experience by sketching a comparison between this Charismatic ritual and an abortion ritual practiced by Buddhists in Japan.

By means of a detailed case study, Chapter Four shows how cultural phenomenology applied to the analysis of bodily experience can serve as a valuable meta-discourse on suffering. This highly disturbing case is one of a healer and patient locked in an agonizing struggle with what they are convinced is a demon, or evil spirit. Based on exhaustive interviews, my method was to present a thorough description of the case to two sets of commentators, one composed of Catholic Charismatic healers, the other of trained secular psychotherapists. I compare the religious accounts centering on whether there was indeed a demon at work with the clinical accounts centering on whether there was a particular disease at work in the affliction. Where cultural phenomenology comes to bear is in addressing the question of how such accounts can have so much in common yet be so different by offering a third account aimed at disclosing the manner in which cultural meaning is objectified out of an indeterminate existential ground.

Chapter Five extends the discussion of healing to a new ethnographic setting, and it also takes aim at a broader social understanding of the relation between ritual healing and identity politics as they interact on three levels in contemporary Navajo society: representation of Navajo identity in relation to the dominant Euro-American society, interaction among religious healing traditions within Navajo society, and transformation of individual experience with respect to dignity and self-worth as a Navajo. I illustrate the first level with an epidemic of hanta virus and a serious drought; the second with respect to the coexistence of several forms of healing; and the third in terms of case studies of Navajo patients who have used these forms of healing. These levels constitute a framework for analyzing the relation between healing and identity politics that is potentially more nuanced than either the position that ritual healing is a futile expression of frustration (what we may call the "opiate of the masses" interpretation) or that ritual healing is a subtle form of political resistance (what we may call the "postmodern liberation of the indigenous voice" interpretation). I suggest that future studies using such a framework begin to distinguish more clearly between a personal politics of collective identity, in which individual actors with clear commitments struggle to assert a shared identity, and a collective politics of personal identity, in which each actor among a group of actors with ambiguous commitments struggles to attain individual identity.

A truly useful model of therapeutic process must be applicable in more than one cultural setting and with more than one form of healing. Chapter Six takes a second pass through the Navajo material in a concerted effort to extend the framework outlined in the preceding chapters, not only shifting focus to a different culture, but also addressing the cultural difference and similarity of therapeutic process in three coexisting forms of healing. Here I examine cases from contemporary Navajo society representing the use of traditional Navajo ceremonies, Native American Church prayer meetings, and Navajo Christian faith healing, all of which are resources in the everyday pursuit of health and well being. I document how these three forms of healing differ with respect to their underlying philosophies and therapeutic principles, while at the same time have a character and appeal that is distinctly Navajo. The four components of therapeutic process serve as a heuristic for making sense of how a broad range of Navajo health concerns are played out in the experience of patients in each of these three forms of healing.

Chapter Seven poses the question of how people experience and make sense of serious illness, in this instance cancer, given the cultural milieu in which they exist. I place particular emphasis on causal reasoning, that is patients' understandings about how their illness came about. A central feature of the chapter is comparison between the ideas expressed by Navajo cancer patients and by a comparable groups of Euro-American cancer patients, between whom there appears an intricate pattern of differences and similarities. I also present an extended discussion of what Navajo patients mean when they attribute their cancer to an exposure to lightning, in which the importance of embodiment becomes strikingly evident. The chapter concludes with a discussion of the methodological difficulties inherent in giving adequate experiential specificity to analysis of issues like causal understanding in the medical systems of different cultures, focusing on the conceptual distinctions between cause and symptom made in different cultures, between understanding disease as entity or process, between biomedical and traditional systems of causal reasoning, and finally between body and mind.

Chapter Eight fleshes out the case of one of the patients who participated in the Navajo cancer study, a young man suffering the effects of a left temporal–parietal brain tumor. The discussion shows how he brought to bear the symbolic resources of his Navajo culture to create meaning for a life plunged into profound existential crisis, interpreting his experiences of losing and regaining the ability to speak and of finding himself inspired with deeply spiritual words of prayer as a call to become a medicine man. In analyzing the case I juxtapose understandings drawn from neurological studies of temporal lobe lesions with interpretation along the lines of cultural phenomenology in order to address two critical issues relevant to whether the latter approach can be ultimately and successfully elaborated. The first is to clarify the relation between embodiment as an experiential understanding of our worldly existence and biology as a form of objectified knowledge about our bodily being. The second is to define the sense in which we are always already in the social world as bodily beings even prior to being able to symbolize or objectify our experience.

Chapter Nine advances the viability of embodiment as a coherent methodological standpoint by elaborating the construct of *somatic modes of attention*, which I define as culturally elaborated ways of attending to and with one's body in surroundings that include the embodied presence of others. Once again I adopt the hermeneutic strategy of beginning with a more abstract, theoretical argument, playing it out in terms of concrete empirical examples, and finally returning to a more general conceptual argument. The empirical examples are in this instance those of revelatory phenomena drawn from a range of religious and non-religious forms of healing, including those of the Catholic Charismatic Renewal, Puerto Rican *espiritismo*, South Asian *siddha* medicine, and North American psychotherapy. This leads back to discussion of the indeterminacy of analytical categories such as intuition, imagination, perception, and sensation as tools in understanding such phenomena and hence to the importance of grasping the essential indeterminacy of existence with reference to the relation between textuality and embodiment.

Finally, Chapter Ten makes another departure in topic while retaining the theoretical concern for the relation between representation and being-in-the-world from the standpoint of embodiment. My principal motivation in making this departure is to meet the rather obvious challenge that if a cultural phenomenology grounded in embodiment is to have a broad relevance for anthropology, it must be able to offer a distinctive interpretive purchase domains of culture beyond ritual healing. As has been increasingly observed by scholars, in contemporary society biotechnology is profoundly implicated in transforming the very bodily conditions for having and inhabiting any world, and doubly so when it incorporates sophisticated computer applications. With this in mind, I examine the National Library of Medicine's Visible Human Project and its use of computer-generated virtual reality to recreate so-called virtual cadavers. The discussion touches on the kinds of beings one encounters in an ethnography of cyberspace, the place of such innovation in the contemporary cultural imaginary, and the concrete experiential implications of the technology's application in anatomy training for medical students and computer-aided surgery, all with an eye on the broader theoretical issue of the relation between representation and being-in-the-world.

In the end, there are probably at least two ways to read this book. One can approach the individual chapters as a series of studies that document the sufferings and strivings of afflicted people trying to become whole and, in many instances, holy. At the same time, one can trace the overlapping trajectories of the studies as steps toward a standpoint for understanding the experiential specificity of human existence, a cultural phenomenology grounded in embodiment. To my way of thinking, the cultural data and details of experience are inseparable from the methodological concepts and theoretical reflections. They constitute a dialogue of concrete and abstract—a dialogue about what it means to be human.

# PART I

*Charismatic Transformations*

# CHAPTER ONE

# The Rhetoric of Transformation in Ritual Healing

Healing at its most human is not an escape into irreality and mystification, but an intensification of the encounter between suffering and hope at the moment in which it finds a voice, where the anguished clash of bare life and raw existence emerges from muteness into articulation. An understanding of healing as an existential process requires description of the processes of treatment and specification of concrete psychological and social effects of therapeutic practices, as well as determination of what counts as an illness in need of treatment in particular cultural contexts, and when it can be said that a cure has been effected. However complex, this task constitutes an essential problem of meaning in anthropology, for it is concerned with the fundamental question of what it means to be a human being, whole and healthy, or distressed and diseased. The interpretive dimension of the problem is highlighted by the fact that many forms of healing are religious in nature, which requires accounting for the role of divine forces and entities (Csordas and Lewton 1998). Given the prevalence of religious healing and the global interrelation of religion and healing, the category of the holy may in its own way be fundamental to our understanding of health and health problems. A complete account of religious healing per se would then have not only to examine the construction of clinical reality with respect to medical motives, but also the construction of sacred reality with respect to religious motives.

To put the issue another way: When we pose the problem of how to understand religious healing, does our enterprise more properly belong under the rubric of comparative religion, or under that of medical science? This question, as commonsensical as it may sound, is perhaps the artifact of Western culture's tendency to compartmentalize experience and reify categories like religion and medicine. Each category spawns its own science, which then assumes that its field of knowledge is analytically distinct from all others. In much that has been written on the subject, the implicit assumption is that although phenomena of religious healing, ecstatic trance, or spirit possession can be acknowledged as religious from the indigenous viewpoint, from the scientific stance they must be viewed in medical or psychiatric terms. In this methodological disposition the relevant questions often have been whether religious experience itself is pathological or therapeutic and whether religious healing can be understood as analogous to psychotherapy.[1]

In actuality, the point of convergence between religion and medicine is not difficult to locate: Both address themselves in one sense or another to suffering (Kleinman 1997) and to salvation (Good 1994). These are indeed broad categories, as suffering can include the social distresses of poverty, oppression, and inequality as well as the painful burden of disease, and salvation as a solution to suffering can be sought in this world or the next as a brief reprieve from pain or a reward for eternity. It is no wonder that such apparently different modes of social action as religion and medicine have evolved to address these profoundly diverse dimensions of humanity's existential condition, nor that they converge in the domain of religious healing. Yet any reader with an anthropological sensibility will suspect that this answer is less than adequate, for "suffering" and "salvation" have the cultural ring of the monotheistic religions (Christianity, Judaism, Islam) in which eschatology and soteriology are centrally at issue. A truly comparative understanding of religious healing must be conceptually grounded in a way that is generalizable beyond these religions and to the least degree possible culturally indebted to them, powerful and influential as they are. The present chapter, and several of those that follow, are intended as steps toward such a comparative understanding.

To begin, the problem of *efficacy* appears repeatedly at the center of debate about religious healing practices. Although other reviewers have chosen to treat the diverse and voluminous literature on this problem (confer Bourguignon 1976; Dow 1986; Moerman 1979; Waldram 2000), my purpose here is to develop an approach that is sensitive to incremental and inconclusive effects that define the lowest threshold of efficacy, in a way that begins to remedy a lack of analytic specificity that hampers any understanding of efficacy. A first step is to be aware of which of three aspects, implicit in most discussions of healing practice, is the focus of analysis. The first is *procedure,* or who does what to whom with respect to medicines administered, prayers recited, objects manipulated, altered states of consciousness induced or evoked. The second aspect of healing practice is what we may call *process,* referring to the nature of participants' experience with respect to encounters with the sacred, episodes of insight, or changes in thought, emotion, attitude, meaning, behavior. Third is *outcome,* or the final disposition of participants both with respect to their expressed level of satisfaction with healing and to change (positive or negative) in symptoms, pathology, or functioning.

Of these three elements, therapeutic procedure has been treated exhaustively in many empirical studies and comparative works. Therapeutic outcome has only recently begun to be treated systematically by anthropologists. However, therapeutic process as defined here has been virtually neglected, and relegated to the status of a "black box." This neglect may originate in a failure to distinguish between prototypical cases for ritual analysis, such as rites of passage (Victor Turner 1969) and ritual healing. What is typically called process in anthropological studies of such rites conforms more to what we are calling procedure. Following this convention, studies of religious healing have been based on descriptions of healing rituals and interviews with ritual specialists, and have included little explicit attention to the phenomenology of the transformative process as lived by participants.[2] In what follows, I will propose a framework for understanding the transformative process using the ethnographic case of Catholic Charismatic healing. In later chapters I will examine the comparative usefulness of this framework in the ethnography of Navajo religious healing.

### Pentecostalism and Healing

The year 1967 was a watershed in the history of religion. In that year a synthesis was made of two forms of Christianity that had previously been as far apart as can be imagined on the spectrum of religious practice and experience. Roman Catholicism is the world's largest and oldest church, intensely liturgical and hierarchical, characterized by a European sensibility, in which the highlights of spirituality have taken the form of highly cultivated monasticism and mysticism. The Pentecostal movement began at the turn of the twentieth century in the United States and is characterized by independent congregations whose participants receive the "Baptism in the Holy Spirit" that fills one with divine power and spiritual gifts, such as glossolalia (speaking in tongues) and faith-healing. The young university-educated Catholics who created "Catholic Pentecostalism" and institutionalized it as the Catholic Charismatic Renewal inaugurated a movement that has become global in scope, a powerful force for evangelism counteracting both conservative Protestant evangelism and radical movements such as liberation theology in the Third World. I have written full length ethnographies of the Catholic Charismatic healing system (Csordas 1994a) and of the use of ritual language to forge a communal way of life in this movement (Csordas 1997). Accordingly, I will give only a brief summary of Charismatic ritual healing practices. I follow this with an examination of two sets of cases, representing discussions from two previous articles in which I developed two successive formulations of an experiential framework for understanding transformative process in religious healing.

Catholic Charismatics participate in the late-twentieth-century shift among Christians from emphasis on suffering and self-mortification as an imitation of Christ to emphasis on the possibility and benefit of divine healing as practiced by Jesus in the gospels (Favazza 1982). The processes of healing and spiritual growth are linked, because illness is typically regarded as an obstacle to spiritual growth. Healing is therefore considered necessary for all persons in the process of spiritual growth, and spiritual growth is in turn conducive to good health. The healing system is holistic in that it aims in principle to integrate all aspects of the person, conceived as a tripartite composite of body, mind, and spirit.

The tripartite concept of the person is the basis for three distinct but interrelated types of healing: *physical healing* of bodily illness, *inner healing* of emotional illness and distress, and *deliverance* from the adverse effects of demons or evil spirits. Physical healing is the simplest in form, in which laying on of hands and, in some instances, anointing with blessed oil accompany prayer. Healing ministers pray for relief from illness, success of medical treatment, lessening of side effects from medication, or release from suffering through death. This is the type of faith healing most widely known in American religious culture and is associated with such popular evangelists as Oral Roberts and Kathryn Kuhlman. Inner healing may be aimed at removing the effects of a particular life trauma, or it may be a review and reinterpretation of an individual's entire life history in light of the "healing presence of Jesus," treating emotional hurts or scars that Charismatics recognize may linger on from an individual's past even after he has received the Holy Spirit. Supplicants are frequently exhorted to forgive others for past wrongs. Vivid imagery often accompanies inner healing, either as a revelation of some repressed experience or as a confirmation that healing is taking place. In deliverance, a supplicant is relieved of oppression by evil spirits. Demons in this instance typically do not have complete control over a person in such a way as to require the formal Church rite of exorcism, but are nevertheless regarded as having a detrimental effect on the person's life, including behavior, personality, and spiritual growth. Evil spirits identified or "discerned" by either the healing minister or the supplicant are dispatched by a "prayer of command" in the name of Jesus Christ (the best general summary by a Charismatic writer is MacNutt 1974). In elaborating this system, Catholic Pentecostals believe that they are serving God's intention to "heal the whole man": body, soul, psyche, and relationships with others. While healing ministers tend to specialize, most recognize the necessity, at times, of using all three forms in varying combinations.

Among these types of healing, physical healing is essentially a descriptive category, while the other three are etiological. That is, physical healing is indicated for specific somatic symptoms and complaints, while the others are indicated when a spiritual, psychological, or demonic cause is discerned to be at the root of the problem. The technique of physical healing consists simply of the laying on of hands accompanied by prayer that the sickness be healed, though in cases such as the mending of broken bones or the reversal of cancer, visualization of the healing process might be included. Physical healing is often the first type of healing encountered by Charismatics, and often occurs in large group settings. In certain cases it may be accompanied by, or even regarded as contingent upon, one of the other types. In spite of its importance, I will not consider physical healing here for the following reasons (but see Csordas 1994a). First, although its structure of meaning is essentially the same, as a form of ritual discourse it has not been elaborated to the same degree as other types of healing, and hence, it is less accessible to interpretation. Second, given the nature of the illness problems addressed, it cannot be adequately analyzed apart from an outcome assessment that documents both biomedical effects and religious expectations (cf. Ness 1980; Young 1976).

In the Healing of Memories an individual's entire life is prayed for in stages, from the moment of conception to the present. Any events or unreconciled relationships that emerge in this review of life history are given special attention; the proceedings are stopped for a period of prayer, and the person is asked to forgive those who have hurt him. The supplicant may be asked to visualize any painful incident that is uncovered. An essential element of the image is the healing presence of Jesus in human form. As one healer explains, "What you are doing when you're praying for someone, you're praying the presence of Christ into the moments of their lives, which may be pleasant moments or painful moments." The healer is aided by supernatural means: He may utter a "prophecy," which is an inspired message from God meant to encourage or admonish the supplicant; or he may receive a "word of knowledge," which is a divinely bestowed intuition into the supplicant's situation that could not have been known by natural means. The healer does most of the talking, guiding the supplicant's reflections while "walking Jesus through" his life. Depending on the individual and the seriousness of his problem, one or several sessions of prayer may be needed. In addition, Charismatics often refer to the Healing of Relationships, which, while not associated with a particular technique of its own, entails a recognition that strains in the interpersonal environment can contribute to the etiology of ills otherwise described as physical, spiritual, emotional, or demonic.

Prayer for Deliverance requires the recognition of a chronic problem, which is interpreted as the presence of evil in a person's life. A distinction is made between *possession,* a rare state in which the demon takes total control of a person's faculties, and *oppression,* in which the effect of the demon is felt in a limited domain of the person's life.[3] Evil spirits have names that are typically those of various sins, habits, or unfavorable behavior traits and tend to appear in clusters. Thus an individual may harbor a sexuality cluster (for example, Lust, Perversion, Masturbation, Adultery) or a falsehood cluster (for example, Falsehood, Lying, Deceit, Exaggeration). The troop of demons is often headed by a "manager spirit," which is analogized to the taproot of a weed—hardest to get out, but if it goes, the lesser ones follow. The healer begins the process of deliverance by "binding" the demon in the name of Christ so that it does not manifest itself and disrupt the proceedings, as it is wont to do (fundamentalist or "classical" Pentecostals disagree on this point—they require some sort of manifestation as a sign that the spirit has departed). The spirit is then commanded to name itself, which it does through the mouth of the supplicant. Some Catholic Pentecostals, however, are so intent on avoiding sensationalism and theatricality in deliverance that they bind the spirit from even this much activity, relying on the healer to identify it by supernatural means, through the spiritual gift of "discernment." The demon is then commanded to depart in the name of our Lord Jesus Christ, an authority that it cannot resist. Nevertheless, if these three things are not done—binding the spirit, addressing it by name, and speaking in the name and authority of Jesus—the demon may cause trouble by refusing to leave, speaking some form of verbal abuse through its host, or physically upsetting the host.[4]

Charismatic ritual healing occurs in a variety of settings. Large healing services originated at the periodic conferences in which movement participants assemble on a national or regional basis both to show their strength and unity and to worship and teach. From the late 1970s to the late 1980s these conference sessions evolved into public healing services in which healing ministers of some reputation attracted Catholics who might not otherwise have participated in the Charismatic Renewal. Healing prayers or petitions for self and others may also occur in a segment of smaller weekly prayer meetings. Following these prayer meetings, prayer for individual supplicants may be conducted in a separate "healing room" by a specially chosen team of healing ministers from within the group, who lay hands on the supplicants and pray for whatever problem they might have. More intensive group healing also occurs in smaller day-long or weekend retreats and "days of renewal." Private sessions may be arranged with an experienced healing minister or healing team, and the most serious and profound healing takes place in such sessions. Some of those who practice in the private setting also have professional training in counseling or psychotherapy and integrate these practices with ritual healing. In addition, private healing prayer sometimes occurs over the telephone. The most highly organized communities also incorporate forms of pastoral counseling into their everyday activities. Finally, healing prayer for oneself or others may be practiced in the solitude of private devotion.

Healers in this tradition—that is, those who "pray with others for healing"—are regarded as having been granted a special charism, or "gift of the Holy Spirit," for this purpose. A priest who is a member of a prayer group is likely to be asked for healing prayers by virtue of his ritual status. Similarly, a person with training in counseling or a conventional form of psychotherapy may be expected by others to play a role in healing prayers. (Catholic Pentecostals do not reject Western medicine out of hand and are willing to make referrals to conventional practitioners when it is deemed necessary.) The gift of healing does not necessarily manifest itself spontaneously or dramatically, and the prospective healer is not required to experience healing himself as part of an initiation. As a group develops, its members may decide to institute a healing team or "ministry" as a group function, in which case members will be asked to pray for guidance as to whether they might feel a call to participate in this work. The healing teams meet regularly to cultivate their gifts and knowledge of healing and may even travel to other localities to workshops in ritual healing. The well-known healers who conduct these workshops often also publish books and articles on healing and sometimes travel to conduct healing services in areas where the gifts of local healers are not as highly developed.

An individual may decide to seek Deliverance or Healing of Memories of his own accord, or on the advice of a relative, a friend, or a pastoral advisor. The following is an account by a healer of how a person presenting a complaint is received in his group's "healing room" following every regular prayer meeting:

> Sometimes what will happen is that an individual will come up to the healing room and they'll be sharing that they're having all these different memories coming back to them and they're having trouble with old things, old sins, old difficulties haunting them; a lot of guilt stuff. And one of the first things that the healing room would ask them is how has this been happening: is it a condemnation thing, where the person can't seem to have any freedom, or is it a thing where it's just memories floating in and out. If it's a floating in and out, it's usually the Lord, if it's the condemnation it's Satan .... Generally they're walking through that time when a lot of memories are coming back to them, and they're being haunted by them, and they're not really feeling free with Christ. So what they do is, either through the healing room, a friend tells them, or somebody tells them, "Your know, I think the Lord's preparing you for the Healing of Memories and you need to check with the Healing Room." Somehow they get the message. Or, they've heard about the healing room, and they say to Father, "Well, I think the Lord might want me to have the Healing of Memories: this is what's happening ... ."

This account suggests that the principal criterion determining resort to ritual psychotherapy is the availability of an interpretation of certain elements of experience as signs indicating a need for healing prayer. It is important to recognize, in a movement where even the most experienced healers have been at work for little more than a decade, that the ability to make such an interpretation is the product of secondary socialization. Unlike the taken-for-granted awareness in some traditional societies that the shaman is a health-care resource, the recruit to Catholic Pentecostalism learns a new and unfamiliar way of interpreting and labeling experience as healthy or in need of healing.[5] The healer then carries the diagnosis a step further, judging on the basis of affective tone (presence or absence of feelings of guilt, condemnation, lack of internal freedom) whether the experiences are being caused by God or Satan. Origin with God indicates the relevance of the Healing of Memories technique, while origin with Satan indicates the use of Deliverance.

The relation of these two forms of healing in practice is, however, somewhat ambiguous, and they often overlap with respect to the type of problem addressed. Some healers specialize only in Deliverance; especially among those influenced by non-Catholic fundamentalist neo-Pentecostals, the Healing of Memories is often regarded as "practicing psychotherapy without a license," or worse, as "non-Scriptural." On the other hand, some healers practice only the Healing of Memories, shying away from the mysterious, little-understood, and hence dangerous, realm of spirits and demons. Still others make use of both techniques; for example, during a certain stage of prayer for the Healing of Memories, the presence of a demon may be discerned and dealt with by Deliverance, following which the Healing of Memories is resumed.

## The Rhetorical Control of Endogenous Processes

To the extent that healing is effective, there are certain elements common to all forms, whether they are religious or biomedical. It is widely agreed that a primary interpersonal aspect of treatment is the emotional support of the suffering individual and reaffirmation of his or her worth in a community or society, while a primary intrapsychic result is the reorganization of the person's taken-for-granted orientation to experience or "assumptive world" (Frank and Frank 1991). As to how these effects are achieved, one view focuses on the impact on a patient of the therapeutic technique or environment, emphasizing "exogenous" processes and mechanisms such as persuasion (Frank and Frank 1991) or suggestion (Calestro 1972), while another focuses on a patient's response to his or her own suffering, emphasizing "endogenous" processes such as sleep and rest, search for insight dreaming, dissociation, or acute psychotic episode (Prince 1976, 1980). These endogenous (and sometimes spontaneous or unconscious) processes can in some instances have a positive, "therapeutic" outcome. Prince accounts for the effect of the exogenous processes by arguing that various forms of psychotherapy, whether associated with the consulting room, with shrines with cults, or with shamans, are in fact techniques for facilitating or manipulating the endogenous processes.

Let us begin with several specific case examples that demonstrate the rhetorical mobilization and manipulation of endogenous processes in healing. The healer who recounted the first two examples was a woman aged 26 who also held a Master's degree in clinical psychology. She was one of the two or three most respected healers in her group, which was the largest prayer group in a large Midwestern city.

### *Example One*

I was praying for a man who had a real history of pain. He was in his mid-50s, and had really committed himself to the Lord about three years ago, but had not really grown a whole lot. He really decided he needed some healing from much pain. We ended up having like about six or seven sessions with this man, and we're still not finished. It's just been going on, it's already gone on a year. Some really specific visions have occurred in that particular situation, and the Lord's working slowly. In one vision, I had a picture of a little boy in an alley—the man had not mentioned anything to me, but I had a picture of a little boy in an alley who was about ten years old, who was frightened, and it was dark.

And I sent it—what I call "sending it back to the Lord": before I test something with the person I usually test it with the Lord, and I ask the Lord is that Him [sending the vision]. If it isn't I command it away; if it is Him, I ask that it just be increased in me. Then I throw it out to the individual who I'm praying with, and I say (what I did with him was just said): "Was there ever a time when you were in a dark alley, do you remember anything like that? Does that mean anything to you as a person?" And he just looked at me amazed and he said, "Yeah, I hadn't thought of it, but . . . ." And he said that when he was about ten years old a man accosted him behind a church in an alley.

He had various sexual problems, and one of the sexual problems had been in a past homosexual relationship, and he had never really understood a lot about that—although he had been married and had several kids. But this [vision] was like the Lord opening a door, a key as to how He wanted to travel through the man's life, really, because that was like the beginning of many of his sexual problems. That same person, I had an image of him in a church, very angry, like shaking his fist, and I asked him what that meant. And he had at one time considered the priesthood, had gone to seminary, and had left the seminary because he was involved with a woman and his confessor said, "You ought to get married." So he left the seminary to get married, but he shook his fist at the Lord at the time, saying that he'd be a priest one way or another. That was another thing that was very significant, because in his particular life he was still facing a lot of guilt about the fact that he had left the priesthood. And that guilt was going over into sexual relationships and problems that he had. It was all bouncing in together, just a lot of meshing. There was a lot of need for Deliverance in that individual's life, too; through different things that he had been involved in his Life, he had just really opened himself up to a lot of demonic activity.

Several points in this account merit attention. First, note the subordination of the entire healing process to the value of spiritual *growth* in the man's life; we shall return to this important element later in the discussion. Second, note again the healer's concern as to whether a spiritual experience originates with God or with Satan. Third, endogenous psychological processes are activated in both healer and supplicant. The process experienced by the healer (vision/visualization) gives form and meaning to that of the supplicant (memory/insight); like the shaman who serves as psychopomp, guiding his client through a journey to the underworld, the Catholic Pentecostal healer by means of gifts of revelation guides the suffering individual through an underworld—not the mythical underworld of shamanism, but the post-Freudian underworld of suppressed memories, remythologized by being subjected to religious techniques.

The ritual healing is not fully accounted for by these observations, however, for the healer noted that in addition to the Healing of Memories the supplicant had need of Deliverance. Note that the supplicant is regarded as having become susceptible to the influence of demons "through different things that he had been involved in in his life"; that is, various actions or experiences in the past had produced in him a spiritually weakened condition that had allowed evil spirits to acquire a foothold from which they harassed or oppressed (but not strictly speaking "possessed") him. These demons were "discerned" or diagnosed by the healer as falling into two "clusters." The first included Lust, Masturbation, and Homosexuality/Adultery ("/" indicates two or more spirits closely intertwined within a cluster); the second included Bitterness/Resentment/Anger, Guilt, Rebellion, Vanity-Pride/Insecurity. In this case it is unclear to what extent the identification of these demonic forces involves an endogenous process (externalization) in the psyche of the supplicant, for it appears that some of the names were discerned by the healer. Nonetheless, it is also the case that the supplicant too must discern, or acknowledge the presence of, an evil spirit. What is significant is the manner in which the clusters of demons are used as signifiers in defining a spiritual-psychodynamic condition, or as spiritual "symptoms" defining a spiritual "syndrome."

It is interesting, given the informant's background as a clinical psychologist, that the clusters of demons should be interpreted as circumscribing particular personality conflicts: the first cluster representing a "bisexual conflict," and the second (described as a "personality cluster") representing a "self-confidence conflict." The healer interpreted the latter conflict as follows:

"[T]hey were apparently working against one another. He had a spirit of Insecurity: the individual felt very insecure in nearly everything he did. And yet, at the same time, there was a spirit of Pride and Vanity. One minute he might be prideful and vain and angry if someone stepped on him and told him something, like he was worthless. And that's a very usual thing."

Although not all Catholic Pentecostal healers bring the same degree of psychological sophistication to their work, the practice of dealing with evil spirits in clusters allows the healers to confront problems on two levels simultaneously: The named spirits can be commanded to depart by the relatively simple religious technique of Deliverance, and the underlying problem represented by the cluster as a whole can be treated through more extensive counseling and prayer. Both are vital media for the rhetoric of transformation that is operative in Catholic Pentecostal ritual psychotherapy.

## Example Two

[There is one] person who walked,[6] and is still walking, with a weight problem, but in the midst of that she had a real need for Deliverance, and she really fought that. Someone else had told her she needed

Deliverance, and she didn't think she did. Then finally she came to me and said, "I think I need Healing of Memories and Deliverance regarding food." She had really fought the battle of food intake. As we began to pray it was pretty obvious that there was a real cluster there, of Gluttony, Lust (meaning lust for food, you know), there was Insecurity—well, the person felt insecure about herself, so she would eat to make herself more secure—the whole cycle of being over-weight was there. But in addition to that there was a cluster of six or seven demons that were feeding onto that … [including] Guilt … .

What happened afterwards wasn't that the person suddenly totally lost weight, okay, because she had these habit patterns built up, but what did happen was a real freedom, a beginning of accepting herself, and a beginning of being aware of what happens when she eats. And she still hasn't lost all her weight, but she's really aware of that free-dom, she's becoming more accepting of herself in different things. Now, one of the things that would be very appropriate in that (and I have done it with other people) would be to suggest that she look at what foods she eats now, and get a reward pattern going for herself in terms of those foods that are good for her vs. those that are not. And that reward would be set up by encouragement through her husband or members of her family, or encouragement through other people.

There's a spirit of Manipulation in that, too, where she manipulated other people, and herself, through comments that she made to herself about food. For example, "Well, I'll just have one piece of cake. Well, one piece of *birthday* cake wouldn't hurt. Do you see anything wrong with a piece of birthday cake, I mean, it's a celebration!?" You know those kind of things that an individual tends to say to themself; they tend to rationalize every ounce of food they eat.

The first element of note in this account is the presence of an initial resistance, or denial of the problem, which is overcome when the person accepts the need for ritual therapy. Second, the healer explicitly acknowl-edges that the ritual therapy addresses the spiritual and psychological dimensions of the psychosomatic problem. The healing does not remove excess weight, but activates endogenous processes including a transforma-tion of attitude into one of positive self-acceptance, and the generation of a degree of insight on the part of the supplicant. Third, the clusters of demons, whose names constitute the principal means of persuasion in this case, include two distinct types: those that label vices, sins, or symptoms, and correspond to names in traditional Christian demonology, such as Lust and Gluttony; and those that label psychological and affective elements such as Guilt and Insecurity, or behavioral traits such as Manipulation. Finally, a fourth point is that although demons are treated ritually and regarded in fact as discrete spiritual entities that act as external agents of distress, the healer shows no hesitancy, in dealing with their effects on the person, to make use of techniques of analysis and treatment common to conventional secular psychotherapy.[7] What is important in this is that the secular techniques are subordinated to the religious meaning. The notion of a "spirit of Manipulation" does not have reference to a type of behavioral disposition as would, for example, the phrase "a spirit of good will" in nonritual discourse; rather, it has reference to a system of moral and cosmological forces that constitutes a radically different universe of discourse.

The following example is excerpted from an account by a man in his early thirties of a healing that he experienced himself. The importance of endogenous process in this account is unmistakable, since the healing occurred entirely without the assistance of a healer. Part A describes one incident from a night in which the man experienced several visionary healing episodes, while Part B describes a healing experience by the same man at a later date. Both are examples of the Healing of Memories.

*Example Three*

*Part A*

[T] hat particular night I had probably four or five [visions] in succession. The whole thing happened in just the space of a few minutes. Another one would hit, and then another one. And one of the biggest things of all that happened was I was actually brought back to the birth experience. It's a known fact, it's a medical fact, that I was born stillborn. There was no life. The doctor has the records and everything. And that whole scene came to my mind—of being born. I could actually see myself coming out of my mom, and the doctor picking me up, and him saying clearly, "That's the second one I've lost today." He put me over there on this table, and they were all over there working on my mom. And it was a Catholic hospital, and there was a nun in there—well, I'm jumping ahead of myself. While I was lying there on the table it was like looking down, seeing yourself there. Very vivid; everything in detail. And the Lord came into the scene again, standing there at my head. He put His hand on my head and He said, "You're going to live because I have things for you to do. I want you to live." And at that I whimpered or jerked or moved, or something. And that nun that was there noticed it, and she ran right over. In fact, her words were, "My God, he's alive!" The doctor came over and he got me going again. And that was a tremendous healing. Because for many, you know, the birth experience is traumatic. It's a tough thing to go through. And we know it's in your mind—you can't remember it, but we know it's there. And how does it affect you when you're stillborn? I think it affects you very negatively, because you hear about it the rest of your life; at home, during the years you are growing up. Whenever the subject comes up about birth in any way. It's always mentioned at the dinner table that "You were stillborn and I had a hard time delivering you;" you know, your mother [says that]. And what that says is that you were very hard on your mom. You need a healing; you made your mother go through pain. And you went through tremendous pain: You even died.

There was a definite feeling of guilt because of hearing her say [this] over the years, [caused] unconsciously—she didn't do it intentionally. But parents do that; I've done it myself, I'm sure, in different circumstances with my own children. But the truth is, the healing was needed, and I was able to get it, and I just had a whole new perspective on it. I think the biggest thing was seeing the Lord, His being visible, and being able to go back to that made me appreciate my mother more. She and I have never gotten along very well—a difficult relationship. Seeing that whole scene really helped that relationship ... .

*Part B*

I would say one of the biggest ones that I ever had was a healing—well, first I can give you some statistics. Among men masturbation is a real problem. Most guys go through their lives with the problem. They think there is something wrong with them or they're odd, or something. They never share it with anybody, and they have very difficult problems with guilt—especially Catholic men—because it was always taught that it was sinful, a very bad thing to get into. And yet, it's a very normal thing. There isn't a man around who hasn't. Well, I went around since I was 14—12, 13, 14—years old up until about a year ago with that problem. And I prayed as soon as I found out that healing was real; and I also read a thing in Father MacNutt's book. He just lightly mentioned that he had seen men healed of that, and alcoholism—he just lumped the things together. Well, I prayed for that healing. I prayed for I guess two weeks, constantly. And then one night it hit—I was healed of it and I knew it. And never again; no urge, no nothing. It was gone. The guilt is the big thing that's gone; no more guilt ... [and] there was that imagining. I won't say imagining—it's real. He was right there and He touched me and he said, "I'm here and it's gone." He's taken it away. That was the greatest healing I've ever had, up to that point. I was just walking around in a cloud. I was elated. It was fantastic. And it was over a year ago, and I've held up. You see, when I was active in other religious activities, Cwsillo and other different things, that problem was still there. When an event was coming up that I was in charge of or have a role to play, I could stay away from it for two or three weeks, but the temptations were awful. They were unmerciful, and I would be fighting them off. Now that's gone. With the healing there is no temptation, nothing. It's just gone. Great healing.

The first relevant point in this account is that the informant, who is himself a healer, prayed for his own healing following exposure to books on the subject; in the episode of Part A it was the writing of Agnes Sanford, the Episcopalian specialist in the Healing of Memories, and in Part B it was the work of Catholic healer Francis MacNutt.[8] Second, Part A is only one episode of a longer evening's experience that included, in addition to a reliving of the birth scene, a healing of resentful memories of a nun who had been his teacher in grade school, and healing of a phobia about feather dust mops that was rooted in memories of a particular mop wielded by the maid in his childhood home. Third, note the prominence of guilt as an affective state in need of healing: In Part A the supplicant gives equal weight to relief from guilt instilled by his mother and the emotional vestiges of his own traumatic experience of near-death, with the improvement in his relationship to his mother in the present seemingly viewed almost as a side-effect of the healing; in Part B the supplicant attributes greater importance to his relief from guilt about masturbation than to his freedom from the urge to masturbate. Fourth, attention should be called to the culture-specific definition in Part B of masturbation as a problem in need of being healed, since in some societies, and in segments of the informant's own society, erotic self-stimulation is regarded as neither sinful nor unhealthy. Finally, note the importance of vivid and specific visualization, including the presence of Jesus as healer, as an endogenous process crucial to the thera-peutic impact of the healing.

The importance of endogenous processes in therapy was suggested to Prince (1980) by his observation of a healing ritual in which there was no healer, performed before the tomb of a Muslim saint in Lucknow, India. Prince's point is strengthened by our own Example Three, in which a heal-ing experience normally guided by a ritual therapist is shown to occur to the supplicant in solitude. Both cases suggest that a comprehensive account of therapeutic effectiveness must find its locus elsewhere than in the trans-ference activated in dyadic patient-therapist interaction. Identification of endogenous processes does not in itself provide an adequate locus for inter-pretation, however, for it must be asked how the processes are activated in therapy, and why different endogenous processes are prevalent in different settings. For example, in the case observed by Prince, dissociation was the principal therapeutic process, while in Catholic Pentecostal healing dissoci-ation plays only a marginal role and is indeed somewhat distrusted. Furthermore, the same processes may be manifested in significantly different forms across cultures and types of therapy. Three such processes were iden-tified in the examples given above: Memory/insight and vision/visualization predominated in the Healing of Memories, while externalization was essen-tial to Deliverance. These basic healing processes are by no means unique to Catholic Pentecostalism. Memory/insight is the key component of psycho-analysis; vision/visualization is the key component of shamanism, and is also found in some forms of humanistic psychotherapy, such as psychosynthesis (Assagioli 1965); externalization is common to other exorcistic forms of cure such as the Thai example discussed by Tambiah (1977), and is also found in contemporary form in Japanese Morita therapy (Kleinman 1980).

The thrust of my argument is that the locus of therapeutic efficacy is in the particular forms and meanings—that is, the discourse—through which the endogenous processes are activated and expressed.[9] Recognizing this

role of discourse resolves the paradox posed by the activation of endogenous processes in the absence of a healer. As suggested by Foucault (1970,
1972), discourse is a semiautonomous process that can be contributed to or
tapped by those conversant with its conventions. Carried forward by its
own structure of implications, discourse itself embodies the therapeutic
efficacy and mystical power of the divine "other."

Understanding the specific nature of this efficacy requires the construction of a hermeneutic of the cultural rhetoric at work in the discourse of
healing.[10] The notion of rhetoric, as against the notions of suggestion, support and nurturance, or placebo effect, contributes a recognition that healing is contingent upon a meaningful and convincing discourse that brings
about a transformation of the phenomenological conditions under which
the patient exists and experiences suffering or distress. It can be shown that
this rhetoric redirects the supplicant's attention to new aspects of his actions
and experiences, or persuades him to attend to accustomed features of
action and experience from new perspectives. Following Schutz (1967), for
whom the particular way people attend to their experiences constitutes the
meaning of those experiences,[11] this redirection of attention amounts to
the creation of meaning for supplicants. To the extent that this new meaning encompasses the person's life experience, healing thus creates for him a
new reality or phenomenological world. As he comes to inhabit this new,
sacred world, the supplicant is healed not in the sense of being restored to
the state in which he existed prior to the onset of illness, but in the sense
of being rhetorically "moved" into a state dissimilar from both pre-illness
and illness reality. The key interpretive task is to show how this reality is
constituted as a transformation of pre-illness and illness realities.

In linking the rhetorical aspect of discourse with the endogenous healing processes, this approach suggests that the transformation brought about
by healing operates on multiple levels. Insofar as endogenous processes take
place on physiological and intrapsychic levels, and rhetoric acts on both the
social level of persuasion and interpersonal influence and the cultural level
of meanings, symbols, and styles of argument, the experience of healing is
an experience of totality. The following discussion will lay the groundwork
for identification of the fundamental components of the rhetoric through
which the endogenous processes are controlled and the healing transformation is achieved.

This phase of the argument begins with the primary community of reference wherein an individual is recognized as ill and provided with appropriate therapy. The primary community of reference plays two crucial roles
in any psychotherapeutic system: It defines the types of problems that
require treatment and establishes the criteria under which it will accept one
of its members as having been cured. The key to the rhetoric of transformation in religious healing is that the definition both of problems and of
cures conform to the agenda of the religious community. It has been found
in various studies that the problems presented are not always psychiatric in
nature (Monfouga-Nicolas 1972; Pressel 1973), and my clinically trained

healer-informant confirms this for Catholic Pentecostals. This does not imply that the religious community necessarily creates problems that it then heals, but rather that it treats spiritual and psychological dimensions of problems not addressed by any other method of healing. With regard to defining a cure, it has been found that in movements whose primary aim is healing, particularly those involving religious spirit possession, healing is to a large extent constituted by accepting membership in the group (Crapanzano 1973; Messing 1959; Monfouga-Nicolas 1972). Among Catholic Pentecostals healing is understood to occur in terms of integration of the healed person into the religious community, although the purpose of the community goes beyond that of healing. Successful and lasting healing is regarded as a continuing process aided by the day-to-day support of fellow Christians.[12]

For Pentecostalism in general, this recognition of the prominent role of the group has led to the conclusion that faith healing treats not pathology, but lifestyle (Pattison 1974). This point, as true as it may be, can easily be misconstrued to suggest that such healing sidesteps real problems by redefining them in a new context or by subordinating them to religious goals. Such an approach would beg the question, in Tambiah's (1977) phrase, of the "performative efficacy" embodied in the rhetoric of ritual healing. Tambiah's account of Thai Buddhist cult of healing through meditation provides a useful parallel in this respect to Catholic Pentecostalism, in spite of the very different notions of the self found in Buddhism and Christianity. Tambiah writes that "in the Buddhist context the individual and personal aspects of illness are assimilated to an enduring cosmic paradigm of theodicy and tranquillity" (1977:123) founded on the notion of *karma* and constituting a "blanket explanation" by means of which patients transcend their present conditions. But, says Tambiah, this is only half the story, since "the experience of illness, pain, and misery is viewed as a real manifestation of one's 'substantial' and corporeal enmeshment" (1977:122), and the teacher (*achan*) is regarded as possessing concrete mystical powers (*iddhi*), which are brought to bear in the meditation discourses that "realistically describe diseases and the techniques for curing them ..." (1977:123). The doctrine holds that the highest spiritual state can be attained only by the person who is in a state of perfect physical and mental health; and the religion offers means of achieving both. A similar rhetoric obtains in Catholic Pentecostalism, where the overall process of healing is subordinated to that of "spiritual growth" of the individual, to the degree that some people believe that everyone should at some point undergo the Healing of Memories, whether or not they have a specific problem. At the same time, the concrete mystical powers invoked in the various healing techniques also address real suffering in the temporal domain. This corresponds to the actions of the biblical Jesus, who while preaching the overwhelming importance of preparing to enter the Kingdom of Heaven, nevertheless employed His spiritual power to bring relief to the afflicted through healing. Thus, although informants acknowledge the primary importance of effects such as changes in attitude, affect, or self-image that pertain to spiritual growth, there is also the feeling of assurance that specific complaints are being directly addressed by healing.

This discussion allows us to make a preliminary formulation, suggesting that the rhetoric of transformation must accomplish three closely related tasks:

1. Predisposition—within the context of the primary community of reference, the supplicant must be persuaded that healing is possible, that the group's claims in this respect are coherent and legitimate.
2. Empowerment—the supplicant must be persuaded that the therapy is efficacious—that he is experiencing the healing effects of spiritual power.
3. Transformation—the supplicant must be persuaded to change—that is, he must accept the cognitive/affective, behavioral transformation that constitutes healing within the religious system.

We shall examine how Catholic Pentecostal healing accomplishes each of these tasks in turn, under the respective headings "Rhetoric of Predisposition," "Rhetoric of Empowerment," and "Rhetoric of Transformation." In the second part of this chapter, I will present additional material that will allow a refinement of this preliminary formulation.

### Rhetoric of Predisposition

Prior to the tasks of empowerment and transformation, there is a level of persuasiveness embedded in the social setting of Charismatic healing that predisposes potential supplicants to the kind of experience that healing makes available. Two contextual features are of immediate relevance. First, Charismatic healing is esoteric in that it is available only to those who have already experienced at least a minimal degree of participation in the movement. Thus, it is unlike therapeutic forms such as the Mexican Spiritualism described by Finkler (1980, 1981) or the Sinhalese exorcism described by Kapferer (1979a, 1979c, 1983), the appeal of which is exoteric in that it is oriented to the health care needs of the general populace. Second, unlike many religious therapies encountered in the ethnopsychiatric literature (Finkler 1980, 1981; Jilek 1974; Messing 1959; Monfouga-Nicolas 1972), participants in Catholic Pentecostalism seldom become involved primarily as a result of seeking healing. Other personal or religious reasons lead most participants to the prayer groups or communities; the prayer meeting is the first ritual setting to which most Catholic Pentecostals are exposed, and only with deepening involvement do they experience the forms of healing described here. These two facts—the esoteric nature of Catholic Pentecostal healing and its secondary role in initiating involvement with the movement—indicate that the rhetoric of healing is derived from the larger field of discourse that defines the movement as a whole. Accordingly, I shall offer here a brief sketch of Catholic Pentecostal ritual language (see Csordas 1997 for a more detailed analysis).

At the core of Catholic Pentecostal ritual discourse is a discrete set of terms that constitute a "vocabulary of motives" (confer Mills 1940) for the movement.[13] These motives are complexes of meaning that orient the action of participants in both ritual and day-to-day settings. The motives are kept in constant circulation in Catholic Pentecostal discourse through their use in the performance of specific genres of ritual language. The vocabulary can be analyzed into several categories as follows: Forms of Relation, such as Authority, Covenant, Promise, Sonship/Brotherhood; Forms of Collectivity, such as Community, People, Army, Kingdom; Qualities, such as Light, Heart, Power, Order; Activities, such as Service, Praise, Warfare, Growth; and Negativities, such as World, Flesh, Devil, Darkness. The genres are modes of speech governed by specific rules of use and identifiable in terms of specific prosodic features. Chief among them are Prayer, Teaching, Sharing (witness or testimony), and Prophecy. A dialectical relation obtains between the vocabulary of motives and the system of sacred genres. Motives orient the action of utterance in the genres. For example, to the extent that an utterance is judged to be inspired, it is perceived as a compelling manifestation of divine "Power." Also, utterances in the genres of ritual language are expected both to originate in and serve the needs of the "Community" of fellow Christians. The vocabulary of motives is in turn influenced by its constant circulation though the system of genres. That is, the meaning of each motive is elaborated, related to the meanings of other motives, and applied to situations of everyday life (Csordas 1980).

Persuasiveness at this level lies in the fact that the techniques of healing are specific applications of motives and genres with which the supplicant is already at least marginally familiar in a larger context. On the psychological plane this amounts to a predisposition to be healed; on a phenomenological level it means that the supplicant is aware that his healing is part of something larger than himself. For Catholic Pentecostals, the possibility of healing is tightly articulated in a web of motives. Thus, being filled with the *Power* of the Holy Spirit does not merely enable one to speak in tongues, but initiates a process of spiritual *Growth,* which in turn makes one increasingly capable of Service to the Community [of Believers]. The *Community* is the instrument through which God's Plan can be realized and His *Kingdom* brought about, and hence must be composed of *Spiritually Mature* members. *Healing* is a function of the *Community* that promotes *Spiritual Maturity,* removing obstacles to further *Growth,* and establishing *Order* in a person's life. Achieving *Growth and Order* is regarded as particularly difficult in the absence of *Community* support.

Awareness of the possibility of Spiritual Gifts and the utterance of ritual genres in a group context prepare the supplicant to be profoundly moved when those resources are brought to bear specifically for his or her benefit. For example, prophecy occurs in a variety of contexts among Catholic Pentecostals. Regarded as a direct message from God, prophecy is uttered in the grammatical first person. It is particularly persuasive and authoritative, both as a concrete manifestation of divine presence, and as a means for guiding and directing groups and individuals. It is seldom used to foretell the future; see Csordas (1997). The rhetorical impact of prophecy is magnified when, within the healing ritual, an individual accustomed to hearing prophecy addressed to groups finds divine attention focused directly on him, through a prophetic message of encouragement, admonition, or exhortation. Likewise, the existence of evil spirits is a taken-for-granted fact of daily life for Charismatics. A feeling of dysphoria or anxiety can be attributed to an evil spirit and informally "delivered" at any time; and in group situations, demonology can serve as a language—a kind of semiological shorthand—for the articulation of interpersonal tension (Csordas 1980a). Again, the rhetorical impact of demonic activity is enhanced when the supplicant in healing realizes that his own life is the locus of such demonic activity. It is further enhanced when referred back to the cosmological setting established by the vocabulary of motives, whence it becomes understandable as a battle in the continuing spiritual warfare between the forces of Light and Darkness.

In sum, the contextual rhetoric of therapeutic ritual creates a predisposition to be healed and an awareness of a larger purpose for one's healing. The importance of cultivating such a predisposition is recognized among Catholic Pentecostals; it is designated in the vocabulary of motives by a term that, blurring the analytic distinction between emic and etic, is identical to that employed by Frank (1973) in his explanation of how faith healing works: Expectant Faith. What the present argument has thus far demonstrated is that, whether it is explicitly recognized on an emic level or posited as an operational category on the etic level, such expectant faith is constituted by a rhetoric specific to the context in which the healing occurs. In the present case, this rhetoric is in turn determined by the esoteric nature of Charismatic healing and the subordination of healing to an overarching discourse composed of a vocabulary of motives and a system of genres.

What endogenous process, if any, is activated by the rhetoric of predisposition in ritual healing? In an indirect way, it is "conversion," the overall process of which is rhetorically controlled by the entire apparatus of ritual language described above. The notion of conversion must be qualified for Catholic Pentecostals, since it is not necessarily a sudden, dramatic, cognitive restructuring; while some participants acknowledge joining in the wake of an intense personal crisis or conversion experience, others regard it as a perfectly natural step at a particular moment in their lives (Csordas 1980a; McGuire 1982). At the same time, the degree of secondary socialization necessary may be relatively insignificant for many Catholics already predisposed to this kind of spirituality. The rhetoric of predisposition in healing is, then, a specific elaboration upon the process of conversion. This is shown in the examples described above. In Example One, three years separated the supplicant's initial involvement and his perception of a need for healing; in Example Two, an initial resistance to healing as recommended by another is eventually overcome; in Example Three, the supplicant is led to healing through motivated reading of the movement's printed literature. Thus, the rhetorical impact at this level is to lay the groundwork for activation of the endogenous processes through which the main work of healing is achieved.

### Rhetoric of Empowerment

Under this heading, we consider those components of ritual therapy by means of which the supplicant is persuaded that he is experiencing the effects of divine power. Power is a key motive for Catholic Pentecostals. The movement is referred to as the Charismatic Renewal precisely because of its ritual use of "charisms," or "Gifts of the Holy Spirit," understood as modes for the expression of spiritual power. The rhetorical impact of the power motive is a function of the way it is grounded in concrete experience. For ritual healing, two principal aspects of empowerment are considered: the role of somatic symbols and physiological process and the interpretation of spontaneous expression of endogenous processes.

One need only recall Mauss's (1950) notion of *les techniques du corps*, in which the human body is simultaneously the primordial object of and tool for cultural action, to understand that the most immediate and concrete means of persuading people of the reality of divine power is to involve their bodies. Symbolically a microcosm, and physiologically the limit of human experience, the body recruited to the cause of symbolic healing invokes a powerful sense of totality, encompassing the whole person. It is in this light that the traditional Pentecostal "laying on of hands" is best understood as a rhetorical technique. This gesture is too often interpreted simply as a magical transference of power. From the perspective elaborated above, however, in which the experience of sacred reality is understood as fundamental to religious healing, it appears as a key symbol with (in Victor Turner's terms) "multivocal" significance. The real import of laying on of hands then emerges from analysis of the meaning communicated by the touch. This analysis begins with a recognition that (in Wittgenstein's terms) the gesture bears a "family resemblance" to the congratulatory pat on the back or the sympathetic hand on the shoulder. Performance of the gesture in the ritual context includes a mimetic amplification of its meaning in two respects. First, it is an imitation of the healing touch of Jesus portrayed in the Bible. Second, it is a metonym of the solidarity of the Christian community; the unity of two bodies touching is the unity of the Church as the Mystical Body of Christ. The importance of touch as a critical rhetorical locus in establishing this unity is suggested in Terrence Turner's observation that "the surface of the body seems everywhere to be treated, not only as the boundary of the individual as a biological and psychological entity but as the frontier of the social self as well" (1980:112). An example in concrete expression is the situation in which a group is "praying over" an individual.

It is necessary for only two or three people to be in physical contact with the supplicant; those surrounding may have their hands on the shoulders of the ones between themselves and the supplicant. There is no evidence that the Catholic Pentecostal concept of power includes its transmissibility *through* other people to its object. In the emic view, it is probably more accurate to say that the spiritual power is called into play primarily by the accompanying prayer, particularly when that prayer is uttered in the spiritually powerful form of glossolalia. It is the appeal to totality embodied in physical union rather than the magical transfer of power wherein lies much of the persuasiveness of the gesture.[14] In this respect also, the gesture carries the connotation of shielding and protecting the distressed supplicant.[15]

Two other strands of meaning can be unwound from this complex symbol, one historical and one genetic. The first has to do with the direct family resemblance between the laying on of hands and the "royal touch" of the Middle Ages, when European monarchs would lay hands on their subjects in a ritual gesture of healing for diseases such as scrofula (Bloch 1973). This practice ended in the eighteenth century, at about the same time that the Wesleyan tradition that spawned Pentecostalism was arriving on the religious scene. In one sense, the Pentecostal laying on of hands can be seen as a Protestantization or democratization of the healing touch attributed to monarchs by virtue of their divine right connection with God. The connection with the royal touch is also evident in the motivational complex defined by the terms *Kingship, Lordship, Authority,* and *Submission.* Catholic Pentecostals repeatedly emphasize the Kingly nature of God, with themselves as builders of and subjects in His Kingdom. The Authority exercised by community leaders, as well as the Authority over evil spirits claimed in Deliverance, are derived directly from this Kingly source. Submission to Authority and to God's Plan and will is a highly articulated motive, present also in phrases such as "yielding to the gift of tongues" and "giving it [a problem] to the Lord." Thus the rhetorical message communicated by allowing hands to be laid on one is that of Submission to divine Authority as well as reception of divine Power. In the gesture of laying on of hands, the supplicant is placed, and places himself, in the hands of the Lord.

The genetic strand of meaning has to do with the experience of touch as intimacy. Movement leaders recognize the possibility for misplaced intimacy in recommendations that healers work in teams, and never alone with a member of the opposite sex. Touch breaks a culturally constructed interpersonal barrier based on a notion of the individual as a discrete, independent entity, on the concept of privacy, and on the injunction "don't touch" in most social settings (Montagu 1978; Shweder and Bourne 1982). Montagu (1978) reviews a substantial body of research that indicates the importance of adequate tactile stimulation in childhood to healthy development, and at the same time indicates the relative lack of such stimulation among American children in cross-cultural perspective. In addition, he cites literature that indicates the therapeutic value of touch in the treatment of skin disorders, asthma, and even schizophrenia. These considerations suggest that, as a *technique du corps,* laying on of hands may have more than symbolic significance, actually helping to compensate for a developmental deficiency. Cross-cultural research correlating the degree of tactile stimulation in childhood with the prevalence of tactile therapies of various intensities, from superficial laying on of hands to full scale body massage (confer Finkler 1980, 1981) would be helpful in clarifying this point. Restricting our view to Catholic Pentecostalism for the present, it will be sufficient to note how the basic tactile imagery is metaphorically amplified in healing. Some Charismatic healers use a "soaking prayer" in which touch is accompanied by effusive verbalization, with language literally playing the role of a ritual substance (confer Gossen 1976) in which the supplicant is immersed.[16] This practice is probably related to the older one of being "washed in the Blood of the Lamb" (the redemptive blood shed by Jesus on the cross), achieved metaphorically as a verbal—to borrow Austin's term—"performative act." Also related to this complex of meaning is the basic notion that in the "Baptism of the Holy Spirit" a person is infused with the Spirit and its Power.

Another among the Catholic Pentecostal *techniques du corps* is a physical sensation that may be a tingling in the hands, but may also be expressed elsewhere in the body, such as in a constricted feeling in the chest. These sensations do not occur in all instances of healing or among all healers; neither are they unique to healing, occurring as well in conjunction with the utterance of prophecy. Yet these sensations have a specific place in the rhetoric of empowerment. Termed *anointings,* they are a sign, not that power is being transferred between individuals, but an affirmation that power is indeed being manifested, and an assurance to both healer and supplicant that the ritual is being correctly performed. They are very likely related on a continuum of degree to the similarly termed *anointings* experienced by the Charismatics' more flamboyant Pentecostal brethren who handle fire or take up serpents. Kane (1982) suggests that these anointings might involve activation of the endogenous opioid (endorphin) mechanism of the central nervous system, with resultant desensitization to pain and fear. Although the middle class Catholic Pentecostals tend to avoid what they regard as histrionics in ritual practice, it is reasonable to suggest that the same physiological mechanisms are available in their enactment of the rhetoric of empowerment.[17]

Moving from *techniques du corps,* the second important aspect of empowerment is the meaning attributed to "spontaneity." In the course of prayer for healing, memories of past events and visual imagery can occur either by intentional evocation or by spontaneous emergence from the preconscious. The rhetoric of empowerment establishes these endogenous processes as manifestations of miraculous power in two principal ways. First, *spontaneity* plays an important role as a motive in the overall system of Catholic Pentecostal discourse. Spontaneity is believed to be a qualitative effect of experiencing the Baptism of the Holy Spirit. Ideally, it characterizes the course of prayer meetings, interaction among individuals (as expressed, for example, in the use of the embrace, or "holy hug," as a greeting), and a person's inner spiritual life when he is "right with the Lord."[18] For the supplicant in healing, the spontaneous insight, visual image, memory, or pronouncement of a demon's name is motivated or oriented as a manifestation of divine power; it is not a human achievement, but a spontaneous gift from God to one of His faithful.

With respect to persuasive force, a cultural rhetoric that so construes endogenous processes as experience of the holy can be fruitfully contrasted with one that does not. In Japanese *Naikan* therapy as described by Murase (1982), clients often experience the unexpected emergence of forgotten memories; but there is no indication that spontaneity per se plays any role in validating the experience, or in bringing about the desired results. Indeed, *Naikan* is for the most part a secular psychotherapy, the rhetoric of which directly invokes traditional cultural values, rather than a ritual psychotherapy that operates by directing the attention of supplicants to their action and experience in such a way as to constitute an essentially sacred world.

Second, the spontaneous activation of endogenous processes is given concrete rhetorical form by defining its results as the fruits of discrete, named "spiritual gifts." A *Word of Knowledge* is a fact about the supplicant's life previously unknown to the healer; for example, the visions experienced by the healer in Example One above are Words of Knowledge about the sources of his distress. "Prophecy" (discussed above) may be uttered as a direct message from God to the supplicant, expressing encouragement, admonition, or exhortation. Two other important spiritual gifts are regarded as divine enhancements of ordinary human qualities. A "Word of Wisdom" is a statement of advice to the supplicant experienced spontaneously as a directive from on high, and regarded by the healer as beyond what he could have arrived at through his own thought processes. *Discernment* is conceived both as a kind of spiritual sixth sense for intuiting the concrete presence of evil and as a spiritually enhanced kind of good judgment in guiding the proceedings and getting at the root of the supplicant's problem. Encompassing these gifts is the gift of "Healing" per se, which is regarded as a divine enhancement of the "ordinary" power of prayer.

The immediate rhetorical impact in the context of a healing session varies among these gifts. A particular compelling vision or prophecy is likely to have a pronounced effect on the session, whereas the healer may not even be aware that advice he has given was the result of a Word of Wisdom until later events prove it to be the case. The main thrust of this battery of spiritual techniques taken as a whole, however, is that it allows the healer to tap the divine reservoir of Knowledge, Wisdom, Discernment—in short, omniscience—about the supplicant. In this way, exercise of the spiritual gifts has the same rhetorical impact as the shaman's "journey": It allows the healer to participate mystically in the inner life of the supplicant, and allows the supplicant to participate in the fruits of endogenous processes experienced by the healer. This mutual participation is not the result of an interpenetration of minds—it is the phenomenological component of co-participation in the social project of generating a discourse that is convincing, a rhetoric that creates the concrete experience of divine power.

There remain two elements of empowerment, both of which can be dealt with briefly. The first is the use of glossolalia (speaking in tongues) in conjunction with healing prayer. Like the laying on of hands, glossolalia is an important multivocal symbol. It is, first, a spiritual gift, and therefore a direct manifestation of mystical power as a divinely inspired language enabling the speaker to praise God with a power lacking in language that is a merely human contrivance. Second, on the level of discourse itself, its rhetorical force stems from its very "senselessness": It shatters the taken-for-granted canons of intelligibility in everyday discourse, thereby creating the possibility for new kinds of meaning (Csordas 1980a). Third, to the degree that one agrees that an essential criterion of mental illness in extreme forms is lack of intelligibility (confer Ingleby 1982), with glossolalia Charismatic healing challenges the forces of chaos on their own ontological turf, claiming that behind its apparent senseless utterance is a divinely motivated and unshakeable moral and cosmological order.

A final element of persuasiveness in the rhetoric of empowerment is the extremely vivid or eidetic quality of some of the visual images experienced by supplicants and healers. This occurrence of eidetic imagery enhances the apperception of spiritual power, as was evident in the statement of the supplicant in Example Three above that the "imaging" was not "imagined," but real. However, clarification of this aspect of empowerment must await further research on eidetic imagery among Catholic Pentecostals, both with respect to visions in healing and visions in prophecy.

### Rhetoric of Transformation

The rhetorical movement of transformation is complete when the supplicant is persuaded to change basic cognitive, affective, and behavioral patterns. In terms of Kapferer's (1979c) related discussion of Sinhalese exorcism, this movement amounts to a reconstruction of self.[19] Healing of Memories and Deliverance have complementary ways of redirecting the supplicant's attention to his action and experience in order to achieve the construction of a self that is healthy, whole, and holy.

The most powerful rhetorical element in the visualizations of the Healing of Memories is the figure of Jesus. The image of Jesus is visually "walked through" the supplicant's entire past life in order to demonstrate concretely that *He was always really there,* although never perceived in the person's former self-conception. In this process the sense of the individual's entire life is transformed by concretely inserting the presence of Jesus. At the same time, since the person's entire life is reviewed within several sessions (in contrast to the lengthy treatment of psychoanalysis), the time frame of a life is collapsed such that it can be experienced as a whole in the present. This concrete combination of divine presence and the construction of a new life (or new past) in the present constitutes the rhetorical key to personal transformation in the

Healing of Memories.[20] The healing is not effected by human hands, but by Jesus Himself, in person. In a religious milieu where the embrace or "holy hug" is a conventional greeting, and where the laying on of hands is a common ritual gesture, it is of significance that the figure of Jesus in these visualizations is often perceived as healing with a touch or embrace. There are few images as symbolically charged as that of being touched by the hand of God. At the same time, the supplicant's role is not altogether passive. In emphasizing the need for forgiveness of those responsible for past emotional hurt, the rhetoric of transformation recruits the supplicant's active participation, and requires an overt commitment to "change his mind."

The basic endogenous process activated in Deliverance is externalization. As noted above, the names of demons from whose influence people are delivered indicate that they are spiritual entities that control particular sins, vices, character flaws, personal weaknesses, or negative affective states. Although the healer sometimes discerns the presence of evil spirits, and constructs the interpretation of their action in clusters, the spirits are usually identified by the supplicant himself. Actually, it is the demon which is thought to identify itself through the mouth of the supplicant, in a kind of inverse prophecy (compare the biblical account of Jesus' encounter with the demon who declared "I am Legion"). In this way, as with the act of forgiveness in the Healing of Memories, the supplicant is recruited to active participation in the therapeutic process. Catholic Pentecostals recognize the power and necessity of uttering these names aloud; we can identify at least three levels on which the supplicant's identification of demons has significant rhetorical impact. On the psychological level, it is an acknowledgment that the person is suffering from these particular problems, and an opening to the subsequent interpretation and counseling by the healer. On the spiritual, or ritual, level, the naming in effect "gets the demon out in the open," where it can be cast out by the relatively simple ritual formula stated above. Since a demon must be commanded by name to depart in the name of Jesus, one of the biggest obstacles to success in the Deliverance ritual is the case when recalcitrant demons refuse to identify themselves. Finally, on a cultural level, by specifically acknowledging the part in his problem played by demons as concrete spiritual entities, the supplicant brings his or her experience to participate in a field of symbols, motives, and meanings that constitute the religious milieu of, and raison d'être for, the Catholic Pentecostal movement.

The transformation brought about in the actual casting out of evil spirits likewise operates on several levels; in fact, it is considerably more important, since in some cases the healer, rather than the supplicant, identifies the demons. On the cultural level, a demon can be understood as a metaphor of self. The cultural logic of metaphoric transformation has been formulated by Fernandez (1974) in his discussion of ritual healing in the Bwiti cult. Translated onto the level of cultural discourse, the self is represented as an "inchoate pronoun," a simple *I* or *me* lacking distinct qualities. The metaphor, in the present case an evil spirit, is predicated onto the inchoate pronoun, giving it a particular cultural value or significance. The metaphor is then symbolically moved along a qualitative continuum (for example, evil–good, weak–strong, darkness–light) that has a particular meaning within the cultural milieu in which the healing takes place. Carried along the vector of this continuum, the inchoate pronoun acquires direction and form. Finally, the self as pronoun acquires qualities defined by the metaphors characteristic of the end of the continuum where it has come to rest. The success of this process on the cultural level persuades the individual to accept the new self-definition as the means for orienting his actions in daily life.

Without an understanding of the complex of symbols and meanings that constitutes the Catholic Pentecostal religious milieu, it would appear that the movement involved in casting out evil spirits is simply a movement out and away from the self, rather than movement along a qualitative continuum. Catholic Pentecostal demonology is much more fully integrated into the entire symbolic system, however, than this interpretation would imply. Juxtaposed to the Catholic Pentecostal terminology described above, the demonology appears almost as its mirror image, a kind of negative vocabulary of motives. Thus, in line with the interpretation put forward by Fernandez (1974), a demon (*Hate,* for example) and a motive (*Love*) can be understood to define opposite ends of a qualitative continuum within which the rhetorical operation is carried out.

The occurrence of demons in clusters extends the transformation to several other levels. First, on the level of message communicated, it calls into play the classic rhetorical device of redundancy. Identifying demons with similar or related names repeats the same general message in several forms, driving home the point that evil is present. This is evident in the grouping of Bitterness/Resentment/Anger in Example One above. Second, the presence of several spirits subordinate to a master demon introduces the notion of cognitive and affective complexity, gives the healer more symbolic substance to work with in the sessions, and increases the potential for drama when spirits are eliminated one by one down to the last, stubborn master demon. Third, as noted in the commentary to Example Two, the names of demons may include both traditional vices and sins (for example, Lust, Gluttony), and behavioral attributes or affective states relevant to a contemporary context (such as Insecurity, Manipulation). In this way, in addition to being linked to the movement's vocabulary of motives, the metaphors of self are linked both to the traditional Catholic belief system and to a vocabulary suitable for the counseling dimension of ritual psychotherapy.

That demonic names refer explicitly to a range of affective states and behavioral attributes is clear in the comment of one healer that there are as many evil spirits as "you could think of psychologically" (that is, as opposed to religiously or spiritually). Thus, there is not only a demon of Fear, but also demons of Fear of the Dark, Fear of Dogs, Fear of Heights, and demons representing all the phobias. The repertoire is enhanced by the introduction of stereotyped elements of folk culture: For example, one healer recalled expelling an "ancestral spirit" from a supplicant of Chinese descent, and others working in a locale of considerable ethnic diversity noted the prevalence of evil spirits such as Superstition among Irish, Legalism among Germans, and Nationalism among Americans. The absolute number of evil spirits appears to correspond to a list of possible adverse effects. These adverse effects are the primary concern of healers, even in the apparently rare cases in which the actual name of the demon is never discerned. Thus the Charismatic demonology can be understood as a semantic illness network similar to that described by Good (1977), where the demons are "images which condense fields of experience, particularly of stressful experience" (1977:39). Furthermore, the appearance of demons in clusters can be understood as a recognition in therapeutic practice of the close semantic links within the network. In all cases, however, through the rhetoric of healing, negative qualities are made to participate in sacred reality by being given the status of spiritual entities—beings whose intent is explicitly evil, but who are subject to divine power and can be commanded to depart, and whose names correspond in a negative sense to the fundamental vocabulary of motives that underlies Catholic Pentecostal discourse.

In sum, in Deliverance, negative metaphors of self drawn from a variety of domains are transformed into positive motives, fundamentally altering the way supplicants attend to their own patterns of cognition, affect, and behavior. In the Healing of Memories, the supplicant's past is transformed by redirecting his attention toward various actions and experiences in such a way as to perceive the role of Jesus in leading him to the present, thereby removing the negative residue of emotionally damaging experiences. In achieving these tasks, the two rituals complement one another in harnessing the endogenous healing processes. In the Healing of Memories, with its activation of memory/insight and vision/visualization, all the action takes place within the person. In Deliverance, which activates externalization, the action occurs between the individual and forces that originate outside his self. Moreover, in the Healing of Memories the goal is forgiveness and reconciliation with one's past, reinterpreting it as part of God's plan that led the person to his present relationship with Jesus within the Charismatic movement. In Deliverance, on the other hand, the attitude is one of active, authoritative grappling with and expulsion of evil influences. In this sense, it might be argued that, taken together, these two rituals offer a balanced approach to healing that deals both with internal, intrapsychic factors and external, social environment factors that contribute to emotional distress.

To the extent that the rituals deal with problems that can be identified as phobias originating in traumatic events of the past, or as obsessive patterns beyond a supplicant's control, they can clearly be understood as forms of ethnopsychiatry. As Kapferer (1979a, 1979c) notes for Sinhalese rites of exorcism, however, they are structurally reminiscent of certain rites of passage. The resemblance may be more than structural in the Catholic Pentecostal case, where some participants hold that everyone should undergo the healing experience as an aid to "spiritual growth," thus deepening their involvement with and commitment to the phenomenological world created by the movement. The vocabulary of motives provides the term that orients the shift in status implied in the notion of a rite of passage, namely spiritual Maturity. Thus we are again faced with simultaneous accounts of the same practices from the perspectives of Health and the Holy. We shall return to the methodological implications of this analytic duality.

## Elements of Charismatic Persuasion and Healing

In the preceding section, the material I discussed was entirely retrospective, and in some instances consisted of third-person accounts by healers about people for whom they had prayed. The discussion that follows is based on a subsequent study of Charismatic healing that I undertook based on the principle that an experientially valid account of therapeutic process would ideally follow a patient through the actual experience of healing (see also Csordas 1994a).[21] I followed both individuals whom I discuss here from beginning to end of their involvement with one of the participating healing ministers. They represent precisely the kind of incremental and inconclusive process that I suggested above as characterizing the lowest limits of therapeutic efficacy in ritual healing, and it is for that reason I chose them as the focus of this analysis. Before beginning this discussion, however, I shall introduce the healer with whom I followed these cases.

Father Felix, an experienced Charismatic healing minister, is a 60-year-old Catholic priest, ordained as a member of a religious order in 1952. He holds a Doctorate of Ministries with concentrations in psychology and counseling and has been an assistant supervisor of a program for priests in Clinical Pastoral Education (CPE). In 1975, as part of an assignment as director of pastoral care at a Catholic medical center, he was asked by the executive administrator to coordinate a Charismatic healing ministry within the hospital. Although he had been aware of Charismatic prayer groups, he had previously taken no interest; thus his involvement in Catholic Pentecostalism began with his consent to become active in Charismatic healing. Since then he has remained active in the healing ministry, leading public healing services and workshops and conducting private healing sessions.

Catholic Pentecostals believe that the power to heal stems from "spiritual gifts" ("charisma" in theological terms) granted by God. As Father Felix continued to work as both a counselor and healing minister, he asked God for "the gift of discernment to be able to know what to pray for. Because a lot of people are coming in. There's a lot of stuff that's unconscious, they can't get in touch with it." *Discernment*—divinely heightened intuition—is understood as a divinely inspired ability to understand people, problems, and situations. Father Felix recounts two incidents in which he felt the granting of this charism was confirmed. In the first, while praying with a parish priest, he spoke about problems that were so uniquely relevant to the priest's situation that the latter thought his parishioners had already spoken to Father Felix about them beforehand. In the second, he discerned that he should ask someone else in the healing group to lead a vocal prayer while he prayed silently with his hand on a priest's back. During the prayer his hand became extremely hot. This heat was also perceived by the supplicant, who later mentioned that he had cancer in his back at the spot where Father Felix had "discerned" that he should place his hand. For Father Felix, the fortuitous placement of his hand was a manifestation of discernment, while the heat was a sign that healing was taking place. Since then, Father Felix has relied strongly on this gift in his healing practice.

Father Felix holds private healing encounters in one of the counseling rooms at the monastery where he resides. The session begins with a period of light talk or counseling, during which the priest typically inquires about changes that may have occurred since the previous session. He then places a straight-backed chair in the center of the small room, asks the supplicant to be seated there, and anoints the person's forehead with holy oil. He stands behind the person with one hand on her head and another on her shoulder, praying silently for approximately five minutes. During this period he often receives "discernment" about the person and the problem. Afterwards he asks the person about any experiences she might have had during the prayer. After this second brief period of conversation and counseling, the session ends, seldom having lasted more than a half-hour.

Father Felix strongly believes in the necessity of "getting to the source" of a problem in order to heal it. From his experience, two important sources of people's problems are evil spirits and previous generations. To eliminate the influence of evil spirits he uses Deliverance prayer, and to eliminate that of previous generations he uses the mass for healing of ancestry. Each of these will be briefly described.

In the Catholic Pentecostal healing system, evil spirits typically are named after emotions or behavioral patterns; Anxiety, Depression, Lust, and Rebellion are all common spirit names. Father Felix agrees with most other Catholic Pentecostal healers interviewed that spirits attack individuals at their most vulnerable points, whether these be the propensity for committing a particular type of sin or the lasting effects of traumatic experience. No one can be completely possessed by Satan unless he makes a conscious decision or pact; all other spiritual afflictions are in the form of oppression or harassment in a particular domain of life experience. Father Felix also allows for human sources of negative emotions in the absence of demonic influence however, and it is a matter for discernment whether a person plagued by depression or lust is in fact under attack by the spirit of Depression or Lust. Among the most common spirits in his experience is Fear-of-Being-Found-Out, which causes such thoughts as "if only people knew the things I did or I think, I'd have no friends." Another very common spirit is Devaluation, akin to Self-Hatred, which causes "low self-image and self-esteem."

Father Felix's typical mode of Deliverance is to pray silently as follows: "By the power of the Word of God, Jesus Christ, and by the power of the Sword of the Spirit I sever forever all negative spiritual, emotional, psychic, or physical negative influences that are bothering my sister [or brother]." Following this general prayer he specifically addresses whatever evil spirits may be present: "You, dark binding forces, I command you in the name of

Jesus Christ to be separate one from the other, to be without communication and to be rendered powerless. You have no more power over this person. He [or she] belongs to Jesus Christ." He then silently commands individual demons by name, as their presence is revealed to him through discernment. He does not necessarily inform supplicants that there are evil spirits involved, but instead waits for a sign in their speech or behavior that confirms his discernment. Yet in withholding this divinely inspired knowledge, he sometimes tells the supplicant that he has discerned things that may hurt if told. In this way he establishes a role both of wise protector and empowered bearer who is in direct contact with the sacred.

In addition to private healing sessions, Father Felix often says a healing-of-ancestry mass in the home of the supplicant. Beforehand, he asks the person to prepare a family tree going back as many generations as possible, noting any important events or health problems, such as suicide, alcoholism, mental illness, or abortion. He then "prays over" the genealogy for discernment about the individuals represented. The principle enacted in this ritual is that illnesses or adverse effects of traumatic experiences can be passed "through the blood line" to successive generations. Part of a person's healing can include the healing through prayer of individual forebears who died without having been healed. In some respects this practice is akin to praying for the souls of the dead, but it goes a step farther in actually trying *to heal* the dead. When this is accomplished, the chain of negative influence is "severed," and the person is freed of the affliction.

This is a brief description of the Charismatic healing ministry as practiced by a single person. While it is well within the bounds of Catholic Pentecostal healing practice as delimited by the research described above, several contextualizing remarks are in order. First, while it is quite common for priests and members of religious orders to practice ritual healing, many Charismatic healing ministers are laypersons. Second, while some healing ministers have had professional training in counseling or psychology, most have had none. With respect to procedure, Father Felix makes less use of guided imagery than do many Charismatic healing ministers, although he encourages spontaneous mental imagery. On the other hand, he makes great use of deliverance, which many healing ministers avoid because of the perceived danger of dealing with powerful evil spirits. Finally, performing the mass for healing of ancestry in supplicants' homes appears to be a practice unique to Father Felix, and it provides him with an opportunity to observe family dynamics in a way that is typically reported only of healers in small-scale traditional societies.

### Case One

Margo is a twenty-seven-year-old woman, third-youngest of nine children, who lives with her parents, three of her sisters, and one sister's three-year-old daughter. She is concurrently under treatment with a psychiatrist (psychopharmacologist) and in therapy with a psychologist, but she has

been frustrated by the failure of both medication and therapy. The diagnostic portion of her interview confirmed panic disorder and major
depression as her principal problems.[22] She and her mother both report
that one of her sisters, who lives at home, suffers from schizophrenic illness.

Margo's illness began in 1985, two years before recourse to the healing
minister. She had dropped out of nursing school after doing less well than
she had hoped and had returned to full-time work as a hospital administrative assistant. She felt overworked and preoccupied by this stressful job.
At the same time she felt that she was "losing" most of her previous friends
as they got married, so that her social life had become "flat." After six
months she "burned out" and took a transfer to a lower-status, less stressful
job. Her first panic attack occurred two months after the transfer.

Difficulties of family life appear to have contributed to the problem. She
regards her parents' marriage as very poor, characterized by frequent loud
arguments. She describes her father as critical, cruel, and authoritarian, to
the point of physically abusive discipline when his children were young.
She is very close to her mother and older sister, but feels a need to distance
herself from emotional overinvolvement and establish an independent life.
She reports developing, one year prior to the onset of her illness, overt hos-
tility and hatred for a previously close sister who had "ruined her own life"
and moved back into the family home after having had a baby with a man
she did not marry. An additional factor in her distress appears to be the
accidental death of a brother some years earlier. Given this constellation of
patterns and events, a major area of intense anxiety for her is relationships
with men. Through psychotherapy she has come to associate this anxiety
with a lack of opportunity to develop a sense of trust for others.

Margo is a practicing Catholic and was involved in Charismatic prayer
groups for a period of months several years prior to her illness, but for no
clear reason she ceased attending. Since the onset of her illness, however,
Margo has frequently attended public healing services and is on the mail-
ing lists of two influential Charismatic healing ministers. At these services
she often experiences "resting in the spirit," a form of motor dissociation
in which a person, at the touch of the healing minister, falls in a peaceful,
relaxing, and rejuvenating swoon as the "Power of the Holy Spirit" over-
comes her. Yet Margo had been disturbed in one of these services when the
healer declared that she was being healed. On inquiry, the healer explained
that her "gift of discernment" revealed that the healing process had already
started. Margo reported being confused and baffled, since "if the healing has
already started, personally I don't feel any different."

Margo called Father Felix to ask for help, and he advised that she attend
his public healing service. At that event she requested prayer for severe
depression, and the priest instructed his assistants to pray for expulsion of a
"spirit of Darkness." He then suggested that she come to him for private
healing sessions. At the initial session he recounted previous situations of
successful healing, and stated that he felt she could be healed quickly. He
"corrected" her idea that prayer would be more successful if she made her

mind blank while he prayed, explaining that she should expect spontaneous mental imagery to emerge from her unconscious during the course of prayer and that God did not need her assistance for the prayer to be successful. He also "corrected" her view that she should cease weekly psychotherapy while undergoing ritual healing.

During the second session, Margo told Father Felix of a disturbing experience she had repeatedly for several months prior to the onset of her illness. As she was drifting off to sleep, Margo "could feel another presence in my room. I could feel someone actually sit down on the end of my bed." She had never mentioned this to her psychiatrist or psychologist, for fear that they would think her crazy. Father Felix agreed that she was right not to have told them, but that he himself was quite familiar with such experiences: It was an evil spirit. This confirmed what she had suspected, and reassured her that it was a phenomenon with which Father Felix could deal.[23]

During the period of silent prayer, perhaps in response to Father Felix's advice to allow thoughts to come to her mind, Margo experienced a series of ideas "coming from all directions." Three issues emerged: the difficulties she experienced in her past administrative job, whether or not to change doctors (she had been told that everything had been tried yet nothing seemed to help), and a disappointing relationship with an older man. The latter situation was one in which the man, who lived in a different city, had courted her for a period of time until she discovered that he was married. She cared for him, but was very angry, and felt conflict about her desire to be with him in spite of a conviction that it would be morally wrong to do so. None of these issues was subsequently discussed with the healer. Father Felix simply told Margo to make note of what came into her mind during prayer because "it would be important" for her.

A final event that unfolded over two sessions had to do with the priest's advice that one can verbally address negative emotions and command them to leave in the name of Jesus Christ. This event was explicitly identified as most significant by Margo in a subsequent interview. She interpreted the advice to mean that the problem is "all in the way that you're thinking." Invoking God indicates that He does not want her to feel as she does, and if she has the strength and faith to say "leave" in His name, the negative emotions of anxiety and depression should go.

During the following session, Father Felix discovered that this technique had not been successful in achieving the goal of changing her attitude. The following key exchange took place:

M: I had thoughts like, you know, I'm slowly going to wither sway. Almost like having some form of cancer. It doesn't leave me. It haunts me. It never leaves me. It won't go away. I can't get rid of it. I don't know how to get rid of it. It's driving me crazy. It's driven me crazy. It's overtaken my whole life. And I . . .

FF: What did I tell you last time? I guess you forgot. About taking authority over these things within yourself. You take authority in the

name of Jesus Christ, and you command them to just get the heck
out. They have to obey.

M: I have said that to myself at different times. Like this whole past
week while I was at mass. I had the tremors and the shakes real bad.
You know, the fears around other people being there, whatever. And I
kept saying that to myself over and over again.

FF: What did you say?

M: I kept saying, you know, "In the name of Christ, leave me, leave
me." Trying to force the way that I thought into another direction,
more positive. And ...

FF: Let me clue you in to something. If you say, for instance, "In the
name of Jesus," right? There's an evil spirit that calls itself "Jesus" ... but
it's a false Jesus. You've got to remember that. Some people get caught
up—it's like conjuring up a spirit, and they're confronting the evil
spirit [that] calls himself "Jesus." So I always use the name "Jesus
Christ" or "Jesus of Nazareth," you know? That Jesus. Oh, yeah,
hundreds of [Spanish-speaking people] call themselves Jesus.

In this interaction (identified as significant by Margo herself), the directive to specify the name Jesus Christ was more than a move by the healer to
cover the technique's lack of success. For the failure to command one's emotions indicates in the logic of the healing system that more than one's emotions are involved. A powerful force must be standing in the way, blocking
the path to healing. In a follow-up interview, Margo acknowledged surprise
at learning both the subtlety of the religious technique and the demonic
cause of her problem. She recalled Father Felix's original invocation of the
"spirit of Darkness" during her first public healing service. She intimated
that she had always "thought [about her problem] along those lines" and
that the idea of evil forces being involved "struck home." With respect to
how this interaction helped her, she responded that it was "to give me
courage and more strength, and more faith. Faith-wise, to know that this is
not of God. And how prayer can build your faith. It can build your strength."

In addition to having a home mass for healing of ancestry, Margo
attended a total of three private sessions with Father Felix. Instead of going
to her fourth session, she kept an appointment with her psychopharmacologist, who decided that since no other treatment had worked, she should
be admitted for electroconvulsive therapy (ECT). She indicated that she
would have resumed the sessions after discharge, but this was precluded by
Father Felix's departure for a long sabbatical.

### Case Two

Ralph is a 25-year-old man who has finished high school and spent a short
period in college. He now lives with his parents and brother, a year his junior, and is under medical and psychiatric treatment for a variety of problems.

The diagnostic portion of our interview confirmed a complex situation revolving around a primary diagnosis of paranoid schizophrenia originating from serious drug abuse; obsessive-compulsive disorder with onset at age 14; probable dysthymic disorder (a mild form of clinical depression); symptoms of agoraphobia, panic disorder, and simple phobia (fear of heights); epilepsy related to a probable brain lesion; and asthma.[24]

In 1983, approximately four years before Ralph tried Charismatic healing, he had a major psychiatric hospitalization following a drug overdose. His inpatient experience was traumatic and appears to have been the occasion on which his principal complaint began: extreme "nervousness" in social situations for fear people are thinking negatively about him, in particular that they are thinking he might be homosexual.[25] Since the advent of these fears, he has been unable to hold a job and finds it nearly intolerable to be in a group of people. Another major source of distress is his brother, who in the past has also been under psychiatric care. He cannot tolerate his brother, who is highly abusive to their parents, so the two have taken turns living with their grandfather in a nearby town. Ralph appears to have a close relationship with his father, but he feels that his mother is critical of him and habitually makes him feel guilty even in small daily events. His primary pleasure comes from listening to recorded music and from writing poetry in a style that he considers similar to that of Kerouac and Ginsberg, though he finds it extremely difficult to write creatively under the influence of his antipsychotic medication.

Ralph's religious background includes exposure to the Charismatic Renewal when he was 16, when he attended a prayer group for about a month with his mother. During this time he had the experience of "Baptism in the Holy Spirit" and became familiar with speaking in tongues and other Charismatic practices. He currently claims not to believe in God, but even so, admits that religious themes consistently emerge in his poetry. The encounter with Father Felix was initiated by Ralph's mother, who thought that, as a psychologist, he could best advise the family about a psychiatrist's recommendation that Ralph submit to electroconvulsive therapy (ECT). Father Felix responded that if Ralph saw him on a regular basis he would not need ETC. Ralph entered the situation expecting counseling for his main problem of social nervousness, and only when the sessions began did he realize they consisted primarily of healing prayer.

Hopes were raised after the first session, during which Ralph experienced warmth emanating from the priest's hands and the sensation of purple rings expanding concentrically in his visual field while his eyes were closed. Father Felix interpreted the vision in terms of Catholic liturgical symbolism, in which purple represents death. He concluded that something negative within Ralph was dying. More important for Ralph, the sense of a benign presence accompanied him for two days after this initial session. This experience encouraged him to attend mass with his grandfather, where he felt his eyes rotating upward in their sockets (nystagmus). One of Ralph's greatest fears is that this occasional phenomenon will occur in public, and its occurrence during the mass prompted him to feel betrayed by God. In subsequent sessions he again experienced heat and color, but the sensations progressively declined in intensity. In addition, although he had prayed silently along with Father Felix during the first few sessions, he ceased this participation in the final ones.

Ralph's post-session interviews reveal his perception of the therapeutic process as unsatisfactory. Two types of comments indicate that the healer at times either overinterpreted or misunderstood Ralph's experience in ways that weakened the rhetorical impact of the healing.

One of Father Felix's overinterpretations occurred when he was trying to convince Ralph that by dwelling on his nervousness he would perpetuate it, just as someone who repeats to himself "don't think about the color green" is in fact thinking about green. During the period of prayer with laying on of hands that followed this conversation, Ralph saw the color green in addition to his usual purple. Father Felix attributed significance to this, pointing out that green is the color of hope in liturgical symbolism. Ralph rejected the interpretation, attributing his vision of green to the suggestion planted by the previous advice, rather than to divine inspiration. In another example, Father Felix asked if he could invite two women from the local Charismatic prayer group to help him in the healing prayer in order to expose Ralph to female influence, which he felt was inadequate in his client's life, and apparently also in response to Ralph's fear of being thought a homosexual. Ralph's response to this therapeutic move was to list a variety of women he knew, rejecting the idea that his exposure to female presence was deficient. Finally, Father Felix attempted to portray as positive Ralph's uncharacteristic attendance at mass and visit to a restaurant with his grandfather. Ralph's response was that he had attended mass only once, and that going to a restaurant never made him as nervous as did being in a group of people.

In addition to these overinterpretations, Father Felix appears to have misunderstood Ralph on a number of occasions. In a segment during which the two discussed whether Ralph's nervousness would prevent him from attending a party, Father Felix stated that he thought mingling with people would be just the thing Ralph needed. In response to Ralph's statement that he was too nervous, the healer said that if you think nervous, you'll be nervous. Ralph objected, "No. I'm not thinking nervous, I am nervous!" In the follow-up interview, he stated specifically that he felt misunderstood, and that with paranoia one cannot simply tell oneself to do something. In another segment Ralph mentioned that his father encouraged him to "be like him" and not care what others think. Father Felix interpreted this as an expression of the father's insecurity, indicating that he did not feel in control of his life and really did care about others' opinions. In the follow-up interview, Ralph took exception to this, arguing that his father had made this statement only once or twice, in the context of encouraging Ralph, and wasn't guilty of "denial." He felt that Father Felix's basic point about people in general was correct, but that he was inaccurate in attributing such denial to his father. Finally, in a session when Ralph stated that no changes had occurred since the previous session, Father Felix turned to the researcher and asked if in fact I could not observe any changes. Ralph interpreted this attempt to solicit impressions of observable behavioral change as an outright contradiction of his report of no internal experience of change. He stated that this made him angry, although it "didn't have anything to do with the praying" as a form of treatment.

In spite of this apparent willingness to separate the religious effects of the prayer from the perceived missteps of the healing minister, successive over-interpretations and misunderstandings appear to have undermined the therapeutic process. Ralph terminated his involvement after five sessions and a healing-of-ancestry mass. Subsequently, Father Felix met in several sessions with the father, praying ostensibly for the second son with the father as "proxy." In private, however, he admitted he was simultaneously praying for the father himself, who he felt had an overly critical and negative manner. The priest felt that the man's manner was somewhat ameliorated through healing, and the family also reported that their second son had become less wrathful and abusive. However, the father soon terminated his sessions with Father Felix as well.

In an interview two months following his termination, Ralph described interactions with a new psychiatrist, who was skeptical of the diagnosis of paranoid schizophrenia and who had successfully hypnotized Ralph into not feeling nervous on a recent date with a woman.[26] He tended to discount the apparent similarity between the peaceful feeling of being prayed over with eyes closed and being placed in light trance and was hopeful about his new course of treatment.

## Incremental Change, Inconclusive Success

Much of the literature on religious healing implies that ritual necessarily and definitively accomplishes, at least in its own terms, what it sets out to do. Far from being definitive, the effects of healing in the two cases presented here are incremental and inconclusive. Both are close to what we could call limiting cases, beyond which the relevance of any idea of efficacy becomes questionable. Even so, the case descriptions indicate that the healing experience was more satisfying for Margo than for Ralph, since he rejected the process and she wished to continue it. This contrast in behavior suggests the need for an interpretive approach sensitive to subtle but important modulations of meaning and experience in the therapeutic process. In the preceding section, I proposed that therapeutic process in ritual healing be analyzed in terms of a model composed of three elements. Following cases prospectively has allowed a refinement of this model, which we can specify as consisting of the disposition of participants, experience of the sacred, negotiation of possibilities or elaboration of alternatives, and actualization of change. We will examine the two cases in light of these four elements of therapeutic process.

## Disposition of Participants

The term "disposition" is fortuitous in that it has the dual meaning of a prevailing mood or tendency and of the act of disposing or arranging in an orderly way. In other words, under this heading we are looking not only at psychological states, such as expectancy or "faith to be healed," but at the disposition of persons within the healing process vis-à-vis social networks and symbolic resources.

While neither client was very active in the Charismatic movement, Margo was more familiar with religious healing through attendance at prayer meetings and public healing services and had no questions about basic religious belief. In addition, her mother was oriented toward Charismatic spirituality and subscribed to the leading Catholic Pentecostal magazine. In spite of strong disappointment at a healing service in which she was told that her healing had already begun, Margo's positive disposition within the process was expressed in her gratitude for having been singled out for one-on-one healing sessions with Father Felix and in her openness to his instructions. She accepted his injunctions both to be open to spontaneous images from her unconscious during prayer and to conceive of her anxiety and depression as "diminishing" from day to day. Her positive disposition was enhanced by the reassurance that an apparition at her bedside, about which she had never told her secular therapist, was not a sign of mental illness but a frequent and fully understandable manifestation of an evil force. Finally, in one session she took the initiative of asking the priest if he had spiritually "picked up" or "discerned" anything particular about her problem while he prayed over her. This anticipation of divine empowerment in fact caught Father Felix by surprise but he was able to summarize several "fears" about which he had been "led" to pray, thus reinforcing Margo's already strong disposition.

Ralph, in contrast, entered the process with ambivalence: He expressed agnosticism but acknowledged a preoccupation with religion that emerged both in his poetry and even occasionally in praying by repeating the name "Jesus." Like Margo, he had been exposed to the practices of the Charismatic Renewal, although he had not been involved for at least eight years and then only briefly. In addition, he had entered healing under the assumption that his sessions with Father Felix would consist not of prayer but of counseling. Nevertheless, his disposition during the process was favorable enough that he prayed along with the priest during the first several sessions. Yet this level of participation diminished with the final result that he discounted the healing process as cultlike.

## Experience of the Sacred

The human capacity to attend to the world as sacred, other, and powerful has been documented repeatedly by phenomenologists of religion (Eliade 1958; Van der Leeuw 1938). Each healing system attends to the human condition differently, elaborating a repertoire of ritual elements that constitute legitimate manifestations of divine power. Within a particular healing system, we are concerned with individual variation in experience of the sacred that may influence the course of therapeutic process.

Margo's experiences of concrete empowerment included periodic "resting in the Spirit" at other healing services before entering the series of sessions with Father Felix. With the priest, instruction to be open to unconscious material resulted in the spontaneous experience of three significant aspects of her problem "rushing at her from all directions." Both the motor dissociation of resting in the Spirit and the spontaneous imagery are examples of concrete embodied experience of the sacred. Ralph's experience of progressively diminishing empowerment began with a distinct experience of abandonment by the transcendent presence that had initially been evoked in the healing prayer. The significance of this event never came to Father Felix's attention during the sessions; hence, he did not have the opportunity of dealing with it in the context of Catholic Pentecostal belief and practice. In short, this experience of the sacred was not incorporated into the therapeutic process for Ralph, and the intensity of his experience of power as presence, heat, and color progressively diminished. It is also possible that the priest's attempt to attribute symbolic meaning to the emergence of green in Ralph's visual field further undermined the evocation of the sacred, since Ralph himself attributed the experience to the power of suggestion rather than to divine power.

The most striking difference between the two cases is that Margo's experiences had more sacred content that pertained immediately to her situation, as opposed to Ralph's vague sense of divine presence, heat, and color that received only minimal interpretation by the healing minister. The observer might surmise that the healer could have worked with this experience either by interpreting it as a mystical companion who could protect the young man from pathological nervousness in social situations or by using it as an experiential wedge into Ralph's agnosticism thereby facilitating greater disposition toward healing. Father Felix might also have taken the occasion to induce behavioral and attitudinal transformation through his stated priority of getting to the "root" of Ralph's problems. Instead, any potential content of Ralph's experience remained unelaborated as insight interpretation, or direction. It is unclear whether this did not occur because the healer was unaware of Ralph's experience of "presence" or because such a strategy would be unacceptable.

Margo's experiences of empowerment were substantially different, rich in biographical meaning (sudden emergence of thoughts about her job doctor, and former boyfriend). For her, the experience was a moment not of abstract but of concrete transcendence. As pointed out by Kapferer, "A ritual fixed in a transcendent moment is empowered to act on contexts external to the performance and to transform them in accordance with the rearrangement or reordering which the transcendent moment of the rite expresses" (1979b:17). Unless the concrete rhetoric within such moments is identified, the phrase "in accordance with" posits no more than an abstract homology between elements of ritual and elements of a distressed life. The concrete experience of the sacred is not an experience of the supernatural but a transformed way of attending to the human world. For Margo but not for Ralph, the link between transcendence and the reordering of life was forged in the biographical content of her transcendent moment.

### Elaboration of Alternatives / Negotiation of Possibilities

A principal task of therapeutic persuasion and healing is to create alternatives by changing the "assumptive world" (Frank 1973) of the afflicted. Different healing systems may conceive the alternatives as new pathways, as a means of becoming unstuck, or of overcoming obstacles, as a way out of trouble, or in terms of a variety of other metaphors. They may use ritual or pragmatic means and may encourage activity or passivity, but the possibilities must be perceived as real and realistic.

The first possibility elaborated for Margo concerned her attitude toward medical treatment. She was persuaded that instead of cooperating with the effects of prescribed medication through a positive attitude, she had been expecting them to fail, and so they had. An extension of this line of thinking was her mother's conjecture that the doctor's unexpected decision to try ECT may have been an effect of the healing prayer.[27]

The second possibility was elaborated through Margo's new understanding of the role of evil spirits, placing "spiritual power" alongside "illness" as conceptual tools for making sense of a frustrating life situation. This alternative was provided along with the reassurance that an unsettling apparition was not a sign of insanity but the manifestation of an evil spirit. She was later persuaded that the technique of "commanding her emotions" was not only a way to invoke divine power but also a way to instill some sense of control over emotions she experienced as uncomfortable and alien. The attribution of the technique's ineffectiveness to interference by an evil spirit not only raised the stakes to a cosmological level, but confirmed her feeling that anxiety and depression were alien to her natural state.

The story for Ralph can be summed up more briefly. There were simply no possibilities generated for him in the healing process. As with Margo, Father Felix offered methods—relaxation, developing a positive attitude, attending social events—but Ralph never perceived them as realistic.

### Actualization of Change

What counts as change, as well as the degree to which that change is seen as significant by participants, cannot be taken for granted in comparative studies of therapeutic process. This insight is all the more important for this discussion where no definitive outcome exists and where our concern is to define minimal elements of efficacy.

The principal evidence for incremental change in Margo's healing is her report of a decision to share her troubles with a younger sister-in-law.

While a reason for this decision did not explicitly emerge in follow-up interviews, it can be suggested that the healer's discourse on "Fear-of-Being-Found-Out" may have planted the idea of seeking support from other rather than attempting to hide her difficulties from them. Attributing her former behavior to a fear that is not only negative but may also represent the activity of an evil spirit is in this instance the key feature of the rhetoric of transformation. Whereas the desire to hide her distress had led to increasing social withdrawal, its linkage to the idea of an evil spirit now motivated Margo to make her distress itself the occasion for social engagement.

Failure to actualize change in Ralph's healing is evident in his explicit rejection of whatever Father Felix offered as evidence of therapeutic change. Attending mass with his grandfather was discounted because it only happened once, going to a restaurant with his grandfather was not significant because he customarily did such things without consequence anyway, and another person's opinion about whether he had changed was discounted both because he had no indication from others that this was so and especially because what mattered to him was that he felt no different. When the researcher asked whether his recent lack of trouble with uncontrolled eye movement was a possible result of healing prayer, Ralph did not reject the possibility outright but greeted it with ambivalence, precluding its classification as an experience of transformation. The healer's perception of positive change in Ralph's father doubtless had minimal effect, since father and son already had a close relationship. Similarly, the parents' report of change in his brother had minimal effect, since strained relations between brothers persisted to the point where they were unable to live in the same house.

In sum, the therapeutic process for Margo was characterized by an initially positive disposition; experiences of divine power with discrete, intelligible content; the elaboration of viable possibilities; and significant, incremental, changes. Ralph exhibited ambivalent disposition, diminishing empowerment, nonrecognition of possibilities, and rejection of change with a strong perception of being misunderstood by the healing minister. In these terms, healing was more successful for Margo than for Ralph, and the analysis thus sheds light on the different modes in which the two terminated their sessions. Ralph left the religious healing process to find apparently greater satisfaction from a psychiatrist/hypnotist, with no sense of continuity from his Charismatic healing encounter. Margo, who was initially demoralized about psychiatry and psychotherapy, left the healing process to try an additional inpatient psychiatric treatment and probably would have continued religious healing if the priest had not left the area for an extended period.

While moving in the right direction, this analysis still does not establish the significance of these transformations in comparison to what clinical thinking would call a cure. What is striking in the examples presented is their incremental character, with no guarantee that they will be permanently integrated into the person's life. The incremental and open-ended process of religious healing may prove to be an essential characteristic that requires some religious cures to be "symbiotic" (Crapanzano 1973): perhaps there is no therapeutic outcome, only therapeutic process. Catholic Pentecostal healing can include the symbiotic goal, encouraging supplicants to incorporate religious meaning and inhabit a religiously defined community. Yet in the sociocultural setting of late-twentieth-century North America, we may readily discern factors that contribute to the fragmentary and inconclusive nature of the healing process that do not pertain in the traditional societies from which the bulk of ethnographic knowledge comes.

First, consider Father Felix's attempt to draw the families of Margo and Ralph into the healing process through the healing-of-ancestry mass. If there is anything unique about Father Felix's healing practice in comparison to that of other Catholic Pentecostal healing ministers, it is his practice of entering the home and mobilizing family support through participation in this event. Most Charismatic healing is based on the model of the individual encounter, and it is not unknown for a woman to be in the healing process to the displeasure of her husband. Even when the healer takes the initiative in mobilizing social support, his authority is not such that he can intervene in the way sometimes described for traditional healers. Margo's father was pointedly absent from her ancestry mass, as was Ralph's brother from his. Ralph's father participated enthusiastically in several private sessions of his own with the priest, but he discontinued them without resolution, simply failing to make another appointment. Thus social support, often cited as one of the hallmarks of ritual healing, is by no means automatic. Support from the family and support from the community of religious believers are not identical or necessarily even compatible. Support from either may be less emphatic than might be expected from the cases commonly reported in the ethnographic literature.

Consider, in addition, the ease with which people may enter and leave the healing process in these examples. In cross-cultural perspective, this kind of mobility among healing resources seems to be a function of both the number of resources available and the exclusivity of each healing form. Finkler (1985) observed a distinction among Mexican Spiritualists between those who were devotees and those who made casual or periodic use of Spiritualist healing; Crapanzano (1973) noted a similar distinction between Hamadsha devotees who experienced a symbiotic cure and others who received a "one-shot" exorcistic cure. As Catholic Pentecostalism has developed over the past two decades, its healing forms have become more accessible to those with only a marginal exposure to the movement.[28] Like Ralph and Margo, they are less likely to become involved in a total "symbiotic cure" and will more likely experience the kinds of incremental transformations documented here. Thus, little understanding will result if research is directed toward definitive therapeutic *outcome*, rather than toward the ambiguities and partial successes (and failures) embedded in therapeutic *process*.

Furthermore, if their diagnoses are correct, Ralph suffers from a serious schizophrenic illness characteristically associated with psychotherapeutic failure, while Margo's problems of depression and panic typically respond well to a variety of psychotherapeutic interventions. Research in traditional societies is often complicated by the fact that the anthropologist does not have comparable diagnostic information; on the other hand, research in contemporary society can be complicated by the fact that the informant *does* have this information. Ralph's rejection of Father Felix's comment, "If you think nervous, you'll be nervous," was based on his conception that clinical paranoia cannot simply be banished by a change of attitude. In contrast, Margo's willingness to tell about her experience of an apparition only to the priest and not to her psychotherapist was based on her concern that she might receive a diagnosis that to her was worse than depression and panic disorder.

One might say that the religious healing encounters of both supplicants were conditioned by previous encounters with mental health professionals, in terms both of knowledge about their conditions and, especially for Margo, of insights gained from previous psychotherapy. This interpretation would represent ethnographic myopia, however. More accurate for both Margo and Ralph, religious healing was an interlude in a history of encounters with the mental health establishment. Herein lies both the clinical and anthropological significance of these cases: Anthropologically, in terms of how the interaction of both religious and clinical meanings shape the illness experience; and clinically, in terms of how the religious encounter may influence the trajectory of the illness. How did Ralph's previous hospitalization and interaction with mental health professionals affect the encounter with Father Felix, and how did the experience with Father Felix influence Ralph's subsequent encounter with the psychiatrist using hypnotherapy? Margo was seeing both a psychiatrist and a psychologist before meeting Father Felix, who suggested that she switch to a Christian psychotherapist while she continued with healing prayer. In the end she appeared committed to both psychiatric treatment and religious healing. But were these independent commitments or did, for example, religious healing influence Margo's willingness to submit to ECT?

The clinician should find this kind of information valuable, but it is not likely to be volunteered by the patient. Like Ralph and Margo, both of whom refused to permit me to contact their physicians, many of those who have recourse to religious healing undoubtedly believe they are better off not informing their physicians unless or until some dramatic change occurs for which they want medical documentation of a miraculous healing. Medical prejudice—real or perceived—against religious conviction may create a critical blind spot in the clinical picture of the large number of people who find religious healing congenial.

We should allow a final word to Father Felix, who was himself disappointed that more noticeable and quicker results had not been achieved in either case. He attributed the difficulty with Ralph both to the supplicant's resistance and to his own failure to include more of a counseling component alongside healing prayer. He also saw Margo's main problem as a negative family environment and her inability to achieve independence from it.

## Conclusion: Therapeutic Process and the Theory of Healing

The method of rhetorical analysis of therapeutic process I have proposed in this chapter treats healing as a discourse that activates and gives meaningful form to endogenous physiological and psychological healing processes in the patient. The net effect is to redirect the patient's attention to various aspects of his or her life in such a way as to create a new meaning for that life and a transformed sense of being a whole and well person. Examining disposition, experience of the sacred, elaboration of alternatives, and actualization of incremental change as elements of therapeutic/transformative process contrasts with studies that emphasize the global role of psychological mechanisms such as suggestion, catharsis, placebo effect, or regression in service of the ego (Calestro 1972; Sargant 1973; Scheff 1979; Torrey 1972). These studies tend to discourage detailed analysis of therapeutic process in the experience of individual persons, since if healing can be accounted for by a nonspecific mechanism, all that need be specified is how that mechanism is triggered. Even when more specificity is given, as in Thomas Scheff's (1979) proposal that a mechanism of distancing is essential to the mechanism of catharsis, analysis tends to discount the nature of distress and the differential effects of healing across individuals. We cannot definitively say, for example, that the technique of commanding her emotions constituted distancing for Margo, and even if we can, the effect may have been more cognitive than cathartic. A similar point applies to invocation of "altered states of consciousness" in explaining the effects of healing.[29] These states cannot be treated like mechanisms such as catharsis or suggestion. Their nature must be defined in cultural as well as psychophysiological terms, and their place within healing systems must be specified.

In staying close to the experiential data, this method also contrasts with other more globally stated conceptions of the healing process. For example, James Dow (1986b) describes the healing process as one in which symbols from the mythic realm are "particularized" in meaning for an individual supplicant. The symbols are then "manipulated" by a healer to mediate or "transact" between the hierarchical levels of society and self. In addition, emotions are "attached" to the symbols to transact between levels of self and soma. In the case of Margo it is certainly possible to label the spirit of Darkness as a transactional symbol to which the healer attaches the emotion of depression; it could just as easily be described as a quality predicated by the healer on an inchoate pronoun (Fernandez 1974), or as a management of meaning by the healer, who is acting as a spiritual broker defining the conditions of the supplicant's participation in the religious group (Kapferer 1976). In short, a model is needed that can specify conditions for change, criteria for a job well done by the healer, and cultural repertoires of significant patient experience.

While Dow states that the healer "persuades" the patient that the mythic symbols are relevant to his or her condition, he does not explain how such persuasion occurs and creates a disposition to be healed. Elements of religious experience are judged by Dow to be "therapeutic preludes," the purpose of which is to establish a therapeutic relationship based on paradox; transcendence (Kapferer 1979c) and experience of the sacred play no part. Finally, the relationship among social, self, and somatic levels is characterized as analogous to that of a "thermostat," such that "it is possible to affect processes in the self and unconscious-somatic systems through the manipulation of symbolic parameters at the social level" (Dow 1986b:63). The thermostat analogy is entirely too mechanistic. What is needed at this stage in the development of a theory of healing is specification of *how* therapeutic process effects transformation in existential states.

An approach grounded in participants' own experience and perceptions of change may arrive at a more pragmatic conceptualization of healing as a cultural process. This should be a goal not only on a conceptual, theoretical level but also on the level of interaction between medical and sacred aspects of complex health-care systems. Having chosen a type of religious healing that is formally and experientially different from psychotherapy, yet sufficiently similar for systematic comparison, I suggest the possibility of a theory of the healing process that will not only include other, more seemingly exotic forms but also permit a rethinking of healing in cosmopolitan biomedicine.

Precisely because this framework aspires to comparative usefulness, a balanced anthropological account must also take into account precise features of the cultural setting in which healing takes place. In this case, both the general fact that the supplicants who seek treatment are practicing Catholics, and the fact of their participation in the Catholic Charismatic movement are relevant. In the first instance, it would be totally erroneous to interpret the presentation of a complaint like masturbation as indicative of sexual unenlightenment; for the Catholic, only two phenomenological categories are relevant: If the action is deliberate it is sinful, and if it is compulsive it is a neurotic symptom. The importance of such cultural relativism at least as a starting point is, of course, one of the main points of ethnopsychiatric research on religious healing and culture-bound syndromes, and hence should not require too great an emphasis.[30] More important is what Von der Heydt (1970), a Catholic psychoanalytic (that is, Jungian) psychotherapist, had observed to be the main principle in work with Catholic patients: "[W]hatever their illness or external problems may be, there is always an underlying conflict connected with their religion and its practice" (1970:76). This is most clearly evident in Example One above, in which the supplicant's first sexually ambivalent experience occurred behind a church, and was followed later in life by a conflicted choice between the married state and the priesthood. It is also evident in Example Three, Part B, in which the supplicant reported making painful attempts to control his habit of masturbation particularly during periods when he was fulfilling special religious duties or functions, and perhaps in Example Two as well, if the supplicant's original resistance to ritual therapy is seen as a refusal to acknowledge a religious dimension to her obesity. Finally, Von der Heydt's insight appears to be confirmed by the prominence of guilt as an element in all three examples.[31]

Can the attribution of problems to evil spirit in Deliverance, and the reconstruction of the past in Healing of Memories, be interpreted as a kind of escape from responsibility for one's own emotional well-being? This is indeed a common attitude of clinical therapists toward religious healing, but it should not be adopted uncritically. For Catholic Charismatic healing, the issue of supplicant participation versus passivity was addressed earlier in this chapter, in noting how in the Healing of Memories the person is encouraged to forgive those who have injured him or her, and how in Deliverance an individual often personally identifies the spirits that are troubling him or her. Personal responsibility is indicated in that the supplicant is subject to counseling that may include recommendations for behavioral change. In addition, the belief that experiences and actions in the past provide the occasion for demons to initiate an influence on the person implies a degree of responsibility in that some of those past actions may be regarded as sinful.

To the extent that the individual *is* relieved of some degree of responsibility, the rhetorical effect may be quite different than that suggested by the notion of escapism. First, in the recognition that the evil represented by the demons is not essential to his own being, the supplicant's sense of his own basic goodness and self-worth is affirmed. Second, he is made to be aware that he is not alone in his suffering, but that his struggle takes place in a larger arena of spiritual forces, namely, the "spiritual warfare" that continues on a cosmic scale between the powers of Light and Darkness. Third, in coming to participate in the phenomenological world created by religious discourse, the supplicant's mode of attention to his experience as a *person* may be altered from the "egocentric" style of Western society to the "sociocentric" style (Shweder and Bourne 1982), in which the individual is subsumed in the religious community, hence by definition relinquishing some degree of responsibility for his emotional well-being to that community. These three points highlight the necessity of accounting for the holy as well as for disease as a substrate of illness and healing. Finally, on the one hand one must weigh the issues of responsibility against Young's (1976) assertion that "exculpation," or transfer of accountability for a person's behavior onto an agency beyond his will, is a fundamental definitional component of any sickness in contrast to other kinds of deviant behavior. In this light *discerning the presence of a demon* must be interpreted in the same category as *diagnosing the presence of a virus.* On the other hand, one must weigh the issue in comparative light against therapies such as the Japanese *Naikan* (Murase 1982), in which feelings of guilt and shame are maximized as part of the therapeutic process. While such a technique may be successful in the Japanese cultural context, it might be proven to be counterproductive in a Western setting.

Two seeming paradoxes in determining the effectiveness of ritual healing lead us back to the questions of experience of the holy as a substrate of illness and healing, and the constitution of clinical and sacred realities through discourse and rhetoric. The first is contained in the conclusion that indigenous and religious therapies must invariably succeed due to the kinds of illness treated and the relative autonomy of these illnesses from biomedically defined disease (Kleinman and Sung 1979), and in the sense that such therapies always meet culturally established expectations and produce predictable results (Young 1976). What seems paradoxical from a scientific, medical point of view is that there can be treatments that in some sense cannot fail, and yet leave so much curative work undone. Catholic Charismatics explicitly say that when a healing is asked for it is invariably granted by the all-benevolent God, *even though* it might not be the one specifically requested. McGuire's (1982) Catholic Charismatic informants reported not only that a supplicant may receive a different kind of healing than requested, but that the healing may occur to someone other than the intended beneficiary of the prayer.[32] The second paradox is that of how criteria of effectiveness can vary so greatly that by one set of standards a patient is declared healed, while by another he remains ill. Whereas for Western medicine relief of symptoms is an important criterion of cure, Kleinman and Sung (1979) report that in Taiwan, even with relief of symptoms, a shaman may judge a supplicant ill until an evil spirit is banished, and a Chinese doctor may make the same judgment if harmony is not restored between yin/yang, hot/cold, or the five body spheres. Likewise, McGuire (1982) observes that Catholic Charismatics may not be regarded as healed if they retain any residue of sin; she goes as far as to suggest a significant overlap between the qualitative continua of health/illness and holiness/sinfulness. Again, in what sense can phenomena constituted in this way be considered to belong to the same universe of discourse as that of clinical medicine?

These paradoxes pertain to the crucial problem of substrates of illness, be they disease in the biomedical sense, disharmony in the traditional Chinese sense, or experience of the holy in the religious sense. The sense of paradox is heightened when analysis confronts methods of healing for which it is impossible to specify not only conditions of possible failure, but even the illness problem being treated and the person benefiting from the therapy. Indeed, in what sense can such practices even be considered to belong to the same category of phenomena as other medical treatments? Does their failure to fail trivialize their medical significance or place the nature of their effectiveness in the residual pigeonhole of placebo effects? Does their extreme indeterminacy in medical terms place them in the domain of some other discipline than medical anthropology, or make them examples of pathology rather than therapy, or simply dump them in the trash heap of the irrational?

A renewed examination of religious modes of healing with the intent of generating interpretations that account equally for their identity as religious and medical phenomena offers, if not a way to resolve these paradoxes, at least a way to confront them—and this is one of the themes I shall return to explicitly in Chapter Four and Chapter Eight. The premise of such an examination is that *disease* and the *holy* are categories on the same phenomenological level, pertaining to ultimate issues of life and death, activating endogenous processes such as those encountered here, and generating fields of interpretive discourse, the intersection of which is discourse about illness. As we have shown for Catholic Charismatic healing, therapeutic rituals can at the same time be viewed as forms of ethnopsychiatry and as rites of passage. Thus it will not do to say that a form of therapy is religious in an "emic" sense but medical in an "etic" sense, as seems to have become the case as anthropological interest in health-related phenomena has developed an identity as the applied or clinical subdiscipline of medical anthropology. We must recognize instead that scientific accounts can be formulated of religious healing *qua* religious as well as in clinical terms. The locus of our paradoxes may then shift from the forms of healing themselves to the methodological inability of comparative religion and medicine to generate mutually intelligible accounts of the same phenomena. Yet if the experiences of disease and of the holy raise some of the same existential questions (confer Comaroff 1982:51–52; Young 1976), and hence are not entirely distinct, there may well be a religious dimension in all forms of healing. In that case, a hermeneutic approach is indicated not only for analysis of overtly religious and folk therapies, but for conventional biomedical healing as well.

None of this should be taken to mean that anthropologists should abandon clinical and applied concerns and take up the interpretation of symbols for its own sake, or subordinate such concerns entirely to the theoretical interests of cultural anthropology broadly conceived. It does mean that medical anthropology and ethnopsychiatry pose fundamental problems of meaning upon which must be brought to bear theoretical as well as applied perspectives, comparative religion as well as medicine, and methods of understanding and interpretation as well as those of explanation and experiment.[33] This is particularly the case given the powerful moral imperative present in the constitution of subjectivity and experience in these domains (Kleinman 1997, 1999). Moreover, the explicit recognition that healing processes possess a kind of creativity (Comaroff 1976; Kleinman 1980) demands that we use methods such as rhetorical analysis that can give an account of that creativity both with respect to therapeutic effect and in relation to other forms of creativity. The study of religious healing is not immune to the methodological implications of a critique of empiricism, such as that made by Byron Good and Mary-Jo Good, who challenge the definition of "meaning as a relationship between language and a reality that lies outside of language," and posit instead that the "meaning of medical discourse is constituted in relationship to socially constructed illness realities" (Good and Good 1982:146; Good 1994).[34] The latter alternative requires a non-empiricist concept of meaning, in which meaning is not *attached* to experience, but is constituted by the way in which a subject *attends* to experience. And experience, in this case—the experience of transformation—is nothing more nor less than the meaningfulness of meaning.

CHAPTER TWO

# Embodiment as a Paradigm for Anthropology

The purpose of this chapter is not to argue that the human body is an important object of anthropological study, but that a paradigm of embodiment can be elaborated for the study of culture and self. By paradigm I mean simply a consistent methodological perspective that encourages reanalysis of existing data and suggests new questions for empirical research. Although I shall argue that a paradigm of embodiment transcends different methodologies, I will not attempt to synthesize the broad multidisciplinary literature on the body.[1] The approach I will develop from the perspective of psychological anthropology leans strongly in the direction of phenomenology. This approach to embodiment begins from the methodological postulate that the body is not an *object* to be studied in relation to culture, but is to be considered as the *subject* of culture, or in other words as the existential ground of culture.[2]

The work of Irving Hallowell is a useful point of departure, since his denomination of the "self as culturally constituted" marked a methodological shift away from concern with personality structure, and remains current in anthropological thought. In his most influential article, Hallowell (1955) articulated two principal concerns, which I will term perception and practice. Perception is a key element in Hallowell's definition of the self as self-awareness, the recognition of oneself as an "object in a world of objects." He saw self-awareness as both necessary to the functioning of society and as a generic aspect of human personality structure. He referred to his outline of a method to study the self as phenomenological "for want of a better term," but I would argue that what he lacked was a more adequately worked out phenomenology. Nevertheless, in directly addressing the problem of perception, Hallowell prefigured an anthropological critique of the distinction between subject and object.

However, although he explicitly recognized the self as a self-objectification and the product of a reflexive mood, Hallowell cast his analysis at the level of the already objectified self. A fully phenomenological account would

EMBODIMENT AS A PARADIGM FOR ANTHROPOLOGY

recognize that while we are capable of becoming objects to ourselves, in daily life this seldom occurs. Such an account would take the decisive step of beginning with the preobjective and prereflective experience of the body, showing that the process of self-objectification is already cultural prior to the analytic distinction between subject and object. Hallowell went only as far as the conventional anthropological concept that the self is constituted in the ontogenetic process of socialization, without taking full cognizance of the constant reconstitution of the self, including the possibilities not only for creative change in some societies, but for varying degrees of self-objectification cross-culturally.

Hallowell's second concern is summarized in the term "behavioral environment," borrowed from the gestalt psychology of Koffka. The proto-phenomenological approach to perception that we have identified accounts for an essential feature of the behavioral environment, namely that it includes not only natural objects but "culturally reified objects," especially supernatural beings and the practices associated with them. The concept thus did more than place the individual in culture, linking behavior to the objective world, but also linked perceptual processes with social constraints and cultural meanings. Thus, the focus of Hallowell's formulation was "orientation" with respect to self, objects, space and time, motivation, and norms. It is in this sense that the term "practice" is relevant to describe Hallowell's concern. If, as Sherry Ortner (1984) has argued, the anthropological conceptualization of practice occurred at a certain theoretical moment, then the concept of behavioral environment is a terminological composite that stands for the context in which practice is carried out, and hence counts as a theoretical stepping-stone between behavior and practice. This is of particular relevance to the present argument because, as we shall see, a theory of practice can best be grounded in the socially informed body.

There are other ways to frame the need for a paradigm of embodiment, of which I shall mention only one. Mauss (1950), in his fragmentary but influential discussion of the person, suggested that all humans have a sense of spiritual and corporal individuality. At the same time, he argued that particular social conditions were associated with qualitative differences among the totemistic personage, the classical persona, and the Christian person.[3] It is of empirical concern to my argument that he saw the development of the individualistic person played out in the arena of sectarian movements of the seventeenth and eighteenth centuries, since the data I analyze below come from their twentieth-century equivalent. It is of methodological concern that he saw the person as associated with the distinction between the world of thought and the material world as promulgated by Descartes and Spinoza, since the paradigm of embodiment has as a principal characteristic the collapse of dualities between mind and body, subject and object. In this light it is of relevance that Mauss himself had already reproduced this duality by elaborating his concept of *la notion du personne* quite independently from that of *les techniques du corps* (1950). Here again we find the themes of perception and practice as domains of the culturally constituted self; but writing nearly two decades before Hallowell, Mauss could not yet treat them together, still less within a consistent paradigm of embodiment.

My plan for outlining such a paradigm begins with a critical examination of two theories of embodiment: Maurice Merleau-Ponty (1962), who elaborates embodiment in the problematic of *perception,* and Pierre Bourdieu (1977, 1984), who situates embodiment in an anthropological discourse of *practice.* My exposition will be hermeneutic in the specific sense of cycling through presentation of methodological concepts and demonstrations of how thinking in terms of embodiment has influenced my own research on healing and ritual language in a contemporary Christian religious movement. I first examine two religious healing services, interpreting multisensory imagery as an embodied cultural process. Then I examine the practice of speaking in tongues or glossolalia as embodied experience within a ritual system and as a cultural operator in the social trajectory of the religious movement. Finally, I return to a general discussion of the implications of embodiment as a methodological paradigm.

## Methodological Orientation to Embodiment

The problematic of both Merleau-Ponty and Bourdieu is formulated in terms of troublesome dualities. For Merleau-Ponty in the domain of perception the principal duality is that of subject–object, while for Bourdieu in the domain of practice it is structure–practice. Both attempt not to mediate but to collapse these dualities, and embodiment is the methodological principle invoked by both. The collapsing of dualities in embodiment requires that the body as a methodological figure must itself be nondualistic, that is, not distinct from or in interaction with an opposed principle of mind. Thus, for Merleau-Ponty the body is a "setting in relation to the world," and consciousness is the body projecting itself into the world; for Bourdieu the socially informed body is the "principle generating and unifying all practices," and consciousness is a form of strategic calculation fused with a system of objective potentialities. I shall briefly elaborate these views as summarized in Merleau-Ponty's concept of the *preobjective* and Bourdieu's concept of the habitus.

## The Perceptual Constitution of Cultural Objects

Merleau-Ponty lays out his position as a critique of empiricism.[4] He examines the constancy hypothesis, which asserts that since perception originates in external stimuli that are registered by our sensory apparatus, there is a "point by point correspondence and constant connection between the stimulus and elementary perception" (1962:7). This is not experientially true, he argues—far from being constant, perception is by nature indeterminate. There is always more than meets the eye, and perception can never outrun itself or exhaust the possibilities of what it perceives. When we make a special effort to see two apparently unequal lines in an optical illusion as really equal, or to see that the triangle is really three lines related by certain geometric properties, we are making an abstraction, not discovering what we really perceive and later name as a triangle or illusion. What we "really" perceive is, in the first case, that one line is longer than the other, and in the second, the triangle. To start from the objective point of view (the triangle as geometric object and the lines of objectively parallel length) and analytically work backward to the perceiving subject does not accurately capture perception as a constituting process.[5]

Merleau-Ponty thus wants our starting point to be the experience of perceiving in all its richness and indeterminacy, because in fact we do not have any objects prior to perception. To the contrary, "Our perception ends in objects ...," which is to say that objects are a secondary product of reflective thinking; on the level of perception we have no objects, we are simply in the world. Merleau-Ponty then wants to ask where perception begins (if it ends in objects), and the answer is, in the body. He wants to step backward from the objective and start with the body in the world. This should also be possible for the study of the self conceived in Hallowell's terms, as an object among other objects.

Since the subject–object distinction is a product of analysis, and objects themselves are end results of perception rather than being given empirically to perception, a concept is necessary to allow us to study the embodied process of perception from beginning to end instead of in reverse. For this purpose Merleau-Ponty offers the concept of the *preobjective*. His project is to "coincide with the act of perception and break with the critical attitude" (1962:238–239) that mistakenly begins with objects. Phenomenology is a descriptive science of existential beginnings, not of already constituted cultural products. If our perception "ends in objects," the goal of a phenomenological anthropology of perception is to capture that moment of transcendence in which perception begins, and, in the midst of arbitrariness and indeterminacy, constitutes and is constituted by culture.

It may be objected that a concept of the preobjective implies that embodied existence is outside or prior to culture. This objection would miss what Merleau-Ponty means by the body as "a certain setting in relation to the world" (1962:303) or a "general power of inhabiting all the environments which the world contains" (1962:311). In fact, the body is in the world from the beginning:

> ... consciousness projects itself into a physical world and has a body, as it projects itself into a cultural world and has its habits: because it cannot be consciousness without playing upon significances given either in the absolute past of nature or in its own personal past, and because any form of lived experience tends toward a certain generality whether that of our habits or that of our bodily functions. (1962:137)

> It is as false to place ourselves in society as an object among other objects, as it is to place society within ourselves as an object of thought, and in both cases the mistake lies in treating the social as an

object. We must return to the social with which we are in contact by
the mere fact of existing, and which we carry about inseparably with
us before any objectification. (1962:362)

By beginning with the preobjective, then, we are not positing a precultural, but a preabstract. The concept offers to cultural analysis the open-
ended human process of taking up and inhabiting the cultural world, in
which our existence transcends but remains grounded in de facto situations.
    Merleau-Ponty gives us the example of a boulder, which is already there
to be encountered, but is not perceived as an obstacle until it is there to be
*surmounted.*[6] Constitution of the cultural object is thus dependent on intentionality (what would make one want to surmount the boulder?), but also
upon the givenness of our upright posture (Straus 1966), which makes
clambering over the boulder a particular way of negotiating it (an option
even if one could walk around it). The anthropological anecdote told by
David Schneider, of the umpire who declares that pitches are neither balls
nor strikes until he calls them such,[7] tells us about an act of bestowing cultural meaning, but it presupposes something about the cultural fact that the
pitches are already there *to be called.* It presupposes objectification of a particular space of the body between the knees and shoulders (the strike zone)
in conjunction with a particular way of moving the arms from the shoulders (swinging the bat). It is the process of this objectification to which
Merleau-Ponty calls our attention.

## Habitus and the Socially Informed Body

Bourdieu's methodological goal for the theory of practice is to delineate a
third order of knowledge beyond both phenomenology[8] and a science of
the objective conditions of possibility of social life. Parallel to Merleau-
Ponty's goal of moving the study of perception from objects to the process
of objectification, Bourdieu's goal is to move beyond analysis of the social
fact as *opus operatum,* to analysis of the *modus operandi* of social life. His strategy is to collapse the dualities of body-mind and sign-significance in the
concept of *habitus.* This concept was introduced by Mauss in his seminal
essay on body techniques, to refer to the sum total of culturally patterned
uses of the body in a society. For Mauss it was a means to organize an otherwise miscellaneous domain of culturally patterned behavior, and received
only a paragraph of elaboration. Even so, Mauss anticipated how a paradigm
of embodiment might mediate fundamental dualities (mind–body, sign–
significance, existence–being)[9] in his statement that the body is simultaneously both the original object upon which the work of culture is carried out,
and the original tool with which that work is achieved (Mauss 1950:372).
It is at once an object of technique, a technical means, and the subjective
origin of technique.
    Bourdieu goes beyond this conception of habitus as a collection of practices,
defining it as a system of perduring dispositions which is the unconscious,

collectively inculcated principle for the generation and structuring of practices and representations (1977:72). This definition holds promise because
it focuses on the psychologically internalized content of the behavioral
environment. For our purposes, it is important that the habitus does not
generate practices unsystematically or at random, because there is a:

> ... Principle generating and unifying all practices, the system of insep-
> arably cognitive and evaluative structures which organizes the vision
> of the world in accordance with the objective structures of a deter-
> minate state of the social world: this principle is nothing other than
> the *socially informed body,* with its tastes and distastes, its compulsions
> and repulsions, with, in a word, all its *senses,* that is to say, not only the
> traditional five senses—which never escape the structuring action of
> social determinisms—but also the sense of necessity and the sense of
> duty, the sense of direction and the sense of reality, the sense of bal-
> ance and the sense of beauty, common sense and the sense of the
> sacred, tactical sense and the sense of responsibility, business sense and
> the sense of propriety, the sense of humor and the sense of absurdity,
> moral sense and the sense of practicality, and so on. (1977:124, empha-
> sis in original)

Bourdieu maintains this groundedness in the body even in discussion of the
"sense of taste" as the cultural operator in his social analysis of aesthetics,
insisting that it is "inseparable from taste in the sense of the capacity to discern the flavors of foods which implies a preference for some of them"
(1984:99).

The locus of Bourdieu's habitus is the conjunction between the objective conditions of life and the totality of aspirations and practices completely compatible with those conditions. Objective conditions do not
cause practices, and neither do practices determine objective conditions:

> The habitus is the universalizing mediation which causes an individ-
> ual agent's practices, without either explicit reason or signifying
> intent, to be none the less "sensible" and "reasonable." That part of
> practices which remains obscure in the eyes of their own producers is
> the aspect by which they are objectively adjusted to other practices
> and to the structures of which the principle of their production is
> itself a product. (1977:79)

In other words, as a universalizing mediation, the habitus has a dual function. In its relation to objective structures it is the principle of generation
of practices (Bourdieu 1977:77), while in its relation to a total repertoire of
social practices, it is their unifying principle (1977:83).[10] With this concept,
Bourdieu offers a social analysis of practice as necessity made into a virtue,
and his image of human activity is Leibniz's magnetic needle that appears
actually to enjoy turning northward (Bourdieu 1977:77; 1984:175).

In this section I have shown that the paradigm of embodiment is super-ordinate to the different empirical interests and divergent methodological propensities of two influential theorists. Thus we have the apparent paradox of positions compatible within the paradigm of embodiment, but articulated in the methodologically incompatible discourses of phenomenology and what we might call "dialectical structuralism."[11] It is natural, however, that contradictions emerge between incipient attempts to forge a paradigm. In the remainder of this chapter I will elaborate a nondualistic paradigm of embodiment for the study of culture. The concepts of the preobjective and habitus will guide analysis in the empirical domain of religious experience and practice.

## Embodied Imagery in Ritual Healing

The healing practices I describe are those of Charismatic Christianity as practiced in contemporary North America. This religion is essentially a form of Pentecostalism, which since the late 1950s has introduced a complex of practices, including faith-healing and speaking in tongues, into established Christian denominations such as Methodist, Episcopalian, and Roman Catholic. Historically, the movement can be said to have originated in the post-World War II search for stability, to have accelerated and acquired a more youthful following during the social turmoil of the 1960s, to have reached a peak of apocalyptic fervor and popular appeal in the 1970s, and to have settled into a socially conservative but theologically enthusiastic niche in the religious ecology of 1980s North America. Its most visible manifestations are in "televangelists" ranging from the Bakkers of the PTL (Praise the Lord) Club to the Roman Catholic Mother Angelica. Less well known are healing services conducted by "spiritually gifted" laypeople or clerics, or networks of intentional communities, non-denominational congregations, and small denominational-based prayer groups. Participants range from lower-middle to professional classes, and except for the slightly younger membership of charismatic intentional communities, the modal age distribution is in the fifties. The data I present in this section include two examples of multisensory imagery[12] in group healing sessions conducted by well-known charismatic evangelists, and one from a private healing session conducted within a Charismatic covenant community.

### *Demons and Self-Objectification*

The first healing session is one led by the Reverend Derek Prince, a leading figure in the practice of Deliverance, or casting out of evil spirits. Reverend Prince typically prays by naming evil spirits of different types, which he then commands to depart their hosts. As the spirits are expelled from those present in the assembly, they produce a physical manifestation as a sign of their departure. Let us first turn our attention to the nature of evil spirits in contemporary Charismatic Christianity, both for ethnographic background, and because the way they are constituted as cultural objects illustrates the importance of Merleau-Ponty's concept of the preobjective.

If we ignore the methodological implications of the dictum that "Our perception ends in objects," we begin with the already constituted object, the Christian evil spirit. It can be described as an intelligent nonmaterial being that is irredeemably evil, is under the domination of Satan, and whose proper abode is Hell. Evil spirits interact with humans by harassing, oppressing, or possessing them. Given this cultural definition, one might hope to reconstruct a demonology similar to the abstract, speculative demonologies of the Middle Ages, and to discover a discourse of interiority/exteriority in which demons transgress body boundaries and are expelled. Indeed, references to spirits being "cast out," and the cultural definition of physical manifestations as "signs" of the spirits "coming out" support the experiential salience of interiority/exteriority, although it may be descriptively as illuminating to say they are being "acted out." These are all late moments in the process of cultural objectification, however. Persons do not perceive a demon inside themselves, they sense a particular thought, behavior, or emotion as outside their control. It is the healer, specialist in cultural objectification, who typically "discerns" whether a supplicant's problem is of demonic origin, and who when faced with a person self-diagnosed as "possessed" is likely instead to attribute that person's presentation to "emotional problems."

To illustrate this demonology in practice, I quote from an edited account by an informant who participated in a healing service conducted by Reverend Prince:

> And as some of [the demons] would come out [from their human hosts], some would come out with a roar. Some would come out with a belch. Some would come out with terrific coughing or choking or twisting of the neck back and forth. There were all kinds of weird and horrible things.... Quite a number of them come out with vomiting. Since there are over 150 kinds of spirits that have been identified, ... maybe 20 of those will come out with vomiting. Ten of them will come with hissing. Two of them will come out with writhing on the floor like a snake. Five of them will come out with rolling of the eyes up to the top of the head. Every spirit of witchcraft ... comes out there with a noise sounding very much like a shriek of a hyena. And it didn't matter whether it was men or women, young or old, whatever.... They all came out with the identical thing.
>
> I'll tell you the story of what happened to me.... He dealt with whole groups [of spirits]. And he got to the group of sexual aberrations. Somewhere along the line, he dealt with the spirit of masturbation.... [He said] "You've known this was a sin, but you did it. You did it deliberately. If you acquired a spirit, now it becomes compulsive and you FEEL that compulsion. If you're Catholic or Lutheran or

Episcopal you may have confessed this sin time again, time again. And you fight it and you don't like it and you hate it and you renounce it and it's still with you. Those are all signs, that whole package. You almost certainly have a spirit. Any of you who have that particular package and think you'd like to be released of the spirit, stand up."

So in that case, I stood up. And there were about 15 or 20 other people. I bet there were a lot more should have, but [chuckles] anyway, there were probably 15 or 20 of us who stood up .... He said, "You foul spirit of masturbation, I'm taking control of you in the name of Jesus and by the power of His precious blood, I cast you out in His holy name." And everyone, their hands went way back. We were standing up. He had asked the group to stand and we went through a prayer of Renunciation and Repentance. So I was standing and quite without thinking of anything, I had no idea what was going to happen. The hands went up like this, the arms this high, and the hands went further than I can do it myself, way back. It didn't hurt. And there was sort of an electric feeling, like a mild electric shock.

Well, he didn't tell us ahead of time what was expected, but that's what happened. Everybody did the same thing. Now I don't know what they felt. But I know what I felt. Something was happening here. And then at a certain point, it all went away and my hands dropped.

The important distinction for our discussion is between demons as cultural objects, and their experiential manifestations as concrete self-objectifications in religious participants. As cultural objects, demons are, in the words of Irving Hallowell, "no more fictitious, in a psychological sense, than is the concept of the self. Consequently, [as] culturally reified objects in the behavioral environment [they] may have functions that can be shown to be directly related to the needs, motivations, and goals of the self" (1955:87). The role of demons in the behavioral environment of Charismatic Christians is two-fold. As a system of representations, the demonology—which this informant estimates to have 150 entries—is a mirror image of the culturally ideal self, representing the range of its negative attributes. In terms of behavioral pragmatics, they are intelligent beings that can be encountered in everyday life and can affect one's thought and behavior.

It is against this cultural background that the manifestations described above can be understood as examples of an embodied process of self-objectification. The preobjective element of this process rests in the fact that participants, like the informant quoted, experience these manifestations as spontaneous and without preordained content. The manifestations are original acts of communication that nevertheless take a limited number of common forms because they emerge from a shared habitus. This character of the preobjective is summarized by Merleau-Ponty:

Anterior to conventional means of expression, which reveal my thoughts to others only because already, for both myself and them, meanings are provided for each sign, and which in this sense do not give rise to genuine communication at all, we must ... recognize a primary process of signification in which the thing expressed does not exist apart from the expression, and in which the signs themselves induce their significance externally .... This incarnate significance is the central phenomenon of which body and mind, sign and significance are abstract moments. (1962:166)

I would suggest that the "thing expressed" that "does not exist apart from the expression" in this case is *not* the cultural object, the evil spirit, for the discourse of spirits is an example of what Merleau-Ponty means by a "conventional means of expression." What is expressed is the transgression or surpassing of a tolerance threshold defined by intensity, generalization, duration, or frequency of distress. There is too much of a particular thought, behavior, or emotion. Self-awareness of this transgression may have already occurred, and self-objectification may have taken place by adopting the conventional demonic idiom. However, the expressive moment that constitutes this form of self-objectification as *healing* is the embodied image that accompanies the casting out of the spirit. This image has a multiple signification: "I have no control over this—it has control over me—I am being released."

This interpretation challenges the common ethnographic description of evil spirits in the language of interiority/exteriority, as transgressors of body boundaries. In Charismatic Christian healing, the language of control/release appears to have as much or greater experiential immediacy. The healer stresses "release" from bondage to the evil spirit over "expulsion" of the spirit that invades and occupies the person. Why this should be is understandable when we are reminded again that the preobjective is not precultural. Control (of one's feelings, actions, thoughts, life course, health, occupation, relationships) is a pervasive theme in the North American cultural context of this healing system. Crawford (1984), for example, offers an ideological analysis of "health" as a symbol that condenses metaphors of self-control and release from pressures. A substantial degree of cultural consistency is evident with the formulation in the Charismatic healing system of problems as loss of control to demonic influence, healing as release from bondage to that influence, and health as surrender to the will of God, whose strength helps restore self-control.

A brief methodological aside is in order to emphasize that analysis in a paradigm of embodiment does not immediately grasp onto transgression of body boundaries as the description of demonic action. Such a description would count as objectivist in the sense that it assumes the demon as already objectified, already a conventional means of expression. Bringing to the fore the rather Foucauldian metaphor of bondage points to the concretely embodied preobjective state of the afflicted rather than to the conventionally expressed invasive action of the demonic object. The metaphor of bondage simultaneously invokes a material/corporal as well as a psychological/spiritual condition addressed by healing.

The analysis of control and release helps us as well to understand certain features of experiential indeterminacy in dealings with evil spirits. There are two loci at which the preobjective perception of demons as emotion, thought, or behavior are indeterminate in practice. First is the threshold of control at which an emotion such as anger becomes the evil spirit of Anger, and the subsequent determination of the degree of purchase that spirit has on a person's life—in order of severity from harassment to oppression to possession. While the degrees of control are thus "objectively" categorized, there are no objective criteria for their determination in practice, since practice operates at the level of preobjective intersubjectivity (empathy and intuition); healers do not "diagnose" but "discern."

Second is the threshold of generalization, where the sufferer's malaise is expressed in multiple characteristics portrayed as clusters of related spirits. Again, although it is established in healing practice that spirit clusters are hierarchically organized around a dominant "manager" or "root" spirit and that certain spirits tend to appear together, in healing with a single person any number of spirits may emerge. In principle, the identification of spirits can be an open-ended excursion through the entire domain of possible spirit names. Again, this domain is culturally predetermined, and both the spontaneous discovery of a series of typically related spirits, and their experiential salience to the supplicant, can be understood in terms of the way dispositions are orchestrated within the habitus.

This orchestration is also the basis for the apparently spontaneous coordination of kinesthetic images culturally defined as manifestations of discrete types of evil spirits in the session narrated above: vomiting, writhing on the floor, hissing, rolling the eyes to the top of the head. Given the ethnographic fact that evil spirits departing a person typically produce a physical manifestation as a sign of their departure, which we can account for in purely cultural or conventional terms, how do we account for the regularized association of particular spirits with particular signs?

Two instances are narrated with enough detail for comment. That the spirit of witchcraft departs with the shriek of a hyena must be understood with respect to the cultural definition of witchcraft as an occult practice connected to Satan, and hence profoundly evil. The bloodcurdling scream is a deeply ingrained somatic component of the experience and symbolism of evil in North America—hence the apparently natural connection between the scream and the spirit. In a group setting such as described by the informant, it makes little difference whether the spirit is first identified and then emerges with a scream, or whether the scream emerges and is subsequently identified as the sign of the spirit; in either case it exemplifies the "arbitrary necessity" (Bourdieu 1977) of evil in the Charismatic Christian habitus.

The narrator's experience with the spirit of masturbation also lends itself to such an interpretation. We begin with the cultural definition of masturbation as a strongly proscribed but compulsive (hence demonic) behavior. The spontaneous collective gesture of arms flung in the air can be understood as a powerful "hands off!" emphasized by strong backward flexion of the hands. That this flexion "does not hurt," although it is farther than one could accomplish "naturally," is consistent with the concept of release from bondage as opposed to punishment for sin. Likewise, the mild electric feeling is understood not as a punitive shock but as an embodiment of spiritual power. Not at issue here is whether most of the men were responding to the cue of one or two others, since the impression of collective spontaneity indicates the immediate, intuitive grasp of the gesture's implicit meaning by them all.

### Image, Emotion, and Bodily Synthesis

The second healing event, described from my own observation, was conducted in the context of a Roman Catholic Charismatic intentional community. The session was led by two visiting Catholic healing evangelists. These healers had recently adopted the currently popular style of the Reverend John Wimber, who, in contrast to the Reverend Prince's emphasis on evil spirits, evokes a diversity of "signs and wonders" in what he refers to as "power evangelism." The signs and wonders are understood as manifestations of divine power intended to prompt the conversion of unbelievers and increase the faith of believers. In addition to the faith healing of physical, emotional, and demonic illness, they include a variety of multisensory images, emotions, and somatic manifestations that indicate the flow of divine power within and among participants. Common elements of the repertoire are rapid fluttering or vibrating of hands and arms, and somatic sensations such as lightness or heaviness, power or love flowing through the body, heat, and tingling. Spontaneous laughter or tears may spread contagiously in waves through the congregation. Many participants "rest in the Spirit," an experience of motor dissociation in which a person is overcome by the power of the Holy Spirit and falls in a semi-swoon, typically experienced as a relaxing and rejuvenating moment in the presence of God. Also common is the "word of knowledge," a form of revelation understood as a divine gift of knowledge about persons or situations not acquired through any channel of human communication, but experienced as a spontaneous thought or image.

The event I observed was a two-day healing conference, to which the leaders brought their own team of experienced healing ministers. The conference consisted of alternating periods of collective prayer, religious song, healing prayer, and lectures. It was stressed that healing and salvation are "almost synonymous," and that the participants should expect healing to occur throughout the sessions, not only during the discrete moments when the healing ministers were praying over them with the laying on of hands. The leaders stated that there was a difference between a gathering for purposes of worship and one for the experience of divine power. "Lots of things will be happening," they said, and the participants should "get their spiritual antennae up" to receive the power. During the proceedings one of them prayed aloud, "More power, Father; release more power."[13]

In the first phase of prayer, the leaders received inspiration through the word of knowledge that God wanted to heal people with back, respiratory, arthritis, and cartilage or tendinitis problems. Such people were asked to come forward for laying on of hands and prayer by the experienced prayer team. In the next phase, all were invited to participate, alternating roles as healing ministers and persons prayed for. The leaders stated that certain among the audience were experiencing a heaviness in their chest and head, a feeling of heat in their faces or lips, or a tingling in their hands. Such people were asked to hold their hands out palms upward in a prayer posture to identify themselves, and those surrounding them were told to lay hands on them in prayer to strengthen the manifestation of divine power and spread the power among themselves. Participants were invited to experience the word of knowledge themselves, and were paired off to pray with whoever responded to the problem they identified.

In contrast to the previously narrated event with Reverend Prince, the multisensory imagery in this instance is a manifestation not of release from evil, but of its cultural inverse, incorporation of divine power. The group leaders' enumeration of the physical accompaniments of divine power that some participants would experience (heaviness, heat, tingling) recapitulates a repertoire acquired from their own experience and from reports of participants in similar events. These somatic images are here being inculcated as *techniques du corps* that will embody dispositions characteristic of the religious milieu. Laughing, crying, and falling can also be objectified as sacred if their spontaneous occurrence is thematized as out of the ordinary, the "otherness," which according to Eliade (1958) is the formal criterion of the sacred.

On the other hand, the leaders' inspired enumeration of predominantly physical ailments is formally similar to Reverend Prince's identification of evil spirits in the psychological domain of negative emotions, thoughts, and behaviors. In a group of two hundred, inspirations that single out culturally common illnesses, or illnesses of particular organ systems, are statistically likely to obtain a response. This is reinforced when the culturally shared knowledge of the body and its ailments is exploited by inviting participants to experience similar inspirations, such that the technique operates communally, rather than unidirectionally from leaders to participants. That this knowledge is not purely conceptual is testified to by the presentation of these revelations in a variety of sensory modes: Participants do not merely draw on a cognitive list of diseases, but are just as likely to visually image a part of the body, or experience pain in their own body. Neither is there a cognitive act of "scanning," either of a list of diseases or of body parts, for the one that "feels right." Inspirations emerge spontaneously, insofar as participants have immediate access to bodily knowledge inculcated as culturally shared dispositions.

That it is a structured form of knowledge, however, is affirmed by the possibility for a poorly formed inspiration to misfire. During the session each participant who had an inspiration was to be approached by the person or persons who recognized their own problem, and they would pray together for that problem. The problems enumerated were specific and localized, specific enough to seem special, but not so specific as to be improbable: the area from the left knee to the lower thigh, left earache, right ear drainage, right deafness, severe lower back pain, alignment problems with ankles, vision (especially right eye), lump near right part of throat, arthritis, left tendon pull, bad hemorrhoids, pregnancy prevented by twisted ovaries, loss of hair due to scalp eczema, grief over lost child, hernia, smoking, chronic stomach acidity, need for counseling. All who articulated a problem appeared to receive responses from the audience except one, a somewhat obese woman with the appearance of not being well adjusted to the collective proceedings. She said that someone was suffering from pain in the right lung, and her overspecific inspiration fell flat. Finally, a young woman approached her for prayer, admitting to me later that she had no "pain in the lung." Instead, she was motivated by a feeling that the woman with the unsuccessful inspiration was the one identified by yet another word of knowledge as being in need of counseling. Since the poorly adjusted woman apparently could not recognize that need, and since the younger woman was herself in training as a counselor, the latter took it upon herself to step forward, forestalling disappointment and offering a supportive interaction.

The interplay of sensory modalities, social interaction, and meaning attribution is illustrated by the experience of another person I was able to follow during the session. He was 30 years of age, married, and working as an assistant store manager. The episode occurred following a period of guided healing prayer during which one of the leader's themes was the need for healing from experiences of rejection. The man was being prayed over with laying on of hands by a friend who had accompanied him and a member of the healing team; the free hand of the latter fluttered continuously during the prayer. The young man broke out laughing, continuing for several minutes until one of the leaders responded by taking all three to the back of the hall, where the prayer could go on more privately. He asked the young man what was happening, crouching at his side while he and his friend both sat and the healing team member stood by their side. The man recounted having responded to the theme of rejection, and secondarily to that of passivity, with the image of a stream flowing over rocks through a broken wall. At the emergence of this image he felt joy and began to laugh. To the leader he stated that this had been a double release for him, both from the sense of not being accepted by others, and in that usually he only laughs inwardly, and was suddenly able to laugh quite openly. His friend then reported an image of a clothes washing machine in action, which was understood as divine "confirmation" that the experience was one of cleansing and freeing from the negative emotion. The attending group leader summarized, saying that God wanted to continue this process, but warning that the young man would be "tested." This follow-up period lasted less than ten minutes.

In this vignette we find the invocation of a culturally common negative affect, taken up by the young man through imagery that is at once visual and kinesthetic. In contrast to the objective compulsion evoked by naming the demon in our earlier example, the leader names an indeterminate affective theme. The religious significance is not that all participants respond to this theme in the same way, but that "God speaks to each individual" in a way concordant with that person's experience. The indeterminacy of a theme like rejection is not the same as ambiguity, in the sense of applicability to any number of diverse situations. In this instance, rejection is indeterminate insofar as one can feel rejected because of a particular event, one can be temperamentally disposed to feeling rejected, or one can be oppressed by an evil spirit of Rejection. Healing does not change the rejecting behavior of others except insofar as they respond differently to the healed person's own behavior; hence the relevance of the leader's statement that the man will be "tested" in the future. For the store manager it is not actual instances of rejection that are treated, but the feeling of being rejected that is replaced by the feeling of joy.

The concreteness of the experience lies in the bodily synthesis of visualization (stream), affect (joy), and kinesthesis (laughing). These expressions, spontaneously coordinated within the North American habitus, do not represent and express an inner experience, but objectify and constitute an embodied healing. The socially informed body deals with the negative emotion in images of breaking through a boundary (water flowing through a broken rock wall), release from repression (ability to laugh openly), cleansing from the sullying effects of the negative emotion (water agitating in a washing machine). Further, it is a particularly male variant of habitus that we see, responding to the emotional combination of rejection and passivity. It thus excludes the kind of experience typical for North American women in devotional settings, such as "I no longer feel rejected because I feel loved by God." Whereas the traditional female variant replaces rejection with acceptance (often passive in the somatic image of being held and nurtured), this male example replaces it with joy (active in the ability to laugh out loud).

As in the case of witticism, which as Bourdieu points out often surprises its author as much as the audience, spontaneous religious images invoke "That part of practices which remains obscure in the eyes of their own producers," the realm of buried possibility in which practices are "objectively adjusted to other practices and to the structures of which the principle of their production is itself the product" (1977:79). Through these embodied images, dispositions of the habitus are manifest in ritual behavior. Because they are shared at a level beneath awareness, they are inevitably misrecognized, and the principle of their production is identified as God instead of as the socially informed body. This conclusion is to be distinguished from Durkheim's functionalist abstraction of the sacred as self-affirmation of social morality and solidarity, as much as it must be distinguished from an incarnational acceptance that "God" inhabits the socially informed body. Instead, it suggests that the lived body is an irreducible principle, the existential ground of culture and the sacred.

## *The Intertwining, Mimesis, and Intersubjectivity*

The third example is a retrospective account that I obtained during an interview with a husband and wife who were leaders of a Charismatic covenant community, but who also had experience in the healing ministry. It emphasizes both the possibility and the compelling nature of multisensory imagery, that is, complex images in more than one sensory modality at a time:

> HL: One time I was praying over a man [for healing]. He had a brain tumor and the doctor had sent him home and said, "Forget it. It's all over." And I had a very strong picture of the tumor actually shrinking. And when he left the tumor hadn't, the tumor was still, see, but I felt, when I had my hand on his head, I felt as if it was like a ball on hand and it got smaller and smaller. And I just, not only through sensory, but through a picture in my mind, I felt it was shrinking. Well, I think it was a week or two later, and [he came back and] said the doctors just don't know what happened. It went away. It was gone.

> TC: Wait a minute, you felt it with your hand, shrinking as well.

> HL: It wasn't shrinking in reality, the growth was still there [inside of his head]. But I sensed it in my hand. I felt it in my hand shrinking. But it wasn't in reality. And I had, and then I had a picture of it shrinking, as well, in my mind.

> TC: It was all the way in his head, or could you feel it from the outside?

> HL: Yeah, he had a lump on his head. I actually felt the lump. And I did sense strongly that he was going to be healed, and I remember sharing with that [i.e., telling him about it]. And he came back and it was gone, totally gone. The doctors were baffled.

Here is a healer struggling to sort out the strands of sensory perception (feeling a lump) from imagery in the tactile and visual modalities (a complex image of the tumor shrinking). This example highlights the intimate connection between touch and sight in a way that appeals directly to the notion of embodiment as the existential ground of culture and self. Certainly, as Walter Ong writes, "the tactile senses combine with sight to register depth and distance when these are presented in the visual field" (1967:1), but what are they registering when presented in the imaginal field? Michael Taussig, in elaborating Walter Benjamin's ideas on Dadism, film, and architecture, also suggests that "tactility, constituting habit, exerts a decisive impact on optical reception" (1992:144). For him, however, this "tactile optics" is closely bound up with mimesis, which "implies *both* copy and substantial connection, *both* visual replication and material transfer" (1992:145, emphasis in original).

In the case of the Charismatic healer the mimetic image is not mere representation, but has a materiality grounded in bodily experience that is at once constitutive of divine power and evidence of efficacy. That materiality is all the more compelling in that it marshals in ritual performance the existential intertwining of the tactile and the visual described by Merleau-Ponty:

> There is a circle of the touched and the touching, the touched takes hold of the touching; there is a circle of the visible and the seeing, the seeing is not without visible existence; there is even an inscription of the touching in the visible, of the seeing in the tangible—and the converse; there is finally a propagation of these changes to all the bodies of the same type and of the same Style which I see and touch—and this by virtue of the fundamental fission or segregation of the sentient and the sensible which, laterally, makes the organs of my body communicate and founds transitivity from one body to another. (1968:143)

The existential richness captured in this passage is in fact a feature of everyday life as embodied existence. The example from ritual healing is one that is performatively exaggerated, but it may be only through such vivid examples that insights about embodiment can begin to be grasped and elaborated. Such examples can lead toward clarification not only of intersensory constitution (in this case with respect to revelatory experience) of the bodily synthesis through imagination, but the imaginal constitution of intersubjectivity (in this case between healer and patient) through mimesis.

## Body and Speech: What Kind of Speaking is Speaking in Tongues?

If embodiment is to attain the status of a paradigm, it should make possible the reinterpretation of data and problems already analyzed from other perspectives; and if this is to be in a strong sense, it should be possible even to construct an embodied account of language, typically the domain of linguistic, semiotic, and textual analyses. With this agenda I turn to the problem of glossolalia, or speaking in tongues, as a cultural and expressive phenomenon. Pentecostal glossolalia (see May 1956 on glossolalia in other traditions) is a form of ritual utterance characterized by its leek of a semantic component. Hence, all syllables are "nonsense syllables." Yet, contemporary Charismatic speakers in tongues may develop distinct phonological–syntactic patterns, and individuals may have more than one glossolalic "prayer language," used in different situations and with different intentions. In addition, they believe that it is at times possible for their apparent gibberish actually to be a natural language (xenoglossia). Despite its semantic indeterminacy and phonological–syntactic variability, glossolalia bears a global meaning as an inspired form of praise to God, and can also be called into play as an experientially profound prayer for divine intercession or guidance. At times it is even understood as the utterance of an inspired message or prophecy from

God. It can be spoken or sung improvisationally, and can be used in private devotion or collective ritual. It is a basic tenet that the expressive powers of glossolalia transcend the inadequacies of natural languages (confer. Csordas 1997).

### Semiosis and Embodiment in the Gestural Constitution of Self

When I first began the study of ritual language, Pentecostal glossolalia was being examined in one of three ways: as a phenomenon of trance or altered state of consciousness (Goodman 1972), as a mechanism of commitment to a fringe religious movement (Gerlach and Hine 1970), or as a ritual speech act within a religious speech community (Samarin 1972). Each of these positions adds to our understanding of the phenomenon, but none exhausts the cultural meaning of glossolalia as a form of utterance that both is and is not language. The question for me became not what social function glossolalia served in religious commitment or as a ritual speech act, or by what mental states it is accompanied, but what can the ritual use of glossolalia tell us about language, culture, the self, and the sacred.

In my own view, the two key facts were that glossolalia took the form of nonsense or gibberish, and that its speakers regarded vernacular language as inadequate for communication with the divine. Glossolalic utterance thus seemed to challenge taken-for-granted canons of vernacular expressivity and intelligibility, and in so doing to call into question conventions of truth, logic, and authority. That glossolalia has this potential for challenge and critique is implicit in contemporary Pentecostal efforts to build the kingdom of God on earth. It is even more strongly borne out by Field's (1982) account of the outlawing of tongues as subversive by British colonial authorities during a post–World War I Watchtower movement in Zambia (then Northern Rhodesia). In the absence of violence—indeed of any overtly political act—on the part of the subjected population, the authorities were totally unnerved, and speaking in tongues became the focus of their repressive campaign.[14] By a semiotic account, then, glossolalia ruptures the world of human meaning, like a wedge forcing an opening in discourse and creating the possibility of creative cultural change, dissolving structures in order to facilitate the emergence of new ones.

The creative potential in glossolalia lies in the phenomenological fact that it is "gibberish," and hence threatening, only to nonparticipants. Yet what is compelling about glossolalia is that it is more than a dramatization of the post-Babel loss of a unified tongue. On the contrary, speaking in tongues is experienced as a redemption of pre-Babel lucidity (Samarin 1979), for despite the existence of distinctly recognizable glossa, the global meaning of glossolalic utterance can be apprehended immediately.

The semiotic interpretation is not incorrect, but additional light is thrown on the creative potential of glossolalia's immediacy when it is viewed as a phenomenon of embodiment. Merleau-Ponty (1962) sees at the root of speech a verbal gesture with immanent meaning, as against a notion of speech as a representation of thought. In this view, speech is coterminous with thought, and we possess words in terms of their articulatory and acoustic style as one of the possible uses of our bodies. Speech does not express or represent thought, since thought is for the most part inchoate until it is spoken (or written). Instead, speech is an act or phonetic gesture in which one takes up an existential position in the world. To follow this line of reasoning does not mean that we are to treat glossolalia only as a gesture, for we must grant its phenomenological reality *as language* for its users. I would argue, with Merleau-Ponty, that *all* language has this gestural or existential meaning, and that glossolalia by its formal characteristic of eliminating the semantic level of linguistic structure highlights precisely the existential reality of intelligent bodies inhabiting a meaningful world. In playing on the gestural characteristic of linguisticality, speaking in tongues is a ritual statement that the speakers inhabit a sacred world, since the gift of ritual language is a gift from God. The stripping away of the semantic dimension in glossolalia is not an absence, but rather the drawing back of a discursive curtain to reveal the grounding of language in natural life, as a bodily act. Glossolalia reveals language as incarnate, and this existential fact is homologous with the religious significance of the Word made Flesh, the unity of human and divine.

The experience of contemporary glossolalists lends support to this position. A common charismatic practice is speaking in tongues to make oneself open to divine guidance. These inspirations frequently take the form of imagery, but also include fully formed verbalizations that seem to emerge spontaneously. Here I would suggest that, just as vernacular speech facilitates and is the embodiment of verbal thought, so glossolalia facilitates and is the embodiment of nonverbal thought. Vernacular speech is "putting it into words"; glossolalic speech is "putting it into images." In glossolalia the physical experience of utterance (*parole*) comes into balance with the intellectual experience of language (*langue*). I would argue not that body and mind merge in glossolalic utterance, but that the utterance takes place at a phenomenological moment prior to distinction between body and mind, a distinction that is in part contingent on the objectifying power of natural language. Preobjective processes of the self emerge, and what is perceived includes both inchoate attributes of self, others, and situations, and what psychoanalysis would call contents of the unconscious. The results do not remain inchoate, however, but are typically taken up into discursive language. The facts that Charismatics typically switch back and forth between glossolalia and the vernacular, and that some of the apparently spontaneous inspirations emerge in verbal form, suggest that speaking in tongues serves the cultural process of self-objectification and is not simply a dreamy state of meditatively emptied consciousness.[15]

Gesture, emotional expression, and language are of a piece in being superimpositions of a human world on a natural or biological world. Because of a "genius for ambiguity which might serve to define man ... Behavior creates meanings which are transcendent in relation to the anatomical apparatus, and yet immanent to the behavior as such, since it communicates itself and is understood" (Merleau-Ponty 1962:189). Thus, a smile for the American and the Japanese is grounded in the same anatomical apparatus, but transcends it by being appropriated or thematized in the one case as friendship and in the other as anger (Ekman 1982). In language, too, this transcendence is both a spontaneous engagement with others and a locus of cultural creativity, since "Speech is the surplus of our existence over natural being" (Merleau-Ponty 1962:197), that is to say, of our existence as persons over mere being as objects or things.

In both these ways (spontaneous engagement and cultural creativity), absence of the semantic component in glossolalia again reveals the gestural meaning of language, such that the sacred becomes concrete in embodied experience. With reference to human engagement, and in comparison with the brain-damaged patient who never feels the need to speak or to whom experience never suggests a question or invites improvisation, Merleau-Ponty quotes Goldstein: "As soon as man uses language to establish a living relation with himself or his fellows, language is no longer an instrument, no longer a means; it is a manifestation, a revelation of intimate being and of the psychic link which unites us to the world and our fellow man" (1962:196). But this element of *communitas* in linguistic utterance is over-shadowed by the fact that once the primordial silence has been shattered by an act of expression, a linguistic and cultural world is constituted. Speech coalesces into constituted languages, the speaking word becomes the already spoken word, and transcendence occurs only in acts of authentic expression such as those of writers, artists, and philosophers. What better way to maximize the gestural element of *communitas,* and what better way to preclude the petrification of *parole* into *langue* than to speak in tongues, always a pure act of expression and never subject to codification. This car-ries us quite a distance beyond the semiotic analysis, which we based on glossolalia's lack of a semantic component and its consequently bold chal-lenge to canons of intelligibility. It suggests that glossolalia offers not only a critique of language, but a positive statement about expressivity, such that its critical force is enhanced by the moral force of its claim to be pure communication, incapable of uttering any "wrong words."

The totalizing aspect of glossolalia does not preclude the possibility noted above for glossolalists to have more than one syntactical phonetic configu-ration or glossa, used in different situations and bearing different expressive and emotional valences. We may see this as a contradiction, or as one of the fruits of indeterminacy and the "genius for ambiguity." Nevertheless, the multiplicity of tongues resonates with Merleau-Ponty's suggestion that ver-bal form may not be as arbitrary as linguistic theory would have it. He sug-gests that the phonetic structures of various languages constitute "several ways for the human body to sing the world's praises and in the last resort to live it" (1962:187). Considered from the perspective of embodiment, it is thus understandable that glossolalia adapts its phonetic contours to the affec-tive contours of different situations; and in an unexpected validation of

Merleau-Ponty's metaphor, I note again that Pentecostal glossolalia is consistently thematized as prayer of praise, and that it is often sung or chanted with improvised harmony and melody lines.

The musical performance of tongues in Charismatic ritual suggests that its temporal structure may be more akin to music than to language, and indeed it has been analogized to scat singing in jazz. The principal difference is that scat is a form of instrumental music in which the voice is the instrument, whereas glossolalia insists on being sung speech. Even when freely improvised, it lacks the temporal contours and resolution of musical form. Because glossolalia lacks the lineality of semantic utterance or music, but also because it highlights the gestural meaning of language as a pure act of expression, it allows language to exist outside time. To the speaker in tongues, temporality becomes eternity, because there is no logical progression, but also because every moment is an existential beginning.

### Embodied Language and Ritual Practice

If embodiment really does advance our understanding of a particular practice, it should also advance our understanding of how practices are related among themselves—this is the contribution of Bourdieu's concept of habitus. "Resting in the Spirit" is one Charismatic practice that on first glance appears quite different from speaking in tongues as a religious experience. In this *technique du corps,* a person is overcome by the power of the Holy Spirit and falls in a state of motor dissociation, while retaining some awareness of the surroundings and subsequent memory of the experience. It is typically characterized as peaceful, relaxing, rejuvenating, healing, and imbued with a sense of divine presence. Among Roman Catholic Charismatics,[16] this practice has incited much more controversy than has speaking in tongues. The principal issue is the "authenticity" of the experience. The fact that this problem never arose with glossolalia can be understood in terms of different uses of the body in the two practices.

In brief, glossolalia cannot be inauthentic as long as it is accompanied by an intention to pray. One cannot have the intention to rest in the spirit, because by definition the experience occurs spontaneously. To be more precise, a person who first begins to speak in tongues is said to "yield to the gift," that is, passively to allow it to be manifest through more or less spontaneous utterance. At the same time, it is said that the neophyte should "step out in faith," actively uttering whatever nonsense syllables he or she can formulate. The combination of active and passive uses of the body in one practice seems to be the concrete operator that allows for experiential communion of human and divine in a speaking body. The ritual status of resting in the spirit is different, emphasizing the subjective passivity of "resting" and the objective passivity of being "overcome." The Protestant term for this practice, "Slaying in the Spirit," even more strongly emphasizes the external force overwhelming a passive or weaker recipient. There is no act of will involved in resting in the spirit, neither is there a willful act of speech—the practice is mute as well as passive. Hence there is the possibility of "inauthenticity" if a person chooses to fall, or falls in conformity to those around him.

This interpretation offers an embodied understanding of the relation between ritual and social life in the Roman Catholic Charismatic movement over its 20-year history. The introduction of resting in the spirit came considerably later than that of speaking in tongues, and corresponded with a social transformation of the movement from a self-perceived vanguard of active renewal in the late 1960s to a source of passive refuge, one conservative movement among others in the Roman Catholic church of the late 1980s. In conjunction with the changed sociopolitical climate across these decades in the United States, the demographic base of the movement has shifted to an older and more conservative group predominantly in their fifties, as well as to a group that includes more working and lower-middle-class people. Thus, the relation between speaking in tongues and resting in the spirit represents the embodiment in ritual practice of differences in generational and class habitus.

The perspective of embodiment can also help us understand the relation between glossolalic prayer and a second form of Charismatic ritual language, prophecy. Prophecy includes a semantic component of the most sacred sort, for the prophetic utterance is understood as a direct message from God. The speaker is not entirely passive, for he or she must "discern" when, where, and whether to utter the inspired words, but the utterance is invariably in the first person, with God as the ostensible speaker. Charismatic prophecy rarely foretells the future, but instead ritually establishes a state of affairs in the world (for example, "You are my people, I am doing a great work among you, Lay down your lives for me"). The gestural nature of prophetic utterance is evident in its content, almost like a verbal pointing. This gestural meaning is made concrete in practice by a direct link with glossolalia, in that prophecy can at times be expressed first in tongues, and subsequently "interpreted" into a vernacular utterance identical to any other prophecy. The difference between prayer and prophecy in tongues is entirely based on tone of voice, volume, and stridency. Thus, through the medium of the body, the relation between glossolalia as prayer and as prophecy is established not as one between activity and passivity, but as one between intimacy (prayer) and authority (prophecy) in the relation between God and humans.

Given that this relationship between glossolalic prayer and vernacular prophecy is grounded in the embodied experience of intimacy and authority, we can understand a further parallel between the two forms in ritual practice. Earlier I described the gestural meaning of glossolalia as a ritual celebration of the open-ended or indeterminate way in which language, gesture, and emotion take up an existential stance in the world. In practice, glossolalic prayer as embodied intimacy is for some individuals free improvisation, but for others it is the redundant repetition of a limited phrase or series of syllables, much in the manner of a *mantra*. Thus, practice follows a continuum between indeterminacy and redundancy. Prophecy as embodied authority follows an inverse continuum between determinacy and redundancy, since in practice it ranges from the unique and creative elaboration of metaphor with explicit rhetorical consequences for mood and motivation, to the highly redundant reproduction of basic meanings through simple prophetic exhortations, the simplest form of verbal pointing.

In conjunction with the way in which ritual activity and passivity have been embodied in the social life of these Charismatic Christians, a movement from intimacy to authority can be seen in the development of charismatic "covenant communities." These intentional communities have cultivated the vanguard mentality of the movement's early days largely through emphasis on prophecy as the authoritative and directive word of God. The increasing reliance on prophecy and the increasingly radical message promulgated have led to a split between two prominent networks of covenant communities, to a self-conception of those communities as a movement distinct from the Catholic Charismatic Renewal as a whole, and finally to proto-schismatic tension between the communities and the Catholic hierarchy. The latter achieved a measure of public visibility in a recent controversy over allegiance of one community to the prophetic authority of another, as opposed to the ecclesiastical authority of the local bishop. The case resulted in litigation by the covenant community in the Vatican, and the bishop's resignation.

From the perspective of embodiment, then, glossolalia asserts the unity of body and mind, establishes a shared human world, and expresses transcendence—as does all language. Thought is not independent of utterance, the human world is constituted in a blend of embodied voices, and every utterance is an initiating utterance, a transcendent beginning. Yet glossolalia does this in a radical way, since the gestural meaning of language predominates. From the perspective of embodiment, the indeterminacy of glossolalia is not only semantic. On a more fundamental level, glossolalia's indeterminacy subsists in its capacity to participate in modes of pure communication and absolute critique, intimacy and authority, activity and passivity, private and collective, a unitary language of pre-Babel and a multiplicity of situationally contoured tongues.[17] Experienced glossolalists do not construe their utterances as childish babble, although the religious theme of childlike simplicity is sometimes invoked to describe a first embarrassed nonsense-utterance. Instead, they see themselves as mature users of a spiritual gift, the purpose of which is to enhance their relationship to the divine.

## Collapsed Dualities: Objectivist Explanations of
## Religious Experience

To the degree that the argument outlined above successfully bridges or integrates domains of perception, practice, and religious experience, I would assert that a paradigm of embodiment does indeed have paradigmatic implications. In the two concluding sections I will elaborate some of these implications. Having concentrated on the domain of religious experience, I will turn first to the critique of explanations grounded in the objectivist mind-body dichotomy, and offer a phenomenological alternative.

Ritual practices are often explained in terms of psychological suggestion or learned behavior on the mental side and physiological mechanisms of trance or catharsis on the physical side. Suggestion and learning are inadequate to account for the phenomena discussed above. In the group setting, the "power of suggestion" takes us no further than the healer's invocation to "release more power, Lord." It accounts for the setting of mood and tone, but not for the structure and efficacy of embodied ritual practices, and not for their character of apparent spontaneity. Neither can learning account for why glossolalia has a particular place in the ritual system (why glossolalia and not some other practice?). Learning may begin to account for its transmission in response to cues, and for its culturally consistent theological meaning, but not for how it can be perceived as *power* in ritual practice.

Likewise, physiological explanations in terms of trance and altered states of consciousness, or catharsis and nervous-emotional discharge, do not take us very far unless we are willing to accept trance and catharsis as ends in themselves rather than as *modus operandi* for the work of culture. For example, the most advanced theory of catharsis, that of Scheff (1979), defines cathartic laughter as the expression of embarrassment. It cannot go beyond this objectivist formulation to account for how such laughter is thematized, or systematically misrecognized, as "joy" in the vignette of the store manager analyzed above, or as "mocking" in other instances in which a demon "refuses to take seriously" attempts of the pious to deliver one of the faithful from its influences.

Part of the inadequacy of these explanations is that they are often derived from research in experimental settings, and research focused on concrete events that does not attempt to transcend those events. These approaches share a weakness outlined by Bourdieu as:

> ... the occasionalist illusion which consists in directly relating practices to properties inscribed in the situation ... the truth of the interaction is never entirely contained in the interaction. This is what social psychology and interactionism or ethnomethodology forget when, reducing the objective structure of the relationship between the assembled individuals to the conjunctural structure of their interaction in a particular situation and group, they seek to explain everything that occurs in an experimental or observed interaction in terms of the experimentally controlled characteristics of the situation, such as the relative spatial positions of the participants or the nature of the channels used. (1977:81–82)

This is true both of the psychological and of the physiological explanations outlined above. The former assume a kind of immediate interpersonal influence, and the latter that ritual interaction operates as a triggering mechanism, as well as that the phenomena of religious experience are results of a stimulus–response pattern operating entirely within the circumscribed ritual event.

In contrast to these positions, to collapse the duality of mind and body yields a phenomenology of perception and self-perception that can pose the question of what is religious about religious experience without falling prey to the fallacies of either empiricism or intellectualism.[18] To explain this approach I must return to my earlier conclusion that certain preobjective phenomena are misrecognized as originating in God instead of in the socially informed body.[19] I would take issue with Durkheim, who identified this misrecognition but adopted a functionalist definition of the sacred as society mystifying and worshiping itself and thereby establishing morality and social solidarity. This was one of the fundamental arguments by which he established the social as a category sui generis, but I believe that in doing so he mistakenly also abolished the sacred as a category sui generis for anthropological theory.

Durkheim's argument was that the way society creates the sacred is by appearing as something radically other and outside the individual, and in the massiveness and mystery of this otherness establishing an absolute moral authority (1965). By restricting the human experience of otherness to the category of the social, however, Durkheim committed a major error of reductionism. Subsequent generations have followed him in this sociological reductionism, in large part precluding an authentically phenomenological and psychocultural theory of religion. Thus Geertz (1973) can posit a definition of religion, and symbolic anthropologists take up the notion that it is a system of symbols, articulated in a system of social relationships. For the psychological anthropologist, it is the next part of Geertz's definition that is of principal concern, that religion acts to establish long-standing moods and motivations. I submit that the theoretical power to get at these moods and motivations may be found among phenomenologists and historians of religion such as Otto (1958), Van der Leeuw (1938), and Eliade (1958). These theorists conceived the sacred in terms of the same "otherness" identified by Durkheim. They differed, however, in regarding this otherness not as a function of society, but as a generic capacity of human nature.

This approach can be applied to the above analyses of embodiment in the Charismatic data, especially the perception of spontaneity as the phenomenological criterion of the divine, and the lack of control as a criterion of the demonic. When a thought or embodied image comes suddenly into consciousness, the Charismatic does not say "I had an insight," but "That wasn't from me, how could I have thought of that. It must be from the Lord." The experience of God does not come from the content of the idea but is constituted by the spontaneous fit of the inspiration with the circumstances. When a bad habit becomes a compulsion, when one can no longer control one's chronically bad temper, the Charismatic does not say "My personality is flawed," but "This is not me, I am under attack by an evil spirit." The demon does not cause the bad habit or the anger but is constituted by the lack of control over these things. The sui generis nature of the sacred is defined not by the capacity to have such experiences, but by the human propensity to thematize them as radically other.

With this conception, the question of what is religious about religious healing can be posed, since the sacred is operationalized by the criterion of the "other." However, since otherness is a characteristic of human consciousness rather than of an objective reality, anything can be perceived as "other," depending on the conditions and configuration of circumstances, so that defining the sacred becomes an ethnographic problem. The paradigmatic significance of embodiment is then to provide the methodological grounds for an empirical (not empiricist) identification of instances of this otherness, and thus for study of the sacred as a modality of human experience.

## Collapsed Dualities: Psychological Anthropology and the Body in the World

In my opening argument I reiterated Hallowell's concern with the subject–object distinction and showed that within the incipient paradigm of embodiment both Merleau-Ponty and Bourdieu require the collapse of such analytic dualities.[20] In the subsequent analyses I attempted to work out some implications of embodiment in the domain of charismatic religious experience. I avoided the assumption that phenomena of perception are mentalistic (subjective) while phenomena of practice are behavioristic (objective) by approaching both within a paradigm that asks how cultural objectifications and objectifications of the self are arrived at in the first place. With Merleau-Ponty I attempted to resist analyzing the objects of religious perception in order to capture the process of objectification, and with Bourdieu to resist constructing models of religious action in order to capture the immanent logic of its production.[21]

The hermeneutic circle of this argument is completed with a return to the subject–object distinction, which in my view frames the central methodological issue of embodiment. Recall that Merleau-Ponty criticized analyzing perception as an intellectual act of grasping external stimuli generated by pre-given objects. His objection was that the object of perception would then have to be either possible or necessary. In fact it is neither— instead, it is *real*. This means that, as Merleau-Ponty has pointed out, "it is given as the infinite sum of an indefinite series of perspectival views in each of which the object is given but in none of which it is given exhaustively" (1964b:15). The critical "but" in this analysis requires the perceptual synthesis of the object to be accomplished by the subject, which is the body as a field of perception and practice (1964b:16). Merleau-Ponty felt that it was necessary to return to this level of real, primordial experience in which the object is present and living, as a starting point for the analysis of language, knowledge, society, and religion. His existential analysis collapses the subject–object duality in order to more precisely pose the question of how the reflective processes of the intellect elaborate these domains of culture from the raw material of perception.

The paradigmatic implications of embodiment extend to how we study perception as such. Beginning with the experiments of Rivers (1901) in the Torres Straits expedition, anthropologists have (1) considered perception strictly as a function of cognition, and seldom with respect to self, emotion, or cultural objects such as supernatural beings; (2) isolated the senses, especially focusing on visual perception, but seldom examining the synthesis and interplay of senses in perceptual life; and (3) focused on contextually abstract experimental tasks, instead of linking the study of perception to that of social practice (confer Cole and Scribner 1974; Bourguignon 1979). Within a paradigm of embodiment, analysis would shift from perceptual categories and questions of classification and differentiation, to perceptual process and questions of objectification and attention/apperception. Looked at in another way, whereas in conventional studies of optical illusions or color perception our questions have been posed in terms of the cultural constitution of perceptual categories, the analyses I have presented raise issues of the perceptual constitution of cultural objects. In taking up a paradigm of embodiment, it is critical to apply the analysis of subject and object to our distinctions between mind and body, between self and other, between cognition and emotion, and between subjectivity and objectivity in the social sciences, particularly psychological anthropology.

First, if we begin with the lived world of perceptual phenomena, our bodies are not objects to us. Quite the contrary, they are an integral part of the perceiving subject. Contrast this with the perspective of Piaget, who argues that "the progress of sensorimotor intelligence leads to the construction of an objective universe in which the subject's own body is an element among others and with which the internal life, localized in the subject's own body, is contrasted" (1967:13). Merleau-Ponty would not deny that we construct an objective universe, nor that development of the capacity to objectify is critical to our makeup, but that the fully developed adult moving about in the world treats his or her body as an object. The slippery moment of Piaget's thought comes in the difference between observing that in reflection the internal life appears localized in the subject's body, and accepting this artifact of consciousness as the end point of development. To do so is to accept the mind-body distinction as given. My argument has been that on the level of perception it is not legitimate to distinguish mind and body. Starting from perception, however, it then becomes relevant (and possible) to ask how our bodies may become objectified through processes of reflection. This contrast is so basic that it gives one pause to think how much psychological anthropology has been influenced by Piaget, and how little by that other professor of child psychology, Merleau-Ponty.[22] The first defines the body as "an element among others in an objective universe," the second as "a setting in relation to the world."

When the body is recognized for what it is in experiential terms, not as an object but as a subject, the mind-body distinction becomes much more uncertain. Psychological anthropology has tended to operate within the mind-body duality, conceptualized as the relation between the subjective mental domain of psychocultural reality and the objective physical domain of biology. The approach I am proposing certainly does not negate the problematic of biology and culture, but by a shift of perspective offers an additional problematic. When both poles of the duality are recast in experiential terms, the dictum of psychological anthropology that all reality is psychological (Bock 1988) no longer carries a mentalistic connotation, but defines culture as embodied from the outset.

If we do not perceive our own bodies as objects, neither do we perceive others as objects. Another person is perceived as another "myself," tearing itself away from being simply a phenomenon in my perceptual field, appropriating my phenomena and conferring on them the dimension of inter-subjective being, and so offering "the task of a true communication" (Merleau-Ponty 1964b:18). As is true of the body, other persons can become objects for us only secondarily, as the result of reflection. Whether or not, and under what conditions, selves do become objectified becomes a question for the anthropology of the self. In addition, the characteristic of being "another myself" is a major part of what distinguishes our experience of the social other from that of the sacred other discussed above, which is in a radical sense "not myself."

Embodiment also has paradigmatic implications for the distinction between cognition and emotion (Rosaldo 1984; Jenkins 1988, 1991). Emotion has attracted growing attention from anthropologists, but has remained conceptually subordinate to cognition. Emotions have been defined as cognitive by making methodological choices to study them through essentially cognitive card-sorting tasks (Lutz 1982), by focusing on the culturally provided schemata for dealing with them (Levy 1973), or by defining them explicitly as interpretations constituted of concepts, beliefs, attitudes, and desires (Solomon 1984). A step toward the present position was taken by Rosaldo (1984), who suggested that emotions are a kind of cognition with a greater "sense of the engagement of the actor's self, ... *embodied* thoughts, thoughts seeped with the apprehension that 'I am involved'" (1984:143, emphasis in original). Although thought and emotion are thus placed on a more even footing, to define emotion as embodied thought preserves the fundamental duality. It precludes the question of how thought in the strict sense is itself embodied, and does not take up the challenge of an authentically "affective" theory of emotion corresponding to the "cognitive" theory (Jenkins 1988, 1991). Rethinking the relation between subject and object also has implications for our conceptions of objectivity as the goal of science. In one of its strongest forms, objectivity is said to be achieved through a process of abstraction whose "aim is to regard the world as centerless, with the viewer as just one of its contents ... . The object is to discount for the features of our pre-reflective outlook that make things appear to us as they do, and thereby to reach an understanding of things as they really are." (Nagel 1979:206, 208). Risking glibness, I would argue that science is not to be run as a discount operation, and that we must start from the pre-reflective if we hope sensibly to pose questions about appearance and reality. The collapse of the subject–object distinction requires us to recognize that if "hard science" deals with hard facts,[23] they are the result of a hardening process, a process of objectification.

Perhaps more immediately compelling to psychological anthropology than this general point about subjectivity and objectivity, Nagel acknowledges that "The problems of personal identity and mind-body arise because certain subjectively apparent facts about the self seem to vanish as one ascends to a more objective standpoint" (1979:210). Before the vanishing point is reached, it is necessary to begin to formulate what Shweder calls a "science of subjectivity," because: "The real world, it seems, is populated with subject-dependent and object-like subjectivity, two types of phenomena for which there is no place in the mutually exclusive and exhaustive realms of the symbol-and-meaning-seeking hermeneuticist and the automated-law-seeking positivist" (1986:178). It is equally in error to seek the objectivist "view from nowhere" and to inordinately privilege subjectivist "inner experience." The most fruitful definition of the real is that quoted above of an indefinite series of perspectival views, none of which exhausts the given objects.[24] Objectivity is not a view from nowhere, but a view from everywhere that the body can take up its position, and in relation to the perspectives of "other myselves." This perspective does not deny that objects are given; as I have emphasized throughout this chapter, the body is in the world from the start. Thus it is not true that contemporary phenomenology denies an "irreducible objective reality" (Nagel 1979:212). Quite differently, phenomenology insists on an *indeterminate* objective reality.

The theme of indeterminacy has arisen several times in this argument with respect to the nature of our analytic categories as well as to the domains of perception and practice.[25] It is not surprising that both theorists we have considered, as a result of the methodological collapse of dualities, recognize an essential principle of indeterminacy within human life. Merleau-Ponty sees in the indeterminacy of perception a transcendence that does not outrun its embodied situation, but which always "asserts more things than it grasps: when I say that I see the ash-tray over there, I suppose as completed an unfolding of experience which could go on ad infinitum, and I commit a whole perceptual future" (1962:361). Bourdieu sees in the indeterminacy of practice that, since no person has conscious mastery of the *modus operandi* that integrates symbolic schemes and practices, the unfolding of his works and actions "always outruns his conscious intentions" (1977:79). This indeterminacy must be squarely faced by embodied accounts of subject-dependent cultural objects that resist isolating the senses from one another, and from social practice, in experimentally restricted settings.

As we have seen in ritual healing and ritual language, embodied selves inhabit a behavioral environment much broader than any single event

If this is the case, then a final paradigmatic implication is that embodiment need not be restricted to a microanalytic application, but as Merleau-Ponty hoped, can be the foundation for analyses of culture and history. Freeing interpretation from the event was critical for Bourdieu, even for his study conducted within a stable traditional society. It is yet more critical with the kind of religious movement I have described, which does not exist in a taken-for-granted world, but is set instead in a contemporary world where the principle of indeterminacy holds sway in a sea of opinion. In this setting, religious practice exploits the preobjective to produce new, sacred objectifications, and exploits the habitus in order to transform the very dispositions of which it is constituted. What is out of the ordinary in such situations, and what therefore can be thematized as sacred, is the evocation in ritual of the preorchestrated dispositions that constitute its sense. The locus of the sacred is the body, for the body is the existential ground of culture.

## Reprise

The argument of this chapter has been that the body is a productive starting point for analyzing culture and self. I have attempted to show that an analysis of perception (the preobjective) and practice (the habitus) grounded in the body leads to collapse of the conventional distinction between subject and object. This collapse allows us to investigate how cultural objects (including selves) are constituted or objectified, not in the processes of ontogenesis and child socialization, but in the ongoing indeterminacy and flux of adult cultural life. To be sure, the empirical examples I have chosen (evil spirits, multisensory imagery, glossolalia, prophecy, and "Resting in the Spirit") come from the specialized domain of ritual practice. Yet if, as I suspect, embodiment has paradigmatic scope, the many analyses of other domains that have begun to be published in the past decade share common features that can be elucidated in future work. This is suggested, as I have argued, by the way embodiment poses additional questions about religious experience and perception beyond those typically asked in psychological anthropology. It is even more strongly suggested by the application of the subject–object analysis to other dualities (mind–body, self–other, cognition–emotion, subjectivity–objectivity) that underlie much of anthropological thought.

# CHAPTER THREE

# A Handmaid's Tale

This chapter has to do with religious rituals directed at the experience o women in North America and Japan who have undergone abortions. In each case, they are rituals aimed at the healing of a particular cultural construction of grief and guilt predicated upon a particular ethnopsychology of the person. I will first present the North American ritual and then contrast it with a parallel ritual in contemporary Japan.

The North American ritual, or more precisely the ritual technique, is disturbing in the way it taps into one of the most emotionally, ethically, and politically provocative issues in contemporary society. It is disturbing in the same sense as is Margaret Atwood's powerful novel, *A Handmaid's Tale,* from which I've borrowed my title. Atwood describes a North American society in the very near and almost-present future in which fundamentalist Christianity has acceded to political power and created a totalitarian state In this psychic, the act of performing abortion is punishable by death and the public exhibition of one's humiliated corpse. Because environmental pollution has decreased the population's fertility to a dangerously low level the Commanders who constitute a ruling elite are assigned Handmaids These fertile young women complement the Commanders' privilege Wives as reproductive servants within their sanctified households.

When I first encountered Atwood's work, I was frankly jolted by the similarity of terminology to that prevalent in some of the Catholic Charismatic "covenant communities" I had been studying. "Household was indeed a specialized term for a Christian living arrangement that included more members than a nuclear family. There was an office o "handmaid," admittedly without reproductive function, but understood a a role in which some women had additional responsibilities for communit service, particularly regarding the well-being of other women, but alway under direct male "headship" or authority. Somewhat ominously, in th leading covenant community, the office of handmaid was itself suspende for a period of several years, presumably because those who held it wer arrogating more authority than was regarded as biblically warranted by th male ruling elite. The ruling elite of these communities, which considered themselves vanguard outposts of a coming kingdom of God (the logical extension of which seemed to me to be Atwood's Republic of Gilead), styled themselves not as Commanders within a religious police state, but in a slightly more bureaucratic vein, as "Coordinators" (Csordas 1997).

The possibility of seeing the Charismatics as "proto-Gileadean" was entranced during my study of their system of ritual healing when I discovered the rite I will describe below. Let me note from the outset that some Catholic Charismatics are quite active in the political opposition to abortion, prompted by the double influence of embracing the conservative position of the Roman Catholic hierarchy and embracing the fundamentalist conservatism of neo-Pentecostalism. Some are additionally active in a campaign to achieve medical recognition of what they call "post-abortion syndrome," a fabricated psychiatric syndrome modeled very closely on the definition of "post-traumatic stress disorder" found in the American Psychiatric Association's Diagnostic and Statistical Manual. Such a disorder is, strictly speaking, a culture-bound disorder in the sense that it is relevant only within a Charismatic culture that defines the experience of abortion as necessarily traumatic.

Leaving that point aside for the present, note that the healing practices we have been discussing among Catholic Charismatics show a remarkable uniformity across regions and locales, at least within North America. This is in part due to a highly developed distribution system for movement publications including books, magazines, and audiotapes, as well as the existence of a class of teachers and healers who travel to workshops, conferences, retreats, and "days of renewal" at which such practices and their rationales are disseminated. Again, the three principal forms of healing are prayer for healing of physical or medical problems, Deliverance or casting out of evil spirits, and inner healing or Healing of Memories.[1] The Healing of Memories is the ritual transformation of the consequences of emotional trauma or "woundedness" by means of prayer. This prayer often includes imaginal processes in the form of guided imagery initiated by the healer or the spontaneous enactment of a scenario by the patient. At times the memory identified as in need of transformation is that of having had an abortion. In Charismatic culture, undergoing an abortion is presumed traumatic to the pregnant woman, entailing the emotional consequences of guilt and the grief of bereavement, and is also presumed to produce a death trauma for the aborted fetus.[2]

Healing of memories for the mother and fetus is described in a book by the highly popular Charismatic Jesuit priests Dennis and Matthew Linn and their collaborator Sheila Fabricant (1985:105–139). Their book treats miscarriages, stillbirths, and abortions as a single class, beginning with a theological discussion emphasizing that while these unbaptized do not necessarily end up in the "limbo" of Catholic lore and can go to heaven, they are in need of healing. The authors go on to a psychological discussion of prenatal research, arguing for the emotional viability, and hence vulnerability of these beings. Then follows a discussion of grief among mothers, which quickly turns to focus on abortion and argues for the commonality of grief and guilt among women who choose abortions.

The authors narrate two cases of praying for such women. The first was a woman who had had one abortion, and had also attempted to abort her now-18-year-old daughter who was having frequent violent outbursts against family members. During a mass offered for the aborted fetus and for "any part of" the living daughter that had died during the abortion attempt, the adult woman collapsed on the floor and experienced all the pains and contractions of labor, following which the healers initiated her symbolically "to give her baby to Jesus and Mary to be cared for." Subsequently the woman claimed that her chronic back pain improved, as did her daughter's violent outbursts, both changes interpreted by the healers as evidence of relief of "the trauma of the abortion." The second case was a woman for whom healing hurt and self-hatred from having an abortion nine years previously caused a variety of other hurts to emerge, including the perinatal effects of grief experienced by her own mother over the death of her father and anger at her relatives who refused to allow the pregnant woman a deathbed visit, as well as the effects of being born with her umbilical cord wrapped around her neck, and of having been physically and sexually abused during childhood.

These examples exhibit an ethnopsychology in which abortion (in a degree greater than miscarriage or stillbirth) is a powerful pathogenic agent, and in which ritual healing is a powerful and occasionally dramatic antidote. The rite often includes specific imaginal techniques. Linn, Linn, and Fabricant describe four steps: (1) the patient visualizes Jesus and Mary holding the child, and the patient holds it with them, asking forgiveness from the deity and the child for any way in which he or she hurt the child, and is instructed to imaginally "see what Jesus or the child says or does in response to you," and with them to forgive anyone else who may have hurt the child; (2) the patient chooses a name for the dead fetus and symbolically baptizes it, with the instruction to "feel the water cleansing and making all things anew," thus granting the fetus the cultural status of a person and, in effect, ritually "undoing" the abortion; (3) the patient prays that the fetus receive divine love, and is instructed to imaginally "place it in the arms of Jesus and Mary and see them do all the things you can't do," and to ask the fetus to become an intercessor for the patient and the patient's family; (4) the patient has a mass offered for the child, and while receiving the Eucharist is instructed to "let Jesus' love and forgiving blood flow through you to the child and to all other deceased members of your family tree" (1985:138–139).

### Person, Gender, and Efficacy

The degree of multisensory vividness that can be attained in what we can call this embodied imaginal performance (see also Csordas 1994a) is evident in the following case narrated by a team of two Charismatic healers (G and H):

G: ... one lady that we had prayed over for an abortion [was so upset that] she turned purple at one point. ... Anyway, we asked the Lord if she could have the vision of her baby, aborted baby. And she physically cupped her hands, arms and hands, as if she was holding a baby. And if you saw her, if you saw any of us, [you'd] probably think we were all nuts. But if you saw her, it looked like she was holding a baby. I mean she was there like this. And talking to it. Of course there was nothing there that anyone could see. But we had just asked the Lord if He would allow her to hold the baby. And the next moment she was holding her baby.

TC: You asked aloud with her or you asked [God] silently whether she could ...

G: No, we asked her first, out loud. And she said she wanted to. Then she wouldn't give it up. So we were quite a while until she was able to let the baby go.

H: And we would just remain silent and just keep praying silently and with our hands on her. So that He [God] would go into her ...

G: Real physical manifestation ...

H: And you could just feel it all around, in the air, of the Lord just loving her.

C: Did she have the physical experience of holding the baby?

G: Oh, yeah.

TC: And what did the purple in her face mean?

G: Well that was before [the imagery sequence]. I just think it was the guilt and the mourning over it.

H: See the thing is she didn't want to come to the acceptance that she had anything to do with the abortion. It was "all her husband's fault." And when she finally came to realize that she had to take a responsibility to ...

G: She started screaming.

H: Then it was kind of scary, ya know. But [we] just loved her through that. And He was there with us. So it was a beautiful experience.

G: And something very interesting on that was, when we deal with the healing for an abortion, we always ask them if they have a sense of what gender the baby is, and if they have any sense of a name ... if they even hear a name or see a name or the Lord places a name in their heart. And I forgot what the name was, but it was a girl. And both ... we dealt with them separately. Both had a sense it was a girl and both came up with the same name. Husband and wife. And they did not consult with each other. Because we saw her first, and then we

ushered her out of the room. There was no communication between
the two. And both sensed that it was a girl, and both came up with
the exact same name. And neither one had talked about this since the
day that the abortion occurred. Never brought it up again. So I mean
there was no possible way that they could have named it . . . that before
the abortion they had even thought it.

I will organize my analysis of this text around the four elements that
I identified in Chapter Two as essential to therapeutic process in ritual
healing. Regarding *disposition,* it is evident that the supplicant must be culturally disposed not only to accept the possibility of divine healing but also
to regard having undergone an abortion as a problem in need of healing.
The healer's presumption that the supplicant's "turning purple" indicated
states of guilt and mourning are part of the taken-for-granted nature of the
latter disposition, apparently never challenged by participants. The presence
of both dispositions is suggested by the apparent fact that the healing was
directed specifically toward the abortion experience and that the woman's
husband was included in a systematic way, separate from his wife. The disposition to maternal attachment enacted in the woman's refusal to relinquish
her imaginal baby is consistent with participation in the healing system.

Nevertheless it is necessary to recognize that the presumption of guilt as
an emotion in the supplicant can, through performance, act as an induction
of guilt. This is especially the case when guilt is regarded not only as an
emotional but an objective state—that is, a state of sin. Characteristically for
Charismatics, there is no explicit discussion of sin and repentance, which
remain implicit in the reference to "taking responsibility for" the action.
In no way does this phrase mean that healing is constituted by "coming to
terms with having made a responsible, though difficult, decision." Instead,
it means that emotional healing requires "acknowledging that by consent-
ing to your husband's demand you too are responsible for a sin," and
accepting divine forgiveness.

Experience of the sacred is actualized by multisensory imagery in several
cultural forms. Gendering and naming the fetus is achieved through revelatory imagery, and the conviction of divine empowerment is reinforced by
the concurrence of husband's and wife's images in the absence of consultation. Divinely granted haptic, kinesthetic, and visual imagery of an exceedingly vivid, eidetic quality is evident in the woman's holding the imaginal
baby and talking to it. The experience of divine presence as a phenomenon
of embodiment is attested by the healers' account that, for their own part,
they could "feel it all around, in the air," and that the supplicant's imagery
sequence was a "real physical manifestation" of divine power entering her.
Finally, although not specifically recounted in this text, it is likely that the
supplicant with her child was led through a complete imaginal performance of baptizing the baby and finally letting it go into the hands of Jesus.

While the imaginal form and eidetic quality of these experiences define
them as sacred, their content achieves the third therapeutic function of

*elaboration of alternatives.* Two such alternatives are implicit in this episode. First is that of actually having a baby, elaborated in the imaginal holding of the baby and its cultural thematic of maternal-child intimacy. Second is that of having the fetus die in a culturally appropriate way, that is, as a baby with definite gender, name, and Christian baptism.

It is the latter alternative that is taken up as part of the *actualization of change,* for part of the efficacy of ritual performance is precisely transforming the fetus into a person. A person in this sense is a cultural representation, or more precisely an objectification of indeterminate self processes (see Chapter Seven and Csordas 1994a:5, 14–15). While both a fetus and a baby are biological entities, whether, and at what point, they are objectified as "persons" varies across cultures. The current North American debate is based on whether the person begins at conception, at birth, or in one of the culturally established "trimesters" between the two. In cross-cultural perspective we see that the issue of personhood extends even beyond birth, however. Among the Northern Cheyenne, children are not participants in the moral community because they lack knowledge or responsibility for their actions, and are therefore considered only "potential" persons (Fogelson 1982; Ann Straus 1977). Among the Mande peoples of Africa, a newborn is not yet a member of the worldly family, remaining unnamed till eight days after birth. The shape of the placenta is examined to determine whether the newborn is in fact not a human person but a *saa* or spirit child (R. Whittemore, personal communication). Among the Dogon, a fetus is conceived as a kind of fish until it has received a series of names and has been circumcised or excised, at which time only it is recognized as truly a boy or girl (Dieterlin 1971:226). For the Tallensi, "it is not until an infant is weaned and has a following sibling (*nyeer*) that it can be said to be on the road to full personhood," a status that is in fact "only attained by degrees over the whole course of a life" (Fortes 1987:261). Among the poorest of Brazil, children are often neither baptized nor named till they are toddlers, and the infant that dies is considered neither a human child nor yet a blessed angel. Instead, "the infant's humanness, its personhood, and its claims on the mother's attention and affections grow over time, slowly, tentatively, and anxiously" (Scheper-Hughes 1990:560).

Such examples could be multiplied, and indeed a paper by Lynn Morgan (1989) does a masterful job of synthesizing the cross-cultural data on the personhood of neonate humans. However in all of these examples, the contrast with the Charismatic practice could not be more striking: Whereas in these instances an already-born infant is *not yet* a person, in Charismatic healing a never-to-be-born fetus is *still* a person. The difference is doubtless grounded in the circumstance that in the former cases, where infant mortality is high, no infant can necessarily be expected to survive, whereas in the middle-class North America of the Charismatics, no infant is ever expected to die. Nevertheless, in all the cases it is the ritual action of naming (and baptizing or its equivalent) that bestows the cultural status of person. Phenomenologically reinforced by imaginal performance, part of the actualization of change in the healing of abortion is creation of a person that can subsequently be prayed for and regarded as being "with Jesus."

This is not all, however, for in this instance actualization of change includes the dual movement of "accepting responsibility" and "letting go." In the healers' account the supplicant's screaming must be categorized as a kind of therapeutic breakthrough that was buffered as they "loved her through that in collaboration with the divine presence." The rather peculiar juxtaposition of "scary" and "beautiful" to describe the situation carries a dual message related both to efficacy to situational dynamics. To redefine a scary situation as a beautiful one is at once to say that what was potentially negative and dangerous was, in fact, highly successful—beauty is synonymous with efficacy. At the same time, it is an acknowledgment that the dynamics of the situation nearly got out of hand but didn't and here, beauty is synonymous with control. Finally, the actualization of "letting go" is the epitome of the Charismatic surrender of control to the deity in exchange for emotional freedom. Here again is a dual meaning. On the one hand, the supplicant "lets go of" the guilt expressed in her cathartic scream, and on the other she "lets go of" her cherished maternal intimacy and the associated grief over its absence by relinquishing the imaginal baby.

In brief summary, in the Charismatic rite for healing abortions we see the rhetorical power of multisensory imaginal performance to create a proto-Gileadean cultural reality for women who participate in the ritual healing system of the Charismatic Renewal. A clear ideological choice is made not to make them feel alright about what they have done but to presume their guilt and absolve them of it through divine forgiveness; not to affirm the pre-personhood of the fetus, but to create a person and bestow upon it an identity by naming/baptizing it and specifying its gender; not to emphasize the termination of the woman's pregnancy but the death trauma of the fetus and to resolve it by commending the unborn soul to the care of the deity.

In her important cultural analysis of the abortion debate in the contemporary United States, Faye Ginsburg (1989) identifies a series of what she calls "interpretive battlegrounds" in the struggle between prochoice and prolife forces. The Charismatic ritual is not a public battleground, but an internal ideological exercise where what is at stake is to intensify the world view that binds the ranks of antiabortion warriors by ritually enacting that world view in a way that displays its doxic qualities. The spontaneous entrainment of multisensory imagery is a product of deeply inculcated dispositions of a patriarchal habitue, and by its spontaneity is a rhetorically powerful display of an ethnopsychological reality. In this capacity the healing ritual goes beyond addressing the issue of fetal personhood to play a powerful role in what Ginsburg calls the "re-negotiation of pregnancy, childbirth, and nurturance … in the construction of female gender identity in American culture" (1989:110). Since the legalization of abortion, motherhood can no longer be presumed to be an ascribed status, the inevitable result of pregnancy conceived as an inevitable process in women's lives.

Instead it becomes an achieved status, the result of a decision that "comes to signify an assertion of a particular construction of female identity," in the face of necessity for rhetorical strategies for reproducing the culture in the absence of its formerly taken-for-granted self-reproduction (1989:109). The ritual undoing of the abortion is just such a strategy, restoring through imaginal performance the inevitability of pregnancy, childbirth, and nurturance. Ginsburg argues that an essential aspect of prolife political action is "the refiguring of a gendered landscape through prayer, demonstration, and efforts to convert others, particularly women in the vulnerable and liminal position of carrying an unwanted pregnancy" (1989:110). The Charismatic healing of abortions extends this refiguring from women who choose to carry an unwanted pregnancy to women who once chose not to carry a pregnancy.

In the example recounted above, the patient was chastised for blaming her husband, an escape from responsibility by citing lack of accountability in the face of the patriarchal authority of the husband. On the one hand, the healer's insistence that she take a share of responsibility for the decision to abort may seem to proffer a degree of empowerment, and the inclusion of the husband in the ritual carries the message that the woman is not abandoned to the emotional consequences of the abortion. On the other hand, insofar as the notions of sin and guilt are inevitably contained within this acceptance of responsibility, the patriarchal logic is enforced wherein the woman is obligated to bear children at all costs, even if her husband abdicates his procreative conscience.

## *Japanese* Mizukoo Kuyo: *Notes toward a Comparison*

In the above discussion of efficacy I situated the Charismatic ritual ethnologically by surveying definitions of the objectification, or coming into being, of persons across a variety of cultures. In this final section I want to return to the same theme with a more precise comparison in mind. (Contemporary Japanese society is the site of a more public ritual practice of postabortion healing.)[3] It is a ritual in which the spirits of aborted fetuses are propitiated through prayer and through representation by stylized statues or tablets. These rites are called *mizuko kuyo,* where *mizuko* refers to fetuses miscarried, stillborn, and aborted, as well as the already-born who succumb to infanticide (LaFleur 1992:16) and *kuyo* is a type ritual based on an offering of simple gifts in thanks to objects or beings that have been in some sense used up, ranging from domestic objects like sewing need to deceased humans (LaFleur 1992:143-146). The *mizuko kuyo* rites appear to be essentially Buddhist in nature, but originated in the social context of the Japanese New Religions since the 1970s (Blacker 1989), and are cited as evidence of the commercialization of contemporary Japanese religion since they are often highly profitable to the temples and organizations that perform them (Picone 1986). In what follows, I will briefly discuss the Japanese Buddhist *mizuko kuyo* in relation the North American Catholic

Charismatic healing of abortions in order to begin to point to the place these overtly similar practices occupy in the cultural configurations of their respective societies.[4]

First let us take care to contextualize the relative social space occupied by these two practices. The American practice is largely a private one that takes place within the membership of a discrete religious movement within Christianity and is a specific instance of the healing system elaborated within that movement. The Japanese practice has a relatively public profile not limited to a particular social group and is an instance of a type of ritual common to a variety of forms of Buddhism. Historically, the Charismatic Renewal and the *mizuko* cult are contemporaneous, products of the post-1960s cultural ferment that spawned the New Age Christian fundamentalism, a renewed interest in Eastern spiritualities in the United States, and the various New Religions and a fluorescence of interest in spirit possession in Japan. Just as the Charismatic Renewal and other forms of neo-Pentecostalism have been associated with the neoconservative Christian right America, some of the Japanese *mizuko* have been observed to have right-wing fundamentalist, naturalist, or Shinto connections.

In the United States, abortion was legalized for the first time in the early 1970s as a result of the Supreme Court decision in *Roe v. Wade,* while in Japan abortion has a deeper history. Both abortion and infanticide were common from the early 1700s to the mid-1800s, when an abortion debate ensued among Buddhist, neo-Shinto, and neo-Confucian positions in the context of a nationalism that demanded population growth and condemned such practices. Only following World War II in 1948 was abortion again legalized. Since that time, it has become the most popular form of birth control in Japan. Just as in the context of the American abortion debate the Charismatic prayer for healing tends to emphasize the aborted rather than the stillborn or miscarried fetus, in the context of the postwar commonality of abortion the aborted fetus has taken precedence as the primary referent of the Japanese term *mizuko.*

In both societies the affective issue addressed by the ritual is guilt, whereas in the United States this is a guilt occurring under the sign of sin, in Japan it is guilt under the sign of necessity. For the Americans abortion is an un-Christian act, and both perpetrator and victim must be ritually brought back into the Christian moral and emotional universe; for the Japanese both the acceptance of abortion as necessary and the acknowledgment of guilt are circumscribed within the Buddhist moral and emotional universe. Both rites are intended to heal the distress experienced by the woman, but the etiology of the illness is somewhat differently construed in the two cases. For Charismatics, any symptoms displayed by the woman are the result of the abortion as psychological trauma compounded by guilt, along with the more or less indirect effects of the restive fetal spirit "crying out" for love and comfort. In Japan such symptoms are attributed to vengeance and resentment on the part of the aborted fetal spirit that is the pained victim of an unnatural, albeit necessary, act.[5] Finally, not only the etiology but the emotional work accomplished by the two rituals is construed differently. As we have seen, for the Charismatics this is a work of forgiveness and of letting go. For the Japanese it is a work of thanks and apology to the fetus, where in cultural context gratitude and guilt are not sharply differentiated. Thus, "[t]here is no great need to determine precisely whether one is addressing a guilt—pre-supposing 'apology' to a *mizuko* or merely expressing 'thanks' to it for having vacated its place in the body of a woman and having moved on, leaving her—and her family—relatively free of its physical presence" (LaFleur 1992:147).

We can now compare the two postabortion healing practices with respect to what they assume and what they produce with regard to the ethno-ontology of the person. The American Charismatic ritual is largely an "imaginal performance" (confer Csordas 1994a) in which the woman may vividly experience holding the imaginal fetus/baby, while the Japanese ritual typically includes the concrete representation of the fetus/baby in the form of a statue. For the Americans, the fetus is a distinct little being that at a certain point is given over to Jesus who is its savior and protector. The Japanese statue *(mizuko jizo),* on the other hand, assimilates the infant and savior in the same representation, a bald and diminutive monklike entity with infantile features sometimes described as "the Bodhisattva who wears a bib." This contrast in the ontological status of the fetus is recapitulated in the respective cultural notions of the coming into being of persons. American Charismatics regard personhood to be definitive at the moment of conception, whereas for Japanese becoming a person is neither a matter of conception nor of birth, but a gradual ontological process wherein "in coming bit by bit into the social world of human beings there is a thickening or densification of being," the inverse of a thinning of being as a person ages into ancestorhood and Buddhahood (LaFleur 1992:33). Thus, for the Charismatics, abortion is the definitive termination of a human life, while in the Japanese vies the aborted fetus can as easily be thought of as returning to a state of prebeing where it may be held till a later date as to a state comparable to that of deceased ancestors.

Given these differences, the intent of the Charismatic ritual is to move rhetorically the dead fetus ahead into a secure post-life union with the deity, whereas the intent of the Japanese ritual is to secure the fetus' good will either as it slips back into its pre-life state or as it advances to the realm of the Buddhas. Charismatics tend to eschew the old Catholic folk notion of a limbo where unbaptized infants must remain separated from the deity (Linn, Linn, and Fabricant 1985), whereas Japanese may embrace a kind of limbo from whence the fetus may return at a later date. In this respect it is instructive to consider the difference in meaning of the ritual symbolism of water and of naming. In the Charismatic ritual imaginal water is used to baptize the fetus, an act that ensures the reunion of the fetus with Jesus. In the Japanese case, water is an essential element in the very definition of the fetus: The term *mizuko* means literally "children of the waters," which in a literal sense refers to the amniotic fluids, while in an ontological sense refers to the ambiguous status of the fetus we have been discussing. Whereas for Charismatics water baptism and return to Jesus is the cultural constitution of the fetus as person, the use of water symbolism in Japan highlights the fluidity of being that characterizes the ontological status of the fetus. Given that in Buddhism impermanence, suffering, and the absence of self are fundamental characteristics of all things, "the fetus as a *mizuko* in the process of sliding from its relative formedness as a human into a state of progressive liquidization is doing no other than following the most basic law of experience" (LaFleur 1992:28). A similar point can be made with respect to naming the aborted fetus. For the American Charismatics, naming is an aspect of baptism that contributes to the objectification of the fetus as person. For the Japanese, while the process of bestowing a posthumous ancestral name (*kaimyo*) is often a part of the ritual, it is often controversial whether it is more appropriate to allow an unnamed fetus to "slip back" into pre-being or to be named and thereby advanced into a state comparable to ancestorhood.

Contemporary civilization has advanced too far into the process of globalization to allow us to presume that the two rituals we have been discussing are necessarily isolated one from the other. Werblowsky (1991) critically refers to claims that there is a movement in the United States that is learning from Japan to fill the lacunae within Christianity, and sarcastically asks whether "in addition to their belief in souls they also believe (in good Japanese fashion) in family trees of souls, in which the souls of even unborn children remain closely related to the ancestors" (1991:327, 328). In this Werblowsky appears to confuse the movement associated with the label of "Zen Catholicism" among progressive Catholic monks with the quite separate and markedly more conservative Catholic Charismatic Renewal. The former is doubtless connected in some degree with the Japan-based Catholic journal of religious studies in which Werblowsky's own article appears. In his own text, however, he implicitly refers to the Catholic Charismatic Renewal, even citing the work by Linn, Linn, and Fabricant. While in addition to Zen Catholicism there is some proselytizing with respect to *mizuko kuyo* on the part of Japanese Buddhists in the West (confer LaFleur 1992:150, 172), if such an influence is present among Charismatics it is certainly less direct than Werblowsky presumes Charismatic healers Linn, Linn, and Fabricant in passing acknowledge awareness of *mizuko kuyo,* citing another Charismatic author who in turn cites an article in *The Wall Street Journal,* of the practice of Japanese women "increasingly going to Buddhist temples where they pay $115 for a ritualized service to get rid of their guilt for the abortion, experienced in recurring bad dreams" (1985:128).

On the other hand, to answer Werblowsky's comment about family trees in the 1980s many Charismatics adopted a form of healing called, variously healing of ancestry or healing the family tree. Along with their more psychological interpretations of guilt and grief, Linn, Linn, and Fabrican (1985) favorably cite this notion, popularized by the British Charismatic psychiatrist Kenneth McCall (1982). They write that the fetus that has not been lovingly accepted by its family and committed to God "will cry out for love and prayer to a living family member," with subsequent psychological impact on parents, on parents' abilities to relate to older children or children yet to be born, and on such children themselves. What is noteworthy here is that McAll's practice was inspired by observing Chinese practices with regard to ancestors and ghosts, implicitly assimilating them to souls in purgatory or limbo, while living and practicing abroad. More significant than whether the Charismatic practice is an instance of either classic cultural diffusion or spurious cultural borrowing, what this suggests is that despite its overt fundamentalist tendencies, the Catholic Charismatic Renewal and contemporary New Religion/Buddhism are mutually participant in the globally prevailing postmodern condition of culture.

## Conclusion

For a society in the throes of moral debate about abortion, where claims are made in terms of moral absolutes, the limits of cultural relativism are tested with the mere observation that "ritual performance creates a cultural reality." In this chapter I have attempted to give an account of the creation of meaning and the nature of therapeutic efficacy in a ritual that rhetorically partakes in this serious cultural debate in contemporary American society, and to contrast it with a parallel ritual in contemporary Japan. The account and the cross-cultural comparison point beyond relativism to the observation that within the limits posed by their own configuration, cultures can create and define the very problems to which they then develop therapeutic solutions. In the end, to cultivate guilt in order to relieve it is doubtless a form of creativity, but this cannot be said without also acknowledging that one of the products of human creativity can be human oppression.

# CHAPTER FOUR

# The Affliction of Martin

To say that "the glass is half empty" or "half full" is to give an existential account of an objective circumstance: The level of fluid is at the midpoint in the glass. To say that an individual is afflicted by "demonic oppression" or "psychopathology" is to give a cultural account of an existential circumstance: A person is suffering. The methodological gulf between the two kinds of accounts is vast. One can appeal to the objective circumstance of fluid in the glass to understand the derivation of optimistic and pessimistic accounts, but how can one define an existential circumstance prior to the elaboration of a cultural account? Equally as important, if one were to develop such an existential language, would it be of value in understanding the cultural logic that distinguishes divergent cultural accounts?

In this chapter I will examine the cultural and existential relation between religious and clinical understandings of human suffering. It is popular to note that what in previous centuries was understood as demonic possession is in the present era understood as psychopathology. Recent work, such as Kenny's (1986) study of multiple personality, suggests that as now, in the nineteenth century the discovery—or perhaps creation—of psychiatric disorders was deeply imbued with cultural meaning. Kenny shows that the experience of individuals who were the prototypical cases of multiple personality in the United States embodied cultural conflicts associated with the nature of personhood, and that their experience was framed alternately as religious and medical. Even more striking is the tragic case documented by Goodman (1981) of the young German woman who died in 1975 after a conflict-laden interaction between religious exorcism from demonic possession and psychiatric medication with anticonvulsant drugs. The fact that the case ended in the court system indicates not only the currency of religious paradigms for understanding distress, but the continued social inability to translate between sacred and psychological interpretations of human reality.

The ethnographic backdrop for our approach to the problem is once again the Catholic Charismatic Renewal in the contemporary United States. We begin with a phenomenological narrative of intense spiritual, physical, and emotional suffering of a young man and the attempt by a Catholic Charismatic religious healer to "deliver" him from the influence of an evil spirit. The narrative is based on conversations with the protagonists over the course of two and a half years. The story and subsequent analysis is one of winks upon winks (Geertz 1984) and of multiple perspectives within divergent cultural accounts. The key informant is the healer, whom we shall call Peggy. As often happens in families of the severely mentally disturbed, she became the one to adopt the role of spokesperson for the afflicted young man, whom we shall call Martin. This situation, combined with the facts of their parallel forms of suffering, and that Martin could not bear to be interviewed himself (was present only at the first interview), made analytic separation of their experiences a near impossible task. Therefore, the narrative must be understood as a text produced in the interviews with Peggy, rather than as a case study of Martin's experience.[1] What follows is a discussion of the cultural logic in two accounts of the case based on commentaries by mental health professionals and Charismatic healing ministers. Finally, a phenomenological analysis of bodily experience provides the grounds for showing how both accounts are cultural objectifications of meaning that is already inherent in basic sensory experience.

## The Affliction of Martin

At the time of our first meeting, Peggy was 42 years old, a mother of three (another child died at birth). She had been married to a successful professional for 20 years. She had long been interested in things spiritual, having practiced yoga since age 14, and having out-of-body experiences and visions of figures of Catholic devotion such as Theresa of Lisieux. She completed 2 years of college in math and chemistry, and early in her marriage worked as a laboratory technician, but had been a housewife for the past 17 years. She and her family are practicing Catholics; her husband is an active layman and plays a role in adult religious training (catechesis) in their Parish. He is aware and somewhat supportive of her activities as a healer, but in practice remains aloof; he never participated in our interview sessions, which Peggy intentionally scheduled for times when he was away from the house, and he never participates in her healing sessions.

Although it is not strictly required among Catholic Charismatics that a healer experience his or her own healing in order to pray for others, Peggy made a surprisingly strong claim of never having felt a need for any healing because her past had been without trauma (this is in spite of the loss of a child at birth) and because her relations with her parents had always been smooth. Nonetheless, she claimed specific experiences to validate her self-definition as a healer, including having heard a voice that said, "You will heal for me."

Peggy already had experienced this ability to heal when, at age 35, she encountered the Catholic Charismatic Renewal. Her husband had heard about a local prayer group and suggested that she might be interested. She attended the movement's standard initiatory program (Life in the Spirit

Seminar) during which she heard God say, "Come follow me," and received the gift of tongues (glossolalia). Following this experience she attended both the Catholic prayer group and a nondenominational Pentecostal group, and reports becoming recognized as a healer in both. In the Catholic group, healing was always prayed for in a collective prayer, and conflict arose when one woman asked Peggy for a private, one-on-one healing session. The group regarded this as potentially dangerous, and as a result Peggy left. She continued her individual healing prayers, but at the time of our first meeting had not been involved in a group for 4 years.

Peggy's methods of healing are somewhat unorthodox vis-à-vis the mainstream of Catholic Pentecostal healing. Although her principal methods (Healing of Memories and Deliverance) are those recognized by Charismatics, she began healing independently of, and prior to, her involvement with the movement. She has not had any guidance from other healers or read any of the movement's substantial literature on healing, but states that her training comes directly from God. She always prays about whether to accept someone as a supplicant in healing, but does not always work through prayer; if the person is not particularly religious, she may not even speak about God in healing.

Her diagnosis of a person's problem is based on her "psychic" gifts: She "becomes" the other person in the sense that she knows their subconscious mind and their past, and she conducts a "mini-scan" of the supplicant by means of visionary "seeing," although she always respects the person's privacy by never probing deeper than she can feel the person wants her to. She also reads a person's "aura" and interprets the meaning of light glowing around each of a person's *cakras* (that is, chakras—this is Peggy's idiosyncratic borrowing from Hinduism and is not typical of Catholic Charismatic thinking). She makes astrological determinations when she deems them appropriate, experiences precognitive dreams, interacts with good and evil spirits, utters words of prophecy, and also gives nutritional counseling. She occasionally consults with non-Christian psychics to assist or confirm some course of action in her own work.

Peggy has a close friend, whom we shall call Randy, a man of about 30 years of age. They share the same spiritual orientation and interests, and often attend visiting evangelists and healers. Randy often spends the evening with Peggy and her family, visiting or watching television. Three years before our interviews, Randy convinced his housemate to see Peggy for spiritual healing. This young man, Martin, was 22 at the time, but his story, by his own and Peggy's account, begins much earlier, when he was 9.

During his childhood, the relation between Martin's parents had not been smooth. He remembers his father as rather cruel, fighting with his mother and being physically and verbally abusive to him. He believes his parents had virtually no sexual relationship. His mother suspected that something was wrong with her husband, but the family doctor did not think it too serious. Then, when Martin was 9, his father committed suicide by shooting himself in the head. Martin heard the shot and found the body. Soon after this event, his mother had a "nervous breakdown" and was committed to a mental hospital; she has been hospitalized several times since.

Martin and his brother (five years his junior) were committed to an orphanage. Later they were assigned to the care of foster parents, adherents of a strict form of evangelical Christianity. Martin's brother eventually rebelled against this environment and moved abroad, severing contact with his mother and Martin as well as with his foster parents. Martin did not rebel overtly, and remembers always being eager to please his foster parents. Peggy believes that the fear of God inspired by their upbringing prevented him from committing suicide during his darkest moment of suffering.

Martin began having sexual fantasies at age 13, but these were apparently not of an obsessive nature. He also reports "seeing colors" at that time. At age 15, he had an experience that he described as feeling like a net descending over his head and enclosing it; at that time he developed a chronic headache that has been with him constantly for the past 10 years. Also at that time, he began experiencing vivid erotic images that were so pervasive and uncontrollable that (in retrospect) he perceived "pornography as a state of being within me." He also began to hear a voice that offered him friendship and companionship, at the same manipulating him and "making deals" with him.

As a freshman in college, Martin developed a terrible guilt about his sexual obsessions, knowing that they were in conflict with the upbringing given by his fundamentalist foster parents. Despite increasing discomfort, he successfully graduated from the nearby state university Phi Beta Kappa in chemistry and mathematics (Peggy made a point of reporting that he has an IQ [intelligence quotient] of 140). Following graduation, he experienced uncontrollable convulsions, but was able to take a laboratory job in biochemistry, and he worked successfully for a year. He experienced increasing difficulty in learning things for his job, with chronic headache and stomach pain accompanying his "fuzzy thinking." A neurologist found no sign of a tumor and recommended biofeedback for his headache, indicating that it was a physical handicap that he must learn to live with. Aspirin and prescription drugs provided no relief. Martin became almost completely disabled and had to leave his job. In the 60 days before he met Peggy, he apparently had almost reached the point of convulsions again. Peggy reports that he "had not slept" during this two-month period and his housemate, Randy, observed that he was "wandering around the house like a zombie."

When Peggy first met Martin, his condition had deteriorated to such an extent that she had him move into a spare room in her house until he could get past the crisis. At this point he had lost considerable weight and was very "dense" in conversation. As he regained some of his physical strength, Peggy began the process of Healing of Memories, retracing the events of Martin's life and praying for the healing of resultant emotional scars or "brokenness."

In the course of Healing of Memories, Peggy felt that Martin successfully overcame a great deal of anger toward his suicidal father, and that his attitude toward his natural mother was also substantially ameliorated (she resides in federally subsidized housing in a neighboring city, living on income from social security and stock investments). Whereas previously he could hardly treat her with civility, he became able to talk to her, to tell her he loved her, and to treat her with compassion. Martin grew stronger physically and emotionally, and regained lost weight. He moved back to his own house, although he still could not work, and spent many of his days at Peggy's house. Given his state at their first encounter, Peggy, Martin, and Randy all agreed that the healing experience literally saved his life. But Martin was not yet healed.

Although he "worked through" a lot of anger toward his father, Martin still felt "more anger than he would like to." (In Healing of Memories, sincere forgiveness of past wrongs is a major component of successful healing.) In addition, he still experienced the full complement of chronic, intractable problems to be described below. Although Peggy at first thought that Healing of Memories, by resolving Martin's "brokenness," would eliminate these problems, she became convinced that they were due to the influence of an evil spirit. In effect, Martin had reached a plateau. Peggy saw him as emotionally "back together" in the sense that, intellectually, he knew what he *wanted* to feel, but was blocked from experiencing it by the evil spirit.

It was God, states Peggy, who told her that she was dealing with an evil spirit or demon. She had thought that she was telepathically "picking up" some of Martin's thoughts, but now realized that it was the spirit speaking. God told her (through inspiration) to "confront" the spirit to determine its identity. When she discovered its presence, it reacted violently with the telepathic message that it was going to "knock Martin out." Peggy tried to get him to sit down, but suddenly he fainted.

The exchange between Peggy and the demon had been silent, and Martin had not anticipated passing out or known anything was "going on." This occurrence was one confirmation of the spirit's presence. Peggy then demanded its identity, and discovered it was a Spirit of Pornography, or alternatively a Spirit of Abnormal Sexuality.[2] This name was "confirmed" in a phone conversation with a local psychic of Peggy's acquaintance, and through several incidents involving Randy and an out-of-town house guest.

Although Peggy thought at first that the spirit was external to Martin, oppressing and harassing him from "outside," she became convinced through its intransigence and degree of influence that it had taken up residence "inside" him, while still not achieving absolute "possession" of his personality. Yet this Spirit of Pornography is itself vassal to a more ominous spirit outside of Martin, but lurking silently nearby. Peggy and Randy succeeded in catching visionary glimpses of this spirit. It appeared as a silent male; tall with dark hair, cloaked, dignified, and exuding a cool power. They discovered its name to be Andronius, but reported no special significance to that name other than that it indicates a high rank in the demonic hierarchy. The role of this spirit was ambiguous, but its presence made Peggy's task as a healer more formidable, and increased the drama of the situation.[3]

Critical to the understanding of therapeutic process in many forms of religious healing is the manner in which life events and symptoms are reinterpreted and made consistent in terms of religious themes (Bourguignon 1976; Monfouga-Nicolas 1972). In the present instance, a double reinterpretation occurred. The first was intrinsic to the Healing of Memories, where Martin was reconciled to the overwhelmingly traumatic events of his early childhood. Themes of brokenness, forgiveness, and the invocation of the healing presence of Jesus within the memory of traumatic events are essential to this reinterpretation. With the lack of success in this healing, the discovery of a demon's presence motivated a second reinterpretation. The evil spirit became a link between the traumatic life events and the genesis of Martin's symptoms.

Specifically, in the Catholic Pentecostal healing system evil spirits are thought to prey upon the vulnerability created by such events. Thus the spirit began hovering around Martin at age 9, after his father's suicide. While in the orphanage, Martin had participated in a séance with some other children, during which he "saw things moving around in the room." While this event had had little relevance to him, it was now recalled and attributed significance as an incident of involvement with the "occult" that allowed the demon to increase its purchase on Martin's "personality." His recollection of "seeing colors" was interpreted by Peggy in terms of the emission of light from the "cakras" as the spirit took over different parts of his body.

Thus, the ostensibly normal sexual fantasies that began at age 13 became obsessive by age 15 as the spirit, which had now entered Martin in earnest, took advantage of the natural erotic drives of an adolescent. Peggy was able to see a "gray mass" engulfing Martin's head, apparently a visionary analogue of the "net" he had felt descending over him at age 15. The voice Martin hears sometimes represented itself as three separate voices, although Martin was convinced that it was really only one spirit. Peggy explained this trickery as an example of typical demonic "deception" (Satan is the Father of Lies), and the three-in-one motif as a "blasphemy" on the part of the spirit in diabolical mockery of the Holy Trinity. The difficulty in separating one's own thoughts from those of the spirit demonstrates how the spirit gradually "inches you away," taking over one's personality in incremental stages.

## The Sensory Organization of Suffering

Given this narrative account of Martin's early life, his crisis, and Peggy's initial attempt to heal him, let us turn to a more detailed description of the nature of his suffering. We shall not frame the description either in terms of pathological symptoms or of demonic manifestations, but instead organize it based on the observation that all of Martin's sensory and cognitive modalities are engaged and enveloped in suffering.

The auditory modality is dominated by a voice experienced as audible; however, Martin does not exactly claim to "hear a voice," instead, he "hears thoughts." He described the way the voice would manipulate him in earlier years. He believed it "knew everything about me." It would make deals with him, such as agreeing not to talk about his father if he did certain things it wanted. It would discuss religion with him; on one occasion, by "outwitting" him in such a discussion, it was able to "get deeper into his gut." It would also make jokes about Martin's foster mother. About his natural mother it would say, "My mother left me" instead of "Your mother left you," in such a way that he tended to lose consciousness of the voice's separateness from himself. The change of grammatical person, "makes you think it's your own thought." In a separate auditory experience not involving the voice, a loud crackling sound occurs when the voice temporarily "releases" its hold.

The pornographic obsession engages Martin's visual modality with compelling eidetic imagery. This imagery can arise spontaneously at any time, even awakening him from sleep. Martin is typically "bombarded" with sexual pictures. The images move rapidly from one scene to another. They often begin with heterosexual pictures but progress to homosexuality, bestiality, sex with children, and sex that includes pain and violence. The pictures are accompanied by almost overwhelming sexual feelings, making resistance (to masturbation) very difficult. These feelings are, in turn, accompanied by equally powerful feelings of acceptance and desire to yield to the influence of the images. It feels "good" and "right" to give in and, at such moments, the voice's presence feels like a kind of friendship and companionship that it is difficult to forego. If there is some weakening of his will in response to the onslaught of pictures, his omnipresent head pain is somewhat diminished. If he begins again to resist, the pain reintensifies. Martin is not always successful in his resistance, and sometimes succumbs to the desire to masturbate. He is concerned that the trajectory of his experience is in the direction of seeing his world exclusively in sexual terms, but continues to struggle against this.

The tactile modality of Martin's experience is dominated by his intractable headache, the pain of which diminishes somewhat if he slackens his resistance to the images. The feeling of a knot in his head accompanies the episode of imagery. This knot remains even when the pain moderates. He often feels a "yanking" and "pulling" in his head, often in response to situations with spiritual content or import, but this experience is only infrequently visible to others in the form of a tic. Periods of maximum affliction are evident by glazing over of Martin's eyes and a thickening or swelling of his eyelids. He also experiences periodic pain in his joints, and varying degrees of pain in his stomach and groin areas. The sensations in his stomach and groin also include yanking or pulling; for example, his testicles are sometimes yanked. These sensations occur at unexpected times, and he often experiences them as sexually stimulating. Martin also feels a yank in his gut any time his father is mentioned. He periodically feels great heat in his body, and his body temperature fluctuates. Sometimes his body feels "like jelly"; there is a "fluid movement" within him, "as if another person were trying to embed its personality" in his body. This configuration of physical sensations is capped by a generalized heavy feeling, as if he was weighted down, and a constant drain of energy. Finally, in one instance during his association with Peggy, he suddenly fell as if violently thrown to the ground.

The gustatory modality is engaged as a terrible taste in Martin's mouth, and at times his food tastes bad. His saliva sometimes thickens, particularly when he first wakes in the morning. Martin's sense of olfaction is not directly affected, but instead he exudes odors that repel others, including extreme halitosis and body odor. These are so strong that, hours after Martin had left after spending a night in a spare room at Peggy's house, her young daughter observed that the room "smelled just like Martin's mouth."

Specific distortions of thought and emotion are also woven into this configuration of affliction. Martin has extreme difficulty maintaining concentration, especially on religious topics. When he accompanies Peggy and her family to Church on Sundays, he finds it difficult to pray. His regular religious reading "blurs out" in front of him. From our own analytical viewpoint, his experience appears to conflate concentration and attention with intensity of feeling and religious faith, for along with these examples Martin included the observation that while he was reading through the Catholic exorcism rite, "I was not sure I was believing it while I was doing it." In addition, only with great force of will can he carry on normal activities such as conversation or computer work. As with his attempts at religious reading, if conversation turns to theological matters his concentration will slip, his eyes glaze over, and his eyelids will visibly thicken.

Among emotional distortions he reports experiences of anger in situations where it is, in retrospect, either inappropriate or a clear overreaction to a minor irritation. He also experiences powerful feelings of anxiety and fear. Martin summarized the overall effect of his tribulation as creating "panic in his body," and the feeling that he was "running from himself."

Martin's sleep patterns are also disrupted. It will be recalled that he had severe insomnia for two months before entering his healing relationship with Peggy. At the time of our interviews, he was still subject to periodic insomnia and was sometimes awakened in the night by a stream of pornographic thoughts and images.

In a phenomenon she described as "psychic mirroring," Peggy began to experience many of the same forms of suffering as did Martin. In general, she feels that through experiencing the pain of others, she can absorb and neutralize that pain. With Martin, however, she acknowledged that the situation "got out of hand," and the demon began to attack Peggy herself at her own "weak points." She reported all the same symptoms as did Martin, with the exception of pain in the joints and halitosis and body odor. Often their bouts with the evil spirit were simultaneous, though their content was not necessarily the same.

Peggy reported beginning to feel some degree of head pain as soon as she heard of Martin, three years before our interviews. The spirit began "jerking her head around" and attacking her in other ways as soon as God told her its identity and she "confronted" it. Later the pain became constant for her and at such an intensity that she had difficulty meditating, and her capacity to remember things was compromised.

The evil spirit also attacked her verbally, screaming, cursing, and calling her "every name in the book," for example screaming "You bitch!" as if from very far away.[4] She stated that this visionary tongue lashing meant that the spirit felt threatened by her because she was the one who uncovered its presence. The demon also tried to deceive and threaten her, saying, for example, "God has abandoned you, you love *me* instead," or "Give me to someone else and I will leave" (demons must be "sent" to Jesus or to hell, never to afflict another person). It may threaten her family with something of a sexual nature or saying it will throw them into the fire. Another rhetorical trick (regarded as typical of demons) is condemnation. If she feels tired in the morning, the spirit might say, "Go have a cup of coffee." If she had one, the spirit would deride her, saying "You know more about nutrition [than to use such an unhealthy substance]."

The evil spirit's sexual assault on Peggy included both visualization and physical sensation such as tugging at her ovaries and sexual probing of her vagina and breasts during the night. Her visualizations were "like a TV show," with a story of sexual episodes accompanying the images. At first these visualizations were somewhat abstract, but they began to include real acquaintances. The visualizations sometimes flowed in a constant stream, coming on almost like a hypnotic state, a state of preoccupation in which thinking clearly was extremely difficult. There was a kind of "pressure" on her eyes that induced her to look at people only sexually. She says, "You feel like you *are* pornography." The visualizations occur with no warning, day or night, sometimes awakening her from sleep. In the midst of one of our interviews, for example, Peggy reported that the vivid image of a woman's vagina had appeared across her field of vision, although she could at the same time still see us.[5]

Peggy reported being especially susceptible to the pornographic imagery when she lost her temper and when she was tired. The spirit also directly caused anger and the urge to utter vulgarities (coprolalia). Although she felt that she "has a temper" from her ethnic background, the spirit intensifies it. The spirit also controlled her facial muscles, drawing them into an expression of anger, and thereby trying to deceive her into feeling angry even when she knew subjectively that she was not.[6] Still, Peggy felt she had more control over her experiences than did Martin, in that she could still find some relief and mystical connection to God through prayer in spite of the intractable pain. She also felt better able to distinguish a thought or emotion of her own from one provoked by the spirit. For example, several times she said she felt anger at my presence, but realized it was the spirit's anger. That is, the spirit perceived a threat, and its anger therefore indicated that we should continue the interviews. On the other hand, Martin was only able to attend the first of our three interviews; the demon had persuaded him that his participation would literally be too painful.

Peggy interpreted her own suffering as essential to Martin's deliverance from the evil spirit. God "told her" that her pain would remain till it departed from Martin; that would be her sign that he was free. She felt that she was only an indirect target for the evil spirit, and that God wanted her to serve in this way as a "barometer" of Martin's suffering so she could understand its dynamic and so speak for Martin. She would not need a Deliverance herself, she felt, but would automatically be freed when he was freed. God also indicated that psychiatrists would be of no help in this process.

Randy also felt that he had a role to play. He only peripherally felt the presence of the evil spirit, sometimes having insomnia when Peggy and Martin had it, sometimes feeling unusual sensations of heat and cold (head hot while feet are cold). Sometimes during a sexual scene in a film he felt an external heat around his head, though this was "not explicitly sexual heat." Aside from these minor experiences, Randy felt that his situation paralleled Martin's in that his own career plans were adrift and at a standstill. He came to regard this apparent stagnation as part of God's plan in the situation so that he could, in a sense, keep Martin company insofar as the latter feels anguish about a life that has for several years been mired in disability. Randy felt that God would allow his own life plans to mature once Martin was healed.

### The Struggle for Integration

Peggy made several tentative attempts to obtain outside help, but felt that a general reluctance on the part of others to deal with problems of demonic origin, as well as the recalcitrance of the spirit itself militated against success. Three times Martin was in the presence of a highly reputed Catholic healer who conceivably could have helped. The first instance was when Peggy and Randy took him to a public healing service in a nearby city, conducted by a nationally known healing priest from New England. Martin became upset and angry; he left without receiving an anointing (with blessed oil) and was vocally angry at Peggy on the drive home. They regarded this as behavior entirely controlled by the evil spirit, whose interests were at great risk in the healing session.

Formal channels within the Church were not effective in obtaining help either. Peggy's parish priests claimed unfamiliarity with demonic phenomena and declined to become engaged in the situation. The local bishop responded that he did not understand such things, and referred them to a Jesuit priest who was respected as a counselor in the diocese. According to Peggy, this priest acknowledged the reality of her healing gifts but was not convinced of the need for Deliverance or exorcism in Martin's case.[7]

Peggy took only one step in the preliminaries to formal exorcism, a mandatory interview with a psychiatrist. It is noteworthy that she and not Martin made this interview. Peggy reports that after listening to her account of the prayer for Healing of Memories applied to the emotional scars of Martin's past, the psychiatrist concluded that Peggy had already done what she herself would do in therapy, and that it would only be a matter of time before Martin improved. Peggy's detailed presentation of symptoms caused by the spirit apparently made little impression on the mental health professional, either with respect to convincing her of the demonic attribution or in suggesting the need for further therapeutic intervention.

However, the psychiatrist reportedly did note that after three years of becoming almost a part of Peggy's family, the difficulty of separation might be creating an obstacle to Martin's improvement. Peggy acknowledged this possibility but also expressed confidence in her ability and motivation (cessation of her own pain) to see the situation to a conclusion. Thus, in isolation and in a painful stalemate with the devil, Peggy, Randy, and Martin faithfully continued to wait for God to determine the moment of Deliverance.

Peggy believed that Martin had great internal strength, and that it was increasing. She believed that, "he will do God's work powerfully at some point in the future." Indeed, this strength made him an important target for Satan. During the second year in which I followed his case, Martin decided to enter training as a Catholic catechist and was baptized a Catholic at Easter of the next year.[8] Peggy reported that after his baptism, Martin began to find temporary relief in the sacraments of the Catholic Church in the sense that the demonic oppression "lightened" from its usual "heaviness." Nevertheless, going to Church was difficult because of the spirit's resistance, and Martin's mind typically became "foggy" just before approaching the altar to receive the Eucharist. However, Martin made a choice to go ahead with his life. He said, "I can remain immobilized or go forward," and he chose to go forward. He enrolled in courses in automotive repairs and computers at the local technical college and soon began to repair cars for Peggy, her husband, and one of his housemates. He also took a course in theology from a priest at the local university, in spite of persistent mental "fogginess" that required him to read and reread his textbooks, and in spite of his persistent pain and uncontrollable sexual imagery. However, according to Peggy, Martin's small successes bolstered his will and strengthened his ego.

In the face of continued tribulation, Peggy claimed to achieve a deeper level of meditation than ever before, in which she found at least temporary refuge. People still came to her for healing, although at one point she had felt so exhausted from her struggle with the evil spirit that she had expressed the intention of retiring. After more than four years of struggle, the encounter with an evil spirit appeared to have settled into a way of life without further dramatic incident. Unable finally to free themselves from the spirit's influence, they concluded that they must, in Peggy's words, "learn to live with the appearance of absolute normality."

Just before Christmas in the year following Martin's baptism, Peggy described the changes that had occurred and the improvements over his previous state as being "on the miracle side." He had completed technical school training in automotive mechanics and continued his computer training. However, most of the basic problems persisted, including slowness, heaviness, dullness, and a weight on his entire body, especially his head pain in other parts of his body, and spontaneous imagery. He no longer experienced fluidity and dissolution of his body because, said Peggy, he now "knows the truth" about his problem. Nevertheless, his problems created difficulty in the school environment.

Particularly bothersome were uncontrollable sexual fantasies about classmates, apparently both male and female. The individuals who attracted his attention were not necessarily ones who "would normally appeal" to him, and he found it "annoying to look at some 300-pounder and want to go to town." Peggy explained that the spirit "sees through his eyes," so that although he does not see anyone who on the "real" level appeals to him, on the "other" level, he wanted to "jump everything that moves." Peggy confirmed that Martin had in college stayed away from girls because of his problem, suggesting that he did not date because of difficulty in "controlling his feelings." He currently had no female friends, and Peggy felt that this would not be advisable unless it was someone of the same spiritual orientation with whom he could be open about his thoughts.

Martin was also afraid that others would be able to perceive the demonic activity through his behavior. In particular, there was a "motion in his eyes" that Peggy could perceive because she is psychic. She said it looked, "almost like another eye behind his eye," but could elaborate no further: She did not know whether the spirit really has eyes of its own or not, but her perception of the hidden eye conforms to her interpretation that the spirit saw through Martin's eyes. She observed, however, that this manifestation occurred only when he was "weak," and often when they were out in public. She qualified this as no more than an exacerbation of a "natural nervousness" associated with going out and felt that other people perceive the demonic manifestations in Martin only as "a little nervousness." He was nevertheless concerned, and she found it necessary to "help keep him focused on the fact that no one can tell."

Peggy herself continued to be afflicted and was concerned about her own anger, worrying particularly that a situation might arise where she would "lose control." Her experience remained linked to Martin's in that whenever he lost a degree of personal control, she was affected. For example, while he was elsewhere working on someone's car, she suddenly felt a tightening in her head, her mind switched to sexual thoughts, and images began to flood in. At the same time, Martin had become frustrated in tightening a bolt and, when he "loses his balance," the evil spirit "releases its personality, which is pornographic." A second example was an instance in which Martin was at school while Peggy had the puppy out in the yard. She suddenly felt a "drop in a sexual place as if my stomach dropped down, *whump.*" She learned later that the same thing had been happening to Martin in his class, a sexual reaction to being surrounded by men and women.

By late summer, despite a continuing "block on his head" and mental "haze," Martin continued to increase his level of social functioning. Peggy's husband assisted with a reference in helping Martin obtain a half-time job working with computer applications of statistics in a division of the local university. In preparation for this, Peggy had spent all of May in "heavy prayer," reciting three rosaries per day, and was "told" by the Virgin Mary that the power of prayer would move the demon. Still it "applied pressure" to prevent Martin from beginning work, causing pain, fear, a turmoil of low self-esteem as it attempted to weaken him in its "emotional grip." Martin and Randy stayed awake praying the entire night before his first day of work. Peggy thought the whole effort would be unsuccessful because the spirit "does have a definite degree of control over him," but she was impressed at Martin's ability to maintain himself with little sleep.

One morning during his first days at work she once again commanded the spirit to depart, and Martin's chest "went *whump*" while she saw it pushed outward from the inside as if there were a "fist from inside, punching." In spite of this demonic harassment, Martin had settled into his work for the two months prior to my final interview with Peggy. His weekly routine included school every morning and work every afternoon. Following work he frequently swam in the faculty pool, which Peggy reported as having a beneficial "cooling" effect given the continued intensity of his suffering. Every night he ate dinner with Peggy and her family, and given that his computer was set up at their house, spent a good deal of his free time there. He had been visiting his mother every Friday for a year until she "slipped" and went off her medication for bipolar depressive disorder. On Sundays he regularly attended mass and communion. Randy, who had formulated a plan to live as a writer and support himself with a low-stress civil service job, typically joined the family for Sunday dinner.

The principal difference Peggy saw in Martin is an increased ability to "discern" his own thoughts and reactions from those of the demon. She elaborated by saying that he at one time had lost sight of his own pain, distancing himself from it and therefore having less sensitivity and feeling in his body. Martin reportedly stated that when the affliction first began he could distinguish between his own self and the alien presence, but that he had lost this ability as the demon progressively merged with his self. He now appeared to be regaining that ability, and could at times say "that is not my thought." He still sensed the pornographic images clearly, but also regarded them as alien. He expressed, according to Peggy, a desire for "freedom" and had become able to utter the prayer of "command" that the spirit depart from him. A principal difficulty that remained is that he had little or no contact with his emotions—no "heart feeling," according to Peggy—for example being unable to distinguish between "love" and "sex." In spite of this isolation from his emotions, he had been learning to conduct his life through an "intellectual" understanding of what is "good or right."

Peggy's interpretation of the situation at this time was that though the evil spirit was rather high-ranking in the demonic hierarchy (as indicated by its mysterious name), the persistence of their prayers had succeeded in denying it access to help from other higher spirits such that it was isolated and (presumably) on the defensive. Martin still at times exuded the "odor of a sickroom," either in his breath or as a body odor, and Peggy could still psychically perceive a gray aura around him indicating "unhealthiness"; she could see as well the spirit "moving behind his eyes." Thus, the problem was still there, but Martin was "moving on despite enormous odds" because "you go on with life."

## Religious and Psychiatric Meanings

In order to generate culturally competent accounts of Martin's affliction from within two distinct healing systems, I prepared a detailed case description based on fieldnotes that covered the period up to Martin's baptism as a Catholic. The case description was given to five Charismatic healing ministers with whom I had worked in a study of therapeutic process in religious healing. All five were recognized as legitimate healers in the local Charismatic movement. The description was also given to five mental health professionals, all of whom have substantial experience in both research and clinical practice. The text examined by these ten individuals was essentially similar to that presented above, such that the reader may formulate his or her own reading of the data in relation to the Charismatic and clinical accounts.

### *The Charismatic Account*

The five Catholic Pentecostal healing ministers included two nuns, a priest, another priest who works as a team with a laywoman, and a laywoman assisted by a team of five other women. Each (referred to hereafter as numbers *1* to *5*) is recognized within the movement, though their reputations range from local to regional to national. One priest has professional training in counseling and psychotherapy; aside from the background in pastoral counseling acquired by priests and nuns as part of their religious training, the knowledge of the others stems solely from their practice of Charismatic healing. Like the mental health professionals, they were told that the purpose of their comments was to "help separate religious and psychiatric meanings" of the case. In accordance with the principles of their healing system, they were asked not for their "diagnostic and dynamic impressions" but for their religious "discernment" of whether or not an evil spirit was present in the case, for an interpretation of what was wrong, and of why the healing seemed not to be effective.

The most striking result of this exercise is the uniform agreement that the healer is a source of the problem, while the issue of presence of an evil spirit appears secondary and is, in fact, the subject of some disagreement among the consultants. These two observations are linked by an important characteristic of the healing system. That is, the knowledge brought to bear on this case by the participating healers is *empirical* knowledge, based on concrete experience in healing encounters and systematized through sharing via media such as conferences and publications.

In evaluating the case description, there was less ambiguity in the healers' application of their empirical knowledge to the practices and experiences of their fellow healer than in their determination of the presence of an evil spirit which, ideally, requires face-to-face interaction with the afflicted person. More than this, however, the healers implicitly assume that if one is gifted and competent to identify evil spirits and to know how to deal with them, then one can use the more or less routine spiritual technique of commanding the demon's departure through invocation of a divine power that is, by definition, greater than that of the demonic. Hence, before elaborating the healing ministers' comments on the issue of demonic presence per se, we must elaborate on their critique of Peggy as a healer.

None of the commentators questions Peggy's motivation to help people as a healer, but the validity of her "calling" to act in that role is explicitly questioned by healers 1–4 and implicitly by 5. They attend closely to the details of the story of her beginning to heal. What appears as an archetypal motif of the prophet's or healer's call, "I heard the voice of God and tried not to pay attention, but the call was insistent and I had to obey," is challenged not on grounds that such calls never occur, but on grounds that it may not be valid in her case. One commentator wants to know *why* she resisted and, it is pointed out, the voice she heard might not have been the voice of God.

Likewise, it is not denied that the "prophecy" she heard at a prayer meeting in confirmation of her healing gift could have occurred; the challenge is that God never "makes" anyone attend an event against their will and that the prophecy she heard at that prayer meeting could have been a message intended for someone else. Even more suspect to the commentators is her statement that she never felt a need for healing in her own life; in the logic of the healing system, not only do healers often experience their own healing in the process of becoming channels of divine power, but the person in general who needs no healing "does not exist."

Peggy's healing practice is also suspect because of her isolation, both in terms of learning and of ongoing support for her work. The healing system is a social system embedded in the Catholic Charismatic movement, which is itself a "movement" of the Holy Spirit; that is, it is understood to be instigated by God. For a healer to say that all her knowledge comes directly from God is then, questionable, not because this is impossible and not because it has never happened, but because the resources for learning through books, tapes, conferences, and the experience(s) of others has been made available by God, and is, therefore, meant to be used.

Again, "gifts" of healing are granted for the sake of "ministering" to and building Christian communities, and are used appropriately only in such contexts. At the same time healing requires guidance, support, and prayer from members of such a community both for success and for the protection of the healing minister who in the nature of the enterprise is exposed to harmful influences. Thus, Peggy's alienation from both Charismatic groups with which she was involved casts further doubt on her validity and bodes ill for the success of her work in the view of the five healers.

Largely as a result of this isolation, healing ministers 1 and 3 explicitly judge Peggy to be "incompetent" as a healer. That is, her marginality does not place her outside the healing system: Her practices are in general recognized by the commentators, and even at some points affirmed, but she is regarded as beyond her abilities, unable to handle the situation because she is ignorant of how things really work. Thus, both healing ministers 1 and 4 draw attention to the "empirical fact" that God never requires a healer to confront a demon to demand its identity. God knows its identity and may reveal it, or one may command in the name of God that the spirit state its name (through the voice of the afflicted person), but one is never left to face a demon on one's own. Without her isolation she would have known that God does not work like that and would not have committed an error of technique. Likewise, although the healer may at times feel another's pain, it simply "does not happen" that such suffering is necessary for a healing to occur or that the healer's release is contingent on the deliverance of the afflicted. Were she not so isolated, she would have known that this is not acceptable; it would have been prevented or she could have obtained help and been saved the consequences of an error of interpretation.

The final criticism of Peggy's healing practice from within the logic of the Charismatic healing system is her use of "occult" practices in combination with, or in place of, healing prayer (healing ministers 1, 4, and 5). Peggy herself explicitly equated her psychic abilities with "what the charismatics would call discernment." In the perception of Charismatic healing ministers, however, a distinction is drawn between the divine gift of "discernment" and "psychic powers," the former is a gift from God, the latter is inspired by Satan. In at least one instance, the commentaries were not in complete agreement about all of Peggy's practices in this respect. Minister 1 grants plausibility to Peggy's feeling that a higher spirit may lurk behind that of Pornography, and that the healer could see it, but suggests that Peggy may be allowing herself to be carried away and should "keep both feet on the ground." Healing minister 4 questions whether Peggy has the spiritual gift of discernment that would allow her to have a visionary glimpse of the spirit, and, more forcefully, argues that unless such knowledge has some clear purpose in the healing, it is a psychic knowledge, the work of the devil.

However, other of Peggy's practices such as astrology, reading of auras, and examining the *cakras* are invariably proscribed and seen as occult, or Satanic, in origin and purpose. Healing minister 4 specified that although some of the *techniques* of the Eastern religions could be used in abstraction from their philosophical underpinnings, acceptance of their basic principles, regarded as contradicting those of Christianity, constitutes involvement in the occult and cannot be tolerated. Therefore Peggy, whether or not she knew it, was inviting demonic influence by the very nature of the practices in which she was engaged.

Given the above critique of Peggy as a healing minister from within the logic of Charismatic healing, one could readily conclude, "No wonder she was unsuccessful in her attempts to heal Martin, and no wonder she was herself susceptible to demonic attack." Yet within the commentaries, the presence of an evil spirit is by no means a yes or no issue, and we must now sort out the healing ministers' own ethnopsychology of the interaction between demonic and psychological forces. Given that most of the healing ministers hesitated to make a definitive judgment about demonic action, their commentaries reveal three areas of divergent interpretation: Relation between demonic action and mental disorder, relation between evil spirits and emotions, and suggestions for therapeutic intervention.

The first area is most clearly exemplified in contrasting healing ministers 1 and 4. The former is a nun with no professional mental health training but extensive experience in Deliverance, and the latter a priest with a doctorate in psychology, but who also has extensive experience in Deliverance. Healing minister 1 concludes that Martin's problem is primarily religious and spiritual and requires Deliverance, though psychiatric follow-up could be necessary or beneficial. She suggests that Martin is "obsessed" by the spirit, which means that it has taken up residence within him while not yet fully "possessing" his personality. Healing minister 4 explicitly labels Martin's symptoms psychotic, specifically schizophrenic and obsessive, using the latter term in its clinical, as opposed to religious, sense. Insofar as demonic symptoms can be identical with those of psychosis, however, and insofar as evil spirits are inherently deceitful and will hence attempt to disguise their identity, their presence is not ruled out. Yet the very fact that the powerful prayer of Deliverance—a divine command that evil spirits must obey—has not been successful is taken as evidence that a psychological problem exists. This problem is labeled *folie à deux,* referring to the enmeshed relationship among the protagonists. Healing minister 4 concludes that *if* there is an evil spirit, Martin and Peggy are both in a position to be manipulated by it, but that the foremost need is for psychiatric evaluation of both Martin and Peggy.

The contrast between the spiritualizing and psychologizing religious perspectives, as we may call them, is best illustrated by comparing the evidence marshaled in these two commentaries. Healing minister 1 cites early involvement in the occult—that is, participation in a childhood séance, difficulty praying, loss of concentration in theological conversations, insomnia, being wakened by pornographic images, inability to tolerate presence of a tested religious healing minister, a voice screaming and cursing which is yet unable to utter the name of Jesus, anger, anxiety, and fear. When entertaining the semidynamic hypothesis that the latter three emotions could be linked to developmental issues from Martin's childhood, she again spiritualizes: The problem may not be with a demonic spirit, but with the restless spirit of Martin's father. Healing minister 4's enumeration of psychotic symptoms includes visualization, the disembodied voice, auditory hallucinations, intractable headache, disabling anxiety, sleep disturbance, yanking and pulling sensation (visceral sensations indicating rage at his father), undifferentiated sexuality, bad taste of food, body odor, and twinges of pain indicating autosuggestion and autohypnosis.

While these two sets of evidence are not in themselves sufficient to contrast two logical styles, several provocative points can be made. First, a great deal of overlap exists between evidence for evil spirits and psychopathology: Visual and auditory phenomena, insomnia, anger, anxiety. However, the more spiritualized account pays much greater attention to content than to form of symptoms. Thus, it is important not only that there are visual phenomena, but that these are pornographic, and not only that there are auditory phenomena, but that the voice screams and curses. This emphasis on content may be related to the apparent emphasis on action over somatic sensation; thus healing minister 1 emphasizes attending a séance, prayer, conversation, and reaction to the presence of another person, while healing minister 4 mentions pain, bad taste, body odor, and yanking and pulling sensations.

Further research may succeed in specifying the different styles of spiritualized and psychologized religious approaches within this healing system, but for the present it will suffice to stress that demonic influence and psychopathology are not mutually exclusive. Thus, between the more clear-cut examples of 1 and 4, healing minister 3 specifically acknowledged a high degree of "spirit activity," while simultaneously suggesting that Martin suffers from schizophrenia. Healing minister 5 summarized the problem as one of imbalance between the demonic/spiritual and psychological/emotional aspects of the case.

More light is shed on the logic of the healing system by consideration of the relation between evil spirit and emotion. In Catholic Pentecostalism it is rare, though not unheard of, for demons to bear names such as Andronius in the present case. Typically, they bear the name of sins (Lust, Gluttony), negative behaviors (Self-Destruction, Rebellion), or negative emotions (Anger, Fear). This leads to a systematic ambiguity in distinguishing where human emotion and behavior leave off and the influence of evil spirits begins. Healing minister 1 does not explicitly label Martin's anger, anxiety, and fear as demons, but does call them "hallmarks of Satan." At the same time she links anger to chronic problems rooted in Martin's childhood. Healing minister 2 indicates that Martin is oppressed by disobedience, rebellion, and rejection. This term is used technically to designate a specific level of demonic influence; recall that healing minister 1 felt Martin was obsessed, indicating a higher degree of influence. However, she says without an actual healing encounter, it cannot be discerned whether these are demons "in the strict sense" of intelligent but evil spiritual entities.

Healing minister 3 identifies homosexuality, guilt, self-hatred, and poor self-image as specifically psychiatric, rather than demonic, problems, but acknowledges that there is some spirit activity that must be dispelled. Healing minister 4, whom we have already identified as the most psychologically oriented, mentions rage, anxiety, and undifferentiated sexuality, but only insofar as they appear as symptoms of psychopathology. Healing minister 4 never rules out the possibility of demonic activity. Healing minister 5 does not name specific emotions or behaviors, but distinguishes demonic/spiritual from psychological/emotional aspects of the case.

Whether or not the emotions and behaviors identified by healing ministers are granted the ontological status of evil spirits or human attributes, they indicate a shared therapeutic style of identifying pragmatic issues dealt within the healing process: Anger, rage, anxiety, fear, disobedience, rebellion, rejection, homosexuality, guilt, self-hatred, poor self-image. These are issues whose concrete content can be explored in the life of the afflicted person, and stand at an intermediate level between Peggy's abstract Andronius and symptomatically superficial Pornography. In the logic of the healing system, when one cannot resolve anger through Healing of Memories or Inner Healing, one may conclude that a spirit of Anger or another spirit is present. When Peggy reached this impasse in her attempt at healing Martin, anger became not symbolically concrete Anger, but a mystically abstract Andronius.

The issue of human versus demonic in attribution of emotion is important for the more orthodox healing ministers in that it determines the preference for further treatment. Healing minister 1 recommends Charismatic Deliverance from evil spirits as the treatment of choice. Healing minister 3 recommends a combination of psychiatric care and Charismatic healing. Healing minister 5 recommends psychotherapy completely removed from a spiritual emphasis, feeling that the situation is already overspiritualized. At the same time, both 3 and 5 suggest that the exorcism prayer within the Catholic sacrament of Baptism should have the effect of releasing Martin from evil spirits if he is properly disposed and has a spiritual relationship with God. Thus the relation between discernment and diagnosis, demon and disease, Deliverance and psychotherapy, remains inherently ambiguous. The healing system allows for interpretation of a situation as demonic influence *instead of* psychopathology, but also allows demonic influence as a *qualification* of psychopathology.

## The Clinical Account

Five mental health professionals were presented with the case materials. These included three psychiatrists and two psychologists; one of the psychologists is female, the others are male. All are university-based and have both clinical and research experience. The initial assumption of this exercise was that the clinical interpretation of Martin's difficulties would be in terms of diagnosable psychopathology. Methodological difficulties exist in that the ethnographic description does not fulfill all the requirements of a definitive diagnostic interview, which can only be conducted in a face-to-face encounter between clinician and patient. Thus, what follows are not true differential diagnoses but diagnostic and dynamic impressions offered by clinical consultants. In this light, it must be stressed that the purpose of the exercise was neither to make a definitive diagnosis of Martin and/or Peggy, nor to compare the different schools of psychotherapy represented. It was instead to elaborate a composite clinical interpretation of an atypical case that could stand in contrast to a composite religious interpretation by the clinicians' counterparts in the Charismatic healing ministry.

In creating a composite clinical understanding from these commentaries, caution must be exercised with respect to disagreements that might occur due to (a) the limitations of the data, and (b) clinicians' adherence to different psychotherapeutic schools. The principal example of the former is the possible diagnosis of schizophrenia. The data allow this diagnosis to be entertained by consultants 1, 2, and 5, and rejected by consultant 4, with consultant 3 refusing any label more specific than psychotic. Based on the written description, both consultants 1 and 5 felt that a diagnostic criterion of schizophrenia, "deterioration in functioning," was present, while consultant 2 saw insufficient deterioration to warrant the diagnosis. Likewise there was disagreement as to whether Martin's visualizations were true hallucinations, pseudohallucinations, or "possible" hallucinations.

An example of the second consideration is the tendency of the psychiatric consultants to comment on disease concepts, while the two psychological consultants emphasized personality disorder or personality structure. This represents neither an inadequacy in the data nor an issue that can be synthesized into a composite interpretation, but a difference in disciplinary emphasis between clinical psychology and psychiatry. Differences among therapeutic schools are reflected in the consultants' various recommendations for family as opposed to individual therapy for the principal actors in the situation.

Given these limitations, the validity of the exercise based on clinical commentaries depends on not overstepping the bounds of the task. Only with such caution can certain differences among consultants be shown to be grounded in commonalities of clinical thought rather than inadequacies in the data. Likewise, even though consultants explicitly stated that the case was unusual, only with the necessary caution can interpretive ambiguities be shown to stem from the atypicality of the ease rather than from different approaches of clinical disciplines or schools.

This situation is further complicated by the fact that differential diagnosis is a complex process of discrimination among disorders that may have certain symptoms in common, or may differ only in time of onset or duration relative to other symptoms. Yet while differential diagnosis in medicine operates primarily by excluding possible in favor of more likely diagnoses, it also allows for the simultaneous and overlapping application of more than one category in the final analysis. This is particularly likely in the case of a person as severely disturbed as Martin. In addition to limitations in the data and atypicality of the situation, diagnosis is complicated by the presence of multiple disorders. Nevertheless, it is possible to examine the clinical logic of the consultants' suggestions based on the American Psychiatric Association's *Diagnostic and Statistical Manual* (DSM-III) (1987).

As discussed above, it appears that "schizophrenic disorder" is a likely candidate as a label for Martin's condition. The commentaries of the mental health professionals also frequently mention depression or severe affective disorder, but the problem here is that major depressive disorder and schizophrenia are mutually exclusive. In addition, the long-term treatment outcome of depression is typically considered better than that of schizophrenia. The DSM-III system, however, includes ways to hedge between these mutually exclusive diagnoses. For example, "depression with mood-congruent psychotic features" would include "delusions or hallucinations whose content is consistent with the themes of either personal inadequacy, guilt, disease, death, nihilism or deserved punishment" (DSM-III:215), a description consistent with Martin's sexual preoccupations and early experience of family trauma. Again, the DSM-III system allows for the possible diagnosis of an admittedly ill-defined schizoaffective disorder if the clinician cannot distinguish between schizophrenia and affective disorder.

In this case the chief concern of differential diagnosis would be to determine whether the affective symptoms preceded the psychotic symptoms; if so one would tend toward depression, if not one would tend toward schizophrenia. The early history of Martin's life is not clear enough to determine this. We know that Martin's father committed suicide, often associated with depression, and that Martin's mother has been diagnosed with bipolar affective disorder (also called manic-depression). The latter fact may suggest a genetic factor predisposing Martin to depression and, at the same time the double loss of parents is precisely the kind of bereavement that has also been associated with depression (Brown and Harris 1978). On the other hand, the abusive family environment that Martin must have endured in his childhood is often suggested to be typical for those who develop schizophrenia at about the same age as the onset of Martin's troubles.

To further complicate the diagnostic picture, some physicians have argued for the existence of a disorder that they call "chronic pain syndrome" (Black 1975) or "learned pain syndrome" (Brena and Chapman 1985). In the sense that Martin's pain has no objectively determined source and that he has learned to associate it with certain situations and thought patterns, this category might appear relevant. This is in accordance with our consultants' mention of somatization (consultant 1) or displacement (consultant 2) of traumatic experience into pain. The neurologist, who originally told Martin that his pain was something he must learn to live with apparently had something like this in mind. It can be assumed, however that in the clinical encounter no mention was made of voices or spontaneous visualizations, for it is possible that at that time Martin regarded them as side-effects of the head pain itself. On the other hand, it must be noted at least in passing that the chronic headache with blurring of consciousness and strange visual manifestations could also suggest migraine (Sacks 1985)

Whether such a thing as chronic pain syndrome exists as a clinical entity is currently a topic of debate among physicians. A subtopic of this debate is the relation between chronic pain and depression (Turner and Romano 1984; Bouckoms et al. 1985; Gupta 1986). Depression here is associated both with losses suffered by Martin in his childhood (confer Brown and Harris 1978) and with chronic entrenched feelings of guilt and anger that are not uncommon among patients with chronic pain. Depression is commonly associated with anxiety, a feature mentioned explicitly by only one consultant, but the evidence for anxiety and panic is complicated by current research suggesting that obsessive-compulsive disorder be included among anxiety disorders (Insel et al. 1985).

In line with the above considerations, the DSM-III also requires obsessive-compulsive disorder to be ruled out in making the differential diagnosis of schizophrenia. The distinction here is that in obsessive-compulsive disorder what appear to be delusions are frequently recognized in a conscious way as irrational and quite unpleasant. It indeed appears that Martin's bizarre and intrusive thoughts are of an obsessive nature and are in addition associated with rigid moral compunctions of a religious nature. Here it is relevant to recall the commentaries' mention of dissociative disorders, and especially the epidemiological observation by consultant 2 that obsessive and dissociative features may co-occur more frequently than previously thought. Following this line, diagnostic logic could lead to a conclusion independent of either depression or schizophrenia as principal diagnoses.

The question of possible personality disorder raised by two of the consultants requires two observations. First, DSM-III requires a diagnostic distinction be made between these disorders and schizophrenia, because they can sometimes include transient psychotic symptoms. In severe paranoid personality disorder, the distinction is based on intensity and severity of paranoid ideation and distortions in communication and perception. The issue of paranoia leads directly to the second observation, in which consultant 5's suggestion of dependent personality structure can be related to the reference by consultant 2 and 3 to *folie à trois,* which is technically defined as a "shared paranoid disorder." This raises a variety of issues simultaneously, all having to do with shared aspects of the situation.

Perhaps Martin is afflicted with a major form of psychopathology, or perhaps the maladaptive characteristics of his personality are exacerbated by the relationships in which he finds himself. In either case, consider the comment by consultant 3 that an apparent predisposition to florid symptoms is exacerbated by the expectations of the system. The question here is whether the beliefs and experiences of the afflicted persons are delusional. In this case, it is critical to distinguish the two. Between consultants 1 and 2, on the one hand, and 3 and 5, on the other, there is some disagreement about whether the religious beliefs themselves are delusional. Consultant 2 points out that beliefs are not delusional if they are shared by a cultural group, but still uses the *folie à trois.* This is not an inconsistency, but rather an implicit recognition that particular experiences may be delusional even though religious or cultural beliefs used to make sense of them are not.

This is an important consideration when we raise the question of whether Peggy herself has a diagnosable psychopathology. Consultants 1 and 2 appear to think not, while 3, 4, and 5 recommend therapy for Peggy as well as Martin, with 3 suggesting the entire healing career may be a "depressive system" and 4 suggesting histrionic personality. Beyond the issues of a shared delusional system and a possible individual diagnosis for Peggy, the consultants seem to share a consensus that Peggy is overly dominant and controlling, and that Martin would benefit by independence from her. The consultants describe this critical aspect of the situation with terms such as "family pathology," "enmeshment," "mutual dependency," "undifferentiated family ego mass," and "interpenetration."

Indeed, Martin's definition of the situation is singularly dependent on Peggy and Randy. They both suggested that because the evil spirit has been harassing Martin since the time of his father's death at age 9, he has nothing with which to compare his present state; that is, he remembers no state of consciousness that was not influenced by the demon and cannot always distinguish his own thoughts from those of the demon. Martin acknowledges that "all he knows" is what Peggy has taught him, thus giving Peggy a powerful role as arbiter and interpreter of his experience. In one instance, for example, when Martin recalled that at an early period in life he felt a psychic connection with his mother, Peggy was quick to add that Martin by himself has no "real psychic power." In addition, and as something of a contradiction, while Peggy claims that Martin has "worked through" a lot of anger and other emotion, she reports that he cannot experience his emotions. This report could be interpreted as indicating either a symptom of schizophrenia—that is, blunted or flat affect—or the emotional dullness of depression. In the present context, however, the fact that he only knows intellectually what he wants to feel but cannot have any emotional experience must be related to the degree of Peggy's control in teaching him what he should be feeling. Both consultants 1 and 3 comment that in effect Randy and Peggy have symbolically agreed to become Martin's parents themselves. Experience of the spirit's "resistance" to outsiders' offers of healing could then be understood as resistance by Martin in order to remain in the dependent relationship.

Interestingly, in the only formal contact with a mental health professional reported in this episode (see above), the psychiatrist suggested to Peggy that the "difficulty of separation might be creating an obstacle to Martin's improvement." Peggy's confidence in her motivation (cessation of her own pain) to see the situation to a conclusion was not shared by the consultants, who noted that Peggy's own emotional needs were apparently being satisfied by the tight bonds among Martin, Randy, and herself. Consultants 1, 2, and 4 explicitly define these bonds as sexual as well as emotional.

Characterization of the trio's bonds as sexual and emotional raises the issue of Peggy's rather aloof husband and the nature of their relationship. The consultants appeared to be in consensus on this issue; and consultant 3 was especially emphatic. He suggested that the entire healing enterprise represented a depressive reaction to the frustrations of the marital relationship, and he perceived a weakness in her example of "intimacy" between husband and wife (for example, weekly two-hour car rides during which they never faced each other but talked while looking ahead out the window).

Peggy admitted that her uncontrollable pornographic fantasies affected her sexual relationship with her husband in the sense that she sometimes became distracted from lovemaking. He was not fully aware of this, however, and they do not talk on a daily basis about her experiences. Peggy herself acknowledged no long-term problem in the relationship, but admitted that early in the three-year trial of Martin's affliction, their relationship was strained by the demands of the situation. She claimed that since then her husband had "grown," accepting the reality of the situation, and waiting and hoping for a resolution. Yet Peggy's approach is never to complain to her husband about her own suffering but to try instead to "be smiling" in his presence.

Meanwhile her husband leads a busy professional life about which she knows little. A newspaper write-up listing his awards and achievements left her surprised that he is so renowned; she never accompanies him on his many travels because she, "doesn't want to leave the kids when the youngest is still ten." In light of data on this topic, the situation suggests not so much an estranged husband, but one who sacrifices intimacy to the demands of professional life and is also somewhat leery of probing too deeply into his wife's spiritual affairs.

What can be concluded from this brief diagnostic exercise is not primarily that the data are inadequate for diagnostic purposes, nor that Martin's situation is too complex to be easily diagnosable. Instead, the exercise suggests that the very diagnostic categories available are fluid, overlapping, and more or less vaguely defined. Martin may have one or a number of serious forms of psychopathology; what appears clear in the diagnostic logic is that the situation is serious, involves psychopathology, and requires psychiatric intervention. Whether Peggy is herself diagnosable hinges in part on whether her religious beliefs and symptoms are related as part of a delusional system, whether her symptoms are a logical consequence of religious beliefs admitted as culturally acceptable, or whether she has made a pathological adaptation to culturally normal religious beliefs. Beyond the attribution of individual psychopathology, the diagnostic logic extends to issues of sexual and emotional relationships characterized by dominance, dependency, lack of fulfillment and frustration, possible shared delusion, spirituality, and isolation. Isolation, along with the persistence and severity of the situation, led consultants 1 and 2 to suggest that as a healer, Peggy may be deviant even in terms of the religious system; consultant 3 observed that Peggy appeared to combine elements of Charismatic Christianity and New Age spirituality. These culturally perceptive comments, in fact, conform to some of the responses of the religious healers noted above.

### Convergence and Divergence

The most striking contrast between commentaries of mental health professionals and Charismatic healing ministers is perhaps not that the latter include the influence of evil spirits, but that they focus on Peggy and technical flaws in the therapeutic process rather than on Martin and his symptoms and pathology. For the healing ministers, Peggy is a "fringe" practitioner of a kind warned against in Catholic Pentecostal teaching and literature on healing: Poorly informed, of questionable competence and legitimacy, lacking the support of a community, and hence "in over her head" in trouble. Partly for this reason, but partly because of the nature of her relationship with Martin, Randy, and her husband, Peggy, too, is afflicted and needs healing, or even psychotherapy, herself.

While the mental health professionals point to the same dynamic and interpersonal issues, they tend to reserve judgment about Peggy out of a desire to respect her religious beliefs. Culturally shared beliefs are not delusional, and

Peggy's bizarre experiences are based on these beliefs; therefore Peggy is probably not diagnosable. Lacking the cultural information about the low degree to which Peggy's beliefs and practices are in fact shared by the most logical reference group, they tend to emphasize Martin's symptoms and pathology. This tendency is, of course, only a relative one, for the mental health professionals certainly identify patterns of enmeshment, *folie à deux,* family pathology, and dependency.

The point is not that the two groups of commentaries reach different conclusions. The point is that different sets of cultural knowledge brought to bear on the problem lead to different emphases and explanations within analyses that are pragmatically similar. Several of the mental health professionals suspected Peggy's lack of orthodoxy. On the other hand, the healers' critique of Peggy's practice as isolated anticipate a clinical judgment of enmeshment among the protagonists. For the therapists, the kind of person who becomes enmeshed may be the kind of person perceived as marginal by her cultural reference group; for the healers, the kind of person who is marginal may be the kind who may become enmeshed.

In order to appreciate fully the interaction of world views in this case we must insert Peggy's religious interpretation between that of the clinicians and the healers. The religious critique of Peggy can be summarized as follows: If there is an evil spirit, she is not handling it properly and, if not, she should acknowledge the need for psychiatric care. There is a chance that both therapy and Deliverance are needed, and she is incompetent in both areas. The fact that the healing ministers pay surprisingly little attention to Martin's own problem is probably because, initially, Peggy's approach and interpretation were not incompatible with theirs and neither was it incompatible with psychotherapeutic interpretations. When, at the outset, Peggy refers to Martin's "brokenness," she has in mind the traumatic early events leading to his father's suicide, Martin's discovery of the body, the loss of his mother, and subsequent assignments to an orphanage and foster parents. For Peggy and other Charismatic healers these early experiences create a vulnerability to demonic influence, while for the clinicians they create a vulnerability to psychiatric disorder. Her focus on resolution of anger and guilt, which includes encouraging concrete steps toward reconciliation between Martin and his mother, is so unobjectionable as to remain unmentioned by both clinicians and healing ministers.

Peggy's subsequent "discovery" of demonic presence was closely associated with recognition that Martin still felt "more anger than he wanted to" toward his father. Indeed, discovery of a spirit is an acknowledgment that a deeper level of unresolved problems exists. Conversely, reaching a therapeutic impasse can be construed as evidence for the presence of an evil spirit—but these are by no means necessarily the same thing. The healing system assumes that a block to change exists within the afflicted person and infers that this block is caused by an evil spirit. Clinical consultant 3 suggested that, at least in this case, the block was the limit of effectiveness for a "transference cure," which could be achieved by a clinically untrained religious healer but which left unresolved deeper conflicts accessible only to a highly skilled psychotherapist.

Whether or not the block is "within" the afflicted person or "between" the person and the healing minister, discovery of a spirit is a rhetorical strategy for transcending the block and bringing the problem out in the open in such a way that it can be challenged with the support of divine power: The lonely and isolated individual is no longer alone because his struggle is now part of a cosmological struggle of universal scope, the spiritual warfare between God and Satan (Csordas 1994a, 1997; also see Dow 1986b; Tambiah 1977). But transcending and bringing out in the open appear to go necessarily hand in hand; transcending alone can be dangerous. For, as noted above, according to some healing ministers Deliverance is easy, because a prayer commanding it to depart, or the exorcism contained within the rite of Baptism, *must* work due to God's inherent power over Satan. It is perhaps important that this is not the only approach to Deliverance, but emerged within the movement as an antidote to practices emphasizing struggle against evil spirits that could last for agonizing hours and include shrieks, vomiting, and writhing on the floor; Peggy may have expected and needed this kind of approach to be convinced of divine empowerment.

Thus, the conventional rhetorical strategy backfired. By positing a source of affliction external to Martin, Peggy made herself vulnerable as well; instead of psychic mirroring of Martin's symptoms, a situation was created in which the evil spirit could attack her directly "at her own weak points." In psychiatric terms, the cool, dignified figure of Andronius became an opaque metaphor of Peggy's uncontrolled countertransference, the impenetrable "humanity" in which the protagonists were enmeshed, and the universal pitfall shared by psychotherapists and exorcists of whatever tradition (Henderson 1982; Good et al. 1982). Thus, the sexual fantasies that began at Martin's puberty, and were exacerbated when the evil spirit took advantage of his unresolved developmental crisis of intimacy, find their parallel in sexual fantasies that reflect Peggy's ambiguous relationship with her husband and her role as housewife/mother, her close spiritual relationship with Randy, and her dominant/dependent relationship with Martin.

## The Existential Ground of Demon and Disease

Comparing medical and sacred realities in this way throws some light on their different properties as systems for organizing experience and some light on the nature of suffering and healing. However, it leaves an essential problem untouched. That is, how is it possible in the first place for such accounts to have so much in common yet be so different; what in fact is the nature of the experience for which they account? We must now turn to this question.

The comparative study of plural healing systems coexisting within an overarching cultural tradition suggests that underlying continuities of process and structure can be found among such systems (Rhodes 1980). In our comparison of North American Catholic Charismatic and psychiatric systems, the principal continuity is the mutual emphasis on residual effects of events in the afflicted person's past, and the principal divergence is the role of spiritual practices and demonic entities. The different ways in which these systems elaborate the implications of these issues may paradoxically contribute to the possibility of their coexistence. That is, insofar as they represent, as it were, intersecting planes in the field of experience, they can be complementary rather than contradictory; an afflicted person is much more likely simultaneously to seek help from a religious healer and a psychotherapist than simultaneously to seek help from a psychoanalyst and a cognitive therapist. Although practitioners in either system may reject the validity of the other (as the healer in the case discussed here rejected any psychiatric interpretation), in principal they are often regarded as complementary and in practice sometimes even actively integrated. At the same time, Christian psychotherapists may reject certain competing forms of therapeutic practice as incompatible not only with each other but with principles of religious healing (Csordas 1990).[9]

We have just now introduced the metaphor of intersecting planes to describe the relation between two readings of Martin's experience. This intersection can be understood in two senses, the cultural and the existential. In the first instance, both readings share a North American cultural proclivity for formulations in strongly *psychological* terms. That is, the interpretations of Martin's experience in both healing systems are predicated on cultural assumptions about emotion, self, and person that begin and end in predominantly psychological understandings. In the second instance, it remains the case that despite our ability to formulate distinct accounts of his experience, there is ultimately only one Martin, in a unique existential situation. I will argue that the common existential ground from which the two accounts are abstracted is his suffering as an *embodied* human being, for whom any distinction among somatic, cognitive, or affective pain is experientially irrelevant.

Specifically, I would argue in favor of bodily experience as the starting point for cultural analysis, the existential ground for divergent cultural elaborations of illness experience and therapeutic intervention. Borrowing our terminology from the existential phenomenology of Merleau-Ponty (1962), we must try to describe Martin's preobjective world of distress and the thematization that creates the possibility for objective entities such as demon or disease to be posited as accounts of that distress. By "preobjective" we refer not in a temporal sense to Martin's experience before he came under the influence of a religious healer but to the manner in which he spontaneously engages the cultural world of everyday life, or on the other hand the degree to which he has lost his hold on that world. Merleau-Ponty would argue that cultural objects such as demons or diseases, no less than natural objects such as rocks or trees, are the end products of a process of abstraction from a perceptual consciousness wherein the sentient human body is an opening on an indeterminate, open-ended, and inexhaustible field: the world.

Critical to our purpose is the understanding that in normal perception one's body is in no sense an object, but always the subject of perception. One does not perceive one's body; one is one's body and perceives *with* it both in the sense that it is a perfectly familiar tool (Mauss 1950) and in the sense that self and body are perfectly coexistent. To perceive a body as an object is thus to have performed a process of abstraction from perceptual experience. When we turn to Martin's situation we are struck first of all with the manner in which all sensory modalities are in crisis. Martin's senses do not give him an immediate grip on the world. In a way he must go through his senses to the world rather than perceiving with them; they are in the way, standing between him and the world such that his perception is not faced with an open horizon but with a wall. This inability to engage the world is thematized in the language of each of the senses, and we must now take a closer look at that language.

The voice Martin hears knew everything about him, would make deals with him, discuss religion, make jokes about his foster mother, cast aspersions on his natural mother, offer friendship and companionship. In short, the voice was thematized as a rather cruel friend, a source of both intimacy and irritation. It is unclear whether, prior to Peggy's identification of the voice with a demonic entity, he perceived the voice as evil, and whether it is legitimate to suggest that Martin was already denying unpleasant thoughts by projecting them onto an alien being. Peggy's reinterpretation of this theme of the cruel friend was that the demon was of the type of the "familiar spirit" (in the sense of the "witch's familiar" and not that of a "family spirit"). Only after the voice was objectified as an evil spirit did it make sense in cultural terms for Martin to say that "it" was able to get deeper into his gut, that it was imitating being three instead of one, or that it "inches you away by merging with your own consciousness." One can only wonder what the consequences might have been if Martin had truly let the utterance "Your mother left you" become "My mother left me" in such a way that he was forced to live through his anger and rage about his feelings of abandonment.

In the visual domain the language used to describe the sexual imagery appears curiously contradictory. Martin is "bombarded" with images in rapid succession, with content of increasing perversion and violence, and the sexual impulse is "almost overpowering." Yet it feels good and right to give in, and giving in is accompanied by feelings of friendship and companionship. Is Martin giving in to purely sexual impulses, or to the anger and rage mentioned above and sexually thematized as bombardment with violent content? Relevant here is Merleau-Ponty's (1962) argument that just as sexuality is an atmosphere permeating our lives as human beings, sexual perception itself is modulated by and fully integrated with the other perceptual functions of our bodies. In the present case, there appears to be a phenomenal interrelation between vision and hearing in Martin's experience of imagery and voice as cruel friendship.

This line of thinking is strengthened by considering the language of touch or bodily sensation in Martin's affliction. Pain occurs primarily in his head, but also in his joints, stomach, and groin. It is described as yanking and pulling, with a knot in the head and an occasional release accompanied by a crackling sound; the yanking and pulling can be sexually stimulating if in the genital area. Any mention of his father will also be answered by a pain in his "gut"; this word also appeared in the interviews in the context of the spirit "getting deeper into his gut." If we bracket the idea that these sensations describe the intentional gripping action of a demon, we can see that they in fact isolate body parts in a way phenomenologically parallel to the experience of religious healers who can sense when one of their followers is healed of a heart problem by a spontaneous painful sensation in the chest. Whereas for the healer each pain is thematized as an index of the outside world to be read in her body, Martin's pains alienate him from the parts of his body in a way more akin to dismemberment. The inner integrity and unity of his body is compromised. In addition, however, the pain is thoroughly integrated with visual and auditory phenomenon. Unlike chronic pain patients whose pain is thought to modulate almost mechanically with varying levels of stress or relaxation, there is a distinct conative dimension to Martin's pain. It modulates directly in relation to his response to the voice and to sexual imagery. If he bears down in resistance he can be assured the pain will increase, and if he gives in it invariable eases.[10]

Description of Martin's bodily sensation includes feelings of heat coursing through his body, his body becoming like jelly, and the sense of a fluid movement through his body. Here is a sense not so much of dismemberment but of *dissolution* of body boundaries. The overall effect is described as creating panic in his body, and the feeling of running from himself. When these sensations are objectified in terms of demonology, the experience is one of feeling an attempt to embed another personality in his body. This must be compared to the image of getting deeper into his gut and the image of being inched away in his interactions with the voice. Each image corresponds to a particular mode of sensory experience, but note especially that the image of being inched away prompted by the more cognitive interaction with the voice is equally as physical as those associated with pain and corporal dissolution.

In addition to the sensations that isolate parts of his body in pain and those that indicate a dissolution of his body as an integrated being in the world, the interviews include a constellation of descriptions including a net descending over Martin's head, visualization of a gray mass around his head, thickening of the saliva, thickening of his eyelids and glazing over of his eyes when his thinking blurs out, a thickness in conversation, feeling heavy, weighted down, or drained of energy. It can be argued that the bad taste in Martin's mouth, along with is halitosis and body odor, be included in this constellation of meanings revolving around heaviness and thickness. The overall theme of these "heaviness" words appears to be that of *immobility*. Note that this immobility can be objectified as the kind of slowness symptomatic of clinical depression, or as a literal manifestation of oppression by an evil spirit. For Martin it was thematized in the recognition that he could "remain immobilized or go forward."

A cultural phenomenology of the existential situation exhibited in this language of the senses can be summarized as a radical narrowing of the horizon of perception and experience. Whereas the unafflicted person in everyday life can ceaselessly continue to explore the world, for Martin the world's horizons have become opaque and impenetrable. The image of dismemberment refers to the inner horizon in which the parts of one's body mutually imply one another or communicate with one another in an experientially undifferentiated and taken-for-granted way, here closed off by the thematization of individual body parts in pain. The image of dissolution refers to the horizon that is the boundary of one's body with the world. In this case, it cannot be said that the horizon is sealed off, but that there is no horizon, no foreground or background of personal reality, neither a direction to explore nor any discrete self to do the exploring. The image of immobility refers to the horizon of action in the world, where one can formulate an unlimited number of open-ended life projects, but which for Martin is sealed off by the total preoccupation with affliction. The voice experienced in the auditory modality participates in all three insofar as what it says is directly connected with pain, makes it difficult for Martin to distinguish his own thoughts from alien ones, and prevents him from engaging in his preferred activities.

The constant barrage of sexual imagery, on the other hand, has its existential significance in opening up an *artificial horizon* of an inexhaustibly sexual world. In this respect the key phrases are that Martin is impelled to see the world exclusively in sexual terms and that he feels "pornography as a state of being" within him. These are not to be distinguished as respectively cognitive and physiological but as alternate phrasings of the same underlying stance toward the world. To recall Merleau-Ponty's (1962) notion of the radical contingency of sexuality as a component of all human experience, it can be said that Martin's coming face to face with this reality is coterminous with the closing of other horizons, such that the truth of sexuality as a state of being within everyone was distorted by appearing as the only transcendent or open-ended modality of experience. Martin's moment of crisis came in the episode of anorexia and insomnia immediately prior to his initial encounter with Peggy. The near absolute collapse of the world and its horizons around him was graphic in his inability to eat, understandable as the inability to allow the world inside himself, and his inability to sleep, understandable as the inability to allow himself an exit from the frozen immediacy of affliction.

Critical to my argument is the recognition that what might appear as distinctly cognitive or affective distortions of Martin's experience are at one with the language of bodily experience. The blurring of Martin's consciousness and his inability to concentrate are closely bound up with other aspects of heaviness and thickness. The special association of these effects

with religious content is tied not only to the preoccupation with an evi
spirit but to religiously motivated sexual guilt that preceded this preoccupation. Feelings of friendship, companionship, goodness, and rightnes
are also associated with yielding to sexual temptation and modulation o
pain. Panic and fear are inseparable from feelings of bodily dissolution
Inappropriate and exaggerated anger, including anger toward his parents
are associated with promptings of the alien voice, but if we are allowec
to apply Peggy's personal report to Martin's parallel experience they arc
also associated with feelings of one's body being manipulated into th
expression and posture of anger. Thus, cognition and affect are not tc
be understood in abstraction from bodily experience. They are equally
components of what Schilder (1950) called the "postural model" that i:
subject to transmutation in a variety of situations, most notably those o
affliction.

A parallel description could be made of Peggy's experience with reference to her own acknowledgment that the evil spirit attacks her at her owr
weak points. In brief, there are three key differences in her experiences
(1) she hears the voice primarily as screaming, cursing, condemning her anc
threatening her family, and concurrently provoking copraphilia in her
(2) sexual images appear in episodes like a television show and have developed from anonymity to inclusion of real people (Martin persisted fo:
some time in the mode of anonymous sexual violence); (3) absence o
halitosis or body odor. The first two features can be understood as concretc
representations of conflicts between her role as Martin's healer and her rolc
within her family, and conflicts over sexual intimacy. The third represent:
the absence of at least one dimension of the heaviness (the olfactory)
which for Martin constitutes an obstructed experiential horizon.

This, then, is our approximate reconstruction of Martin's preobjectivc
experience of distress and his initial thematization of that distress, prior tc
objectification of his experience in the accounts of religious healing or psychiatry. My argument is that each system presupposes this experience, anc
that its account is in that precise sense an abstraction (see Figure 4.1). Eacf
account thematizes preobjective experience according to its own principles. In the religious system, the relevant principle is *moral,* and can be statec

| Religious Account | Psychiatric Account |
|---|---|
| Moral prinicple: Good/Evil | Empirical principle: Body/Mind |
| Origin: Occasion | Origin: Cause |
| Entity: Demon | Entity: Disease |
| Evidence: Manifestation | Evidence: Symptom? Syndrome |
| First-person Process: | Third-person Process: |
|   Oppression/Struggle | Disorder/Somatization |

**Figure 4.1**   Cultural Accounts as Objectifications of Experience

as the contradiction between good and evil. In the psychiatric system, the relevant principle is empirical, and can be stated as the dichotomy between body and mind. Based on these principles, the systems posit either a demon or a disease as an objective entity.

The nature of these cultural objects is directly related to variations in the definition of the person in the two systems. The Catholic Pentecostal person is a tripartite composite of body-mind-spirit, in contrast to the conventional contemporary Western mind-body. The domain of spirit is equally as empirical as mind and body, and equally susceptible to both positive and negative influences. Evil is ontologically real and is embodied in active, intentional beings—that is, the evil spirits. Thus, the term "entity" directs analytic attention to the ontological claim by healers and clinicians that demons and diseases are empirically real things in the world. In their respective systems, the demon is a spiritual substrate of distress, and the disease is a biological substrate of distress.

In recent years, scholars have questioned the status of disease as an empirical entity (Campbell 1976:50–51) and reinterpreted it as a symbolic or conceptual form in terms of which clinicians organizes their interpretation and their patients' experience of affliction and distress (Kleinman 1980, 1983). In this view, the substrate is the phenomenology of affliction, or the experience of illness, and the status of the disease as an entity is made problematic. It is in this sense, and on this level, that the diagnostic logic of disorders and the logic of discernment in the religious healing system are generated in the two sets of commentaries presented above.

The categories of disease and demon organize the understanding of how the distressful condition comes about quite differently. A disease has an underlying *cause* in the strict sense, understood as some kind of infection, degeneration, trauma, genetic abnormality, biochemical imbalance, and so on. A demon, by contrast, has an underlying *occasion* or circumstance by means of which it can gain a purchase on a person through certain vulnerabilities. The occasion can be a traumatic event or the existence of sin. Sin in turn can be the personal sin of the afflicted person, a sinful environment to which the person was exposed, or the general cosmological condition of original sin that permeates the world. This is specifically an occasion and not a cause, for the evil spirit that is thus "allowed in" is, properly speaking, itself the cause of the problem. The evil spirit then accounts for a variety of manifestations that constitute the person's affliction.

This disjunction accounts for the different way the two categories name the problem and the way they are posited as entities. A disease is more than a summary label for a constellation of symptoms, as a demon is more than a summary label for a constellation of manifestations. A disease names a third-person process that has a specifiable course, natural history, or range of predictable outcomes. A demon typically names a behavioral trait or affective state and posits it as a first-person process, endowing it with intentionality and, hence, precluding the possibility of either a completely circumscribed set of symptoms or a completely specifiable natural history. It is precisely by endowing the behavioral trait or affective state with intentionality that the religious system establishes the demonic entity as a cause rather than as something that is caused. In the psychiatric system the equivalent traits and states are objectified not as ontologically real entities but at the more specific descriptive or attribute level of symptoms.

Thus, although the phenomena of preobjective experience are treated or thematized by both systems as a kind of evidence for the posited objective entity, the epistemological status of this evidence is different in each instance. On the religious side a vision of light is a *manifestation* of a demon possessing one's *cakras;* pain is the manifestation of a being who will punish one for resistance to its will; and dull-mindedness is a manifestation of a being intent on interfering with one's performance of the work of God. On the psychiatric side peculiar gustatory sensations are *symptoms* of temporal lobe epilepsy; insomnia, weight loss, and poor concentration are symptoms of depression; and hearing thoughts and experiencing visual imagery are symptoms of atypical psychosis. Given this formulation it can be suggested that one of the difficulties in Peggy's attempts at healing was precisely a preoccupation with phenomena as evidence, and a consequent inability to deal adequately with the task of healing. In her isolation from like-minded individuals she became so intent on proving her diagnosis that she cultivated the very phenomena she hoped to eliminate.

However, the relation among manifestations of a demon need not be as systematic as those among symptoms of a disorder. As discussed in the preceding chapters, specific spirits are sometimes identified by specific manifestations (see also Csordas 1994a), but my point is somewhat different. In a psychosomatic model, emotions can be understood as causing physical distress. Thus, in Martin's case, clinical consultant 1 suggests that "chronic entrenched feelings of guilt and anger" are directly associated with experience of chronic pain. Charismatic healing ministers' familiarity with popular psychology includes the concept of psychosomatic distress, and in practice they tend to integrate it into their work. However, positing an evil spirit preempts the direct connection between pain and affect: The spirit causes them both, or different spirits cause them. I suggest that Peggy's strict adherence to the logic of demonic causality prevented her from seeing the interrelation of features of Martin's distress in any other way. The manifestations have no inherent relation among themselves as do symptoms; they are related only as items in a list of problems caused by the demon.

Once the entity of demon or disease is objectified, it in turn becomes the trope by means of which experience is organized, interpreted, and thematized. This leads to rather different consequences in the two accounts. A demon is posited as an *oppression of the afflicted* with the intent to achieve control of a person's soul, initiating a powerful existential *struggle.* Negative experiences are thematized as forms of oppression. In the strongest formulation of this logic, there would appear to be no compelling reason to look for a relationship of causation or influence between thought and emotion on the one hand and sensory disturbance on the other. Suffering is cumulative, each form being just one more way that the person is hurt, one more channel of demonic harm, one more area of life under siege.

The process of psychiatric disease, however, is posited not as oppression but as disorder. Experiences thematized under the trope of disorder are those which can influence one another, rebound upon one another, and especially mask one another *via* mechanisms such as dissociation, obsession, and somatization. Of particular interest with respect to lived body experience is the concept of somatization, which in psychiatry and anthropology is variously defined as presentation of physical symptoms in the absence of organic pathology, amplification of organic physical symptoms beyond physiological expectations, presentation of somatic symptoms as an alternate expression of personal or social problems, and a mechanism by which emotions give rise to somatic signs and symptoms (Kirmayer 1984). In the present case, although organic pathology in the form of temporal lobe epilepsy or biological (hereditary) depression are not ruled out, somatization can be understood as a transmutation of cognition and affect.

The relationships among evil spirits are markedly different than the relationships among illnesses. Differential diagnosis is precisely a process of differentiation, whereas the discernment of evil spirits is additive. Temporal lobe epilepsy may be ruled out in favor of schizophrenia in Martin's case, meaning that symptoms originally suggesting epilepsy will appear in a different configuration and carry different connotations with regard to the expected course of the illness. The healer does not rule out the presence of particular evil spirits, for discernment of a spirit's presence carries with it an apodictic certainty. It is almost never a question of reorganizing the manifestations in a more satisfactory way under the name of a different demon, although the presence of additional spirits may be discovered.

It is, however, common for evil spirits to gather in clusters and "work together" and, in addition, to be under the hierarchical coordination of a single "master" or "manager" spirit. By itself, this clustering might seem analogous to the patterning of symptoms into a syndrome, but to make this analogy would be to err from the analysis that demon is to manifestation as disease is to symptom. A more accurate parallel is as follows. Insofar as the differential diagnostician is left with more than one apparently confirmed diagnosis, the diseases are superimposed and understood as complicating one another, but very likely they will be analyzed into primary and secondary diagnoses, such as schizophrenia with secondary anxiety and somatization. Similarly, the Catholic Charismatic healer might discern a principal spirit of Self-Destruction, with attendant spirits of Rebellion, Hatred, and Anger.

We have already suggested that positing a demon is in one sense a rhetorical strategy, and there are indeed a variety of intriguing analyses of sickness as rhetorical process (Frankenberg 1986; Chesebro 1982). Can it be said that making a diagnosis is a rhetorical strategy in the same sense or on the same level of analysis as discovering a demon? Superficially they have in common that they name the problem, and we can concur with arguments that naming may both offer a sense of control and the reassurance of knowing what is wrong, and may limit the choices for treatment and shape the course of an illness. From the perspective of labeling theory, it can also be argued that demon and disease both insinuate themselves into the very being of a person, not only accounting for symptoms but transforming a person's identity and experience of self. What makes these parallels superficial is that the way demon and disease name a problem and the way they exist as entities bespeak two different culturally constituted ways of organizing experience in a therapeutic process.

A more significant sense of the parallelism between demon and disease as rhetorical strategies can be made clear by being specific about what the Charismatic discernment of evil spirits is *not*: itself a symptom of psychopathology. James Henderson (1982) discusses "demonological neuroses" as they appear for psychiatry, giving a case of Freud's along with one of his own. He argues that the phenomenon can be understood in terms of internal object relations theory and as an indication of psychodynamic processes of introjection and incorporation. However, the cases he discusses are ones in which the presence of a demon is the patient's presenting complaint, and hence part of the patient's pathology. In the case of Martin, and in most situations of Deliverance among Catholic Charismatics, the presence of an evil spirit is not given but discovered or discerned by the healer. Even when the evil spirit names itself through the voice of the afflicted person, it usually does so only on direct questioning by the healer.

Catholic healers themselves have encountered cases like those discussed by Henderson. One healer told of a man who had contacted several priests in vain in the belief that he was being tortured by evil spirits. After spending several sessions with this man himself, the healer concluded that he could be of no help. He indicated that the man probably had serious emotional problems rather than demonic oppression, and suggested to him that the reason he had gone from priest to priest was that none would validate his self-attribution of demonization. We see here that care must be taken to distinguish between evil spirits as a symptom of psychopathology and as the religious equivalent of a diagnostic category. Whereas in Martin's case it may be legitimate to describe the voices he hears in terms of ego introjection, the evil spirit must itself be described in terms of externalization in roughly the same sense as is a disease.

Yet where these two entities diverge the most is precisely in their rhetorical properties or possibilities. The fact that an evil spirit is a first-person process with an intentional history rather than a third-person process with a natural history means that it can be questioned and commanded. Hence, it can be manipulated in its intimate relations with the afflicted person. Moreover, the form of this intervention is the same whatever spirits might be discerned to be present, and the healing is culminated when the evil spirit is ritually commanded to depart. The psychiatrist does not command schizophrenia or depression in the same way as the healer commands an evil spirit, but intervenes in it as in an event or against a thing. If the psychiatric patient acknowledges the presence of a disease it is something he "has" rather than something vicious that is attacking him, or something he already "is" ("I guess I'm crazy") rather than something that is not him but wants to possess him.[11] Moreover, because each disease implies a different natural history, it also implies a different treatment; the psychiatrist is much less comfortable in saying that psychotherapy is appropriate for all psychiatric disease than the healer is in saying that Deliverance prayer is appropriate for all instances of demonic oppression.

Again, because a demon is a first-person entity, it can play an immediate rhetorical role as an actor in the healing process, although some Charismatic healing ministers refrain from informing supplicants that they have discerned a demonic presence, preferring to cast it out silently and thus avoid disruption and histrionic display. On the other hand, because a disease is essentially a third-person entity it can still more easily be thought of as treatable without, for example, the psychiatric patient ever knowing that it is called schizophrenia. Yet even here, some advocates of "psychoeducational" programs regard naming and understanding the illness as essential to its treatment. Moreover, in a rhetorical sense, diseases can sometimes be granted at least a metaphorical intentionality, as when cancer is described as a "vicious killer" or an "invader." There is a profound qualitative gap, however, between understanding the hearing of voices as a symptom and as an intentional verbalization. Martin sometimes experienced what seemed to be three different voices. However, he believed that in fact a single spirit was "mimicking" being three. Peggy's religious interpretation was that first of all this was exactly the kind of deception that is typical of spirit behavior, and second that the three-in-one illusion was an intentional blasphemy on the part of the spirit, in diabolical imitation of the godhead of trinitarian Christianity.

If the potential multiplication of Martin's voices can be understood as the potential for dissociation and fragmentation of self, then Peggy's rationale for keeping them unified appears as a kind of spiritual damage control. This is especially the case given the ominous presence of Andronius, the master spirit. Already beyond control, to grant a multiplicity of voices and identities to the spirit would surely have added to the sense of danger in the situation. On the other hand, in more typical cases there may be a rhetorical advantage to having clusters of spirits present, both in that it allows a more complex interpretation of what may be a very complex personal situation, and in that it allows a feeling of incremental progress if demons can be expelled one by one over the course of several healing sessions (Chapter One, Csordas 1994a). There appears to be nothing directly parallel to this in psychiatric treatment, which is not to say that psychiatric diseases do not have rhetorical properties of their own. Certainly doctors and patients both can construct elaborate discourses ("let me tell you about my schizophrenia ...") about a disease in such a way as to influence the course of an illness.

The fact that demon and disease are constituted differently and hence have different properties does not in itself determine their relation should they both be applied in a particular case. Demon and disease can be completely redundant, accounting for exactly the same constellation of symptoms but applying at different ontological levels: Healing minister 4 asserted that schizophrenia and the effects of demons can be identical, and only with the spiritual gift of discernment can they be distinguished. They can overlap, including either variations in features or variant interpretations of the same features, as was evident in the comparison between healing ministers 1 and 4, and in the reference by 5 to spiritual and psychological aspects of the case. Demon and disease could be judged to coexist as mutually complicating conditions, or be mutually exclusive as strict alternatives. In the commentaries of the Charismatic healing ministers, the principal trope can be either oppression or disorder, either evil spirit or psychiatric diagnosis.

## Conclusion

The cultural comparison I have elaborated highlights the pragmatic merit of conceiving not only demons, but also diagnostic categories or diseases as interpretive forms rather than as ontological entities. To see psychiatric diagnosis as an interpretive or hermeneutic process (Good and Good 1980) is essential to the development of methods for parallel analyses of religious and medical accounts of distress, wherein convergences and divergences of presupposition and interpretation can be detailed systematically.

The phenomenological description of Martin's affliction as an embodied totality provides the basis for a critique of both these accounts. Disorder and oppression are each processes of an objective entity, either disease or demon. In proposing that affect and cognition cause bodily sensations through somatization, or that an internal mechanism transmutes them into bodily signs, the clinical view misses the unity of somatic and psychic experience that we have demonstrated in the case study. In this way it is subject to the same criticisms that can be made of any type of mechanistic empiricism (Merleau-Ponty 1962). On the other hand, in proposing that all sensory, somatic, cognitive, and affective manifestation are caused by a demon, the notion of oppression admits that somatic and psychic experience are all of a piece by placing them on a par. This notion, however, makes the error of attributing the unity to an abstract constituting consciousness, namely the evil spirit, instead of to the essential unity of the human being in which every modality of perception is conditioned by every other. In this way it is subject to the same criticisms that can be made of any type of rationalist intellectualism (Merleau-Ponty 1962).

I have argued that the paradigm of embodiment is useful in comparing different cultural accounts of experience by providing a description of the existential common ground from which those accounts are abstracted. In spite of this advantage, does phenomenological description of embodied experience offer merely another objectification of the same order as demon, disease, or emotion? My necessarily brief answer will be to show how the paradigm of embodiment helps to reveal the embedded themes that are elaborated as cultural objects by following Martin's experience through his eventual return to a moderate level of social functioning.

Let us return to the images of dissolution, dismemberment, and immobility that we found to be themes of Martin's lived body experience. During the final period in which I followed this case, Peggy reported that feelings of fluidity and dissolving no longer characterized Martin's experience, while most other problems remained. It would appear that the reintegration of body image was his critical achievement. Judging by the language of panic and self-alienation in which he described it, this had been the most distressing dimension of Martin's affliction.

Certainly, in a society where the ethnopsychology of the ego ideal is radically individualistic, an integrated body image could be expected to be critical to acceptable daily functioning. From the perspective of embodiment, a bit of data that may otherwise appear as minor emerges as prominent in Martin's move toward engagement in the world of daily life: He had begun swimming almost every day. In her idiom, Peggy interpreted this as beneficial primarily in terms of "cooling" him from the heat of his oppression. In the phenomenological idiom we can suggest that the flow of water over his skin helped redefine body boundaries against dissolution, that the coordinated muscular action helped redefine bodily integrity against dismemberment, and that continuous motion helped redefine the ability to act against immobility. Yet, Martin had no more than reached another plateau, and insofar as he continued to suffer with only the outward appearance of normality, the evil spirit appears as a condensed symbol of his affliction. From a perspective outside the religious definition of reality, a demon from which one can be delivered may be a metaphor of disease; a demon from which one cannot be delivered is a metaphor of chronicity.

Beyond the questions of metaphor, translation, or equivalence of meaning, analysis of religious and psychiatric meanings in this case suggests the fruitfulness of a cultural phenomenology in comparing radically different accounts of experience. A return to the phenomena of preobjective experience reveals the common ground from which such accounts are built, through alternative thematizations that lead to the positing of cultural objects such as demons and diseases. I have attempted to describe the existential ground presupposed by religious and clinical reflection, and in so doing argued that to explain religious phenomena of affliction solely in medical terms is to merely put one view of the world in place of another.

# PART II

*Navajo Transformations*

CHAPTER FIVE

# Ritual Healing and the Politics of Identity in Contemporary Navajo Society

My point of departure is the intersection of three heavily traveled conceptual highways that wind across American anthropology. The first is ritual healing, which has preoccupied anthropology as religion, as performance, as therapy, and as a window on broader cultural processes (Csordas and Kleinman 1996; Dow 1986a; Kleinman 1980; Levi-Strauss 1966). Second is identity politics—that is, the deployment of representation and mobilization of community within plural societies in the name of gender, sexual orientation, ethnicity, race, or religion—which has in recent years captured the attention of both cultural anthropology and interdisciplinary cultural studies (Calhoun 1994; Friedman 1992; Giddens 1990; Lash and Friedman 1992). Third, Navajo society remains one of the most heavily documented, most frequently drawn on for ethnographic examples, and most irritated by the persistent probing of anthropologists of all stripes (Farella 1984; Kluckhohn and Leighton 1946; Lamphere 1977; Witherspoon 1977). In this article, I elaborate the relation between ritual healing and identity politics in contemporary Navajo society by presenting a conceptual framework that can potentially be applied across a wider range of societies.

What is the purpose of asking about the relation between ritual healing and identity politics? Doing so allows me to address in specific fashion the perennial issue of the relation between religion and politics, both of which are forms of power but with ostensibly different motives and modes of operation (Fogelson and Adams 1977). It allows me to address the parallel issues of the individual in relation to the collective and of microsocial in relation to macrosocial processes. Stated strongly, ritual healing *is* a form of identity politics, as suggested by Rudolph Virchow's famous dictum that politics is nothing but medicine on a grand scale. Stated somewhat less strongly, ritual healing is a window onto larger cultural processes, as in the notion of cultural performance (Geertz 1973; Singer 1972).

Why ask this question in the particular context of Navajo society? The Navajo setting requires us to confront an empirical situation that further undermines the increasingly shaky distinction between tradition and modernity. To be precise, on the one hand typical accounts treat ritual healing as "traditional" and backward-looking in values and goals even though it is practiced in postmodern settings, including that of Navajo society (see the literature reviewed in Csordas and Lewton 1998). On the other hand, scholars often discuss identity politics in terms of modernity, diaspora, post-colonialism, and globalization (Calhoun 1994, Lash and Friedman 1992), whereas "tradition" itself is a central orienting concept in everyday life for Navajos struggling for sovereignty as a fourth world nation.

In this article, my purpose is to show that the relation between ritual healing and identity politics in Navajo society is played out on three different levels of social generality. On the broadest level, healing articulates Navajo identity in relation to the dominant Anglo-American society. Here I will be concerned with cultural representation of events in which identity is at stake in the public sphere defined in part by the news media. On an intermediate level within Navajo society, healing and identity are closely interrelated in the interactions among three relatively distinct forms of healing. Specifically, I will highlight negotiation among participants in these healing forms around issues of competition and cooperation. Finally, on the individual level, healing frames the relation between personal and collective identity in terms of dignity and self-worth as a Navajo. Here, my focus is on the behavioral and experiential transformation of patients and their immediate social relations. Basing my presentation on this threefold analytic framework, I will return in the conclusion to the above-mentioned series of conceptual relations between religion and politics, tradition and modernity, individual and collective, microsocial and macrosocial.[1]

## Navajoland in the Nineties

The Navajo (Diné) are an Athabaskan people who, along with the kindred Apache peoples, migrated south from Alaska and Canada to what is now the U.S. Southwest approximately 500 years ago, roughly the same time as Spaniards were migrating north from Mexico into the same region. The contemporary Navajo Nation comprises more than 17.5 million acres (roughly the size of West Virginia) in the Four Corners region where New Mexico, Arizona, Utah, and Colorado meet. It lies immediately to the east of Grand Canyon National Park and completely surrounds the Hopi Indian reservation. The reservation and its boundaries are an institution of the U.S. federal government, established by an imposed treaty in 1868 as the condition for the Navajos' release from captivity at Bosque Redondo near Fort Sumner in eastern New Mexico. The collective trauma of the Long Walk—their forced march into collective exile from their homeland, following military defeat by U.S. government troops using a scorched-earth policy at the command of the infamous Colonel Kit Carson—is critical to contemporary Navajos' sense of identity as a people. Today, the Navajo reservation is divided into five federal administrative districts or agencies as well as into 110 indigenously recognized localities or chapters. Each chapter sends a delegate to the Navajo tribal council established in the 1930s. The chief executive of the Navajo tribal government is a president chosen in a general election for a four-year term.

According to the 1990 U.S. census, the population of the Navajo Nation was 155,276, of whom 96 percent were American Indian. Although precise figures are not available, as many as 50,000 Navajos may live in other regions of the United States, many maintaining close ties to their homeland, for a total of roughly 200,000 Navajos. These figures make the Navajo, along with Cherokee and Sioux, among the largest Indian tribes in the United States. Given the size and geographical expanse of Navajoland, it is not surprising that there exists a degree of regional cultural variation among Navajos. This variation corresponds to differences in microecological zones within Navajoland and, more recently, to development of semi-urban administrative and commercial centers. In addition, residents of some areas of western Navajoland are relatively isolated either by the stark physical landscape around Black Mesa or by the interposition of Hopiland; while in areas of eastern Navajoland, residents' contact with non-Navajos has been quite common due to the checkerboard pattern of land holdings. Regional variations doubtless are becoming less salient as more paved roads have decreased isolation over the past 20 years. There are, nevertheless, slight dialectal differences in lexicon, accent, and the construction of certain expressions, and there appears to be some variation in the distribution of ceremonial knowledge among traditional Navajos.

Navajo society is traditionally organized around a system of exogamous matrilineal clans. There is common agreement on the identity of the four original clans said to have been created by the deity Changing Woman, but the system is quite complex and several versions of clan classification are extant. Several clans are regarded as extinct, and a good number are adopted clans representing groups of foreigners who at various historical moments were incorporated into Navajo society. Traditional subsistence is based on a combination of farming (primarily corn) and livestock raising (primarily sheep). The Navajo undertake farming and livestock production in varying combinations, depending on their ecological zones within Navajoland. In the twentieth century, these have been supplemented by wage labor, first in railroad construction and the mining of coal and uranium, and more recently in service occupations in the vast bureaucracies of the federal Bureau of Indian Affairs, the federal Indian Health Service, and the Navajo tribal government. Many of the debates over tribal sovereignty in Navajoland have to do with tribal control of services rather than issues of legal jurisdiction, although the topic of whether the Navajo Nation will open itself to the casino industry has recently entered public debate. In general, although Navajos remain an economically poor people, their land, natural resources, population, and cultural and linguistic base place them as relatively well off in comparison to many other Indian tribes in the United States.

## Healing and the Representation of Social Self and Other

Popular awareness of Navajo society outside Navajoland is based largely on media reports of disputes between Hopi and Navajo tribes over reservation boundaries that were never of their own making. Popular awareness is also based on fictional accounts of the Navajo, including the detective novels of Tony Hillerman whose heroes are a traditional Navajo policeman (who wants to be modern) and an acculturated policeman (who wants to be a medicine man). Perhaps even more influential (especially for travelers to the southwest) have been the marvelous woolen rugs and silver jewelry sold at the reservation trading posts and roadside stands near Monument Valley and the Grand Canyon. Popular awareness of the dominant "Anglo" society among Navajos comes from direct interaction in the four reservation border towns of Gallup, Farmington, Flagstaff, and Page, as well as in the four more distant cities of Albuquerque, Phoenix, Salt Lake, and Denver.[2] On the reservation, Navajos encounter tourists, missionaries, and employees of the Indian Health Service, Bureau of Indian Affairs, and Navajo tribal bureaucracy. Awareness of Anglo culture is also developed during military service, which is not uncommon among Navajo youth, and via cultural and technological innovation. Over the past generation, the Navajo have readily adopted the pickup truck and, with the dramatic extension of paved roads on the reservation, have enthusiastically put it to the service of their love of travel and visiting. They have also adopted television and, more recently, computers and faxes. Traces of the postmodern pervasiveness of electronic media are apparent in the image of the traditional chanter who takes appointments for ceremonies by cellular phone from his pickup truck.

All of the preceding points of contact are occasional sites of cultural activity that could be defined as identity politics, whether expressed overtly as the literal politics of tribal sovereignty or covertly in the form of humorous stories about the curious ways of white people.[3] In this section, I will examine two events in recent Navajo history that bring healing to the fore as an articulation of the relations between Navajo and Anglo-American societies.

### *Reflections on a Mystery Illness*

In May 1993, the news media reported the sudden outbreak of a mysterious and deadly illness in the southwestern Untied States, centered in the eastern area of the Navajo reservation. This illness typically began with flu-like symptoms and, within scarcely more than 24 hours, progressed to total respiratory collapse. By mid-August, the Centers for Disease Control had recorded 30 cases, 20 of which had resulted in death (Centers for Disease Control 1993:612). Significantly, the first patients were all Navajo. The CDC sent an emergency team onto the reservation to try to identify the source and vector of the mystery illness, and they set up a hotline that reported possible leads and kept a tally of new cases. Along with the federal investigators, an army of national and international media personnel invaded Navajoland, prying into isolated communities and poking microphones into bemused or embittered faces. From there, the situation deteriorated.

Early media reports referred to the unidentified illness with names like "reservation flu" and "Navajo flu" (see Bales 1994). Some Navajos had difficulty getting service in restaurants, and tourists were observed driving across the reservation wearing surgical masks (Grady 1993). In early June, a front-page story in the *Washington Post* reported that school officials in Los Angeles had canceled the visit of a class of Navajo third-graders who had flown there to meet suburban Californian pen pals (Pressley 1993). Navajos were profoundly insulted at the apparent implication that they were a disease-ridden people or, worse, that they were somehow responsible for the outbreak. At the least, transformation of the epidemic into a global media event focused negative attention on the reservation and its people. In the frantic ten-day period before the illness was identified, the intensity and contrast between on-reservation and off-reservation opinions were particularly vivid to me because the opinions were being voiced in the weeks before my wife and I were to leave for a summer of fieldwork in Navajoland with our then six-month-old twin children. Friends in the university community, including our family pediatrician, expressed serious concerns about the wisdom of our departure before knowing the nature and degree of contagiousness of the mystery illness. Friends on the reservation, including Anglo physicians, were nonplused by the episode, pointing out that life there was proceeding much as usual, that the outbreak seemed to be quite localized, and that, in any case, fatalities occurred every day for a variety of causes among which this was only one more.

With our appraisal of the situation suspended between these poles of panic and complacency we set out, hoping that by the time we reached Albuquerque the mystery would be resolved. We checked the CDC telephone hotline at every night's stop along the highway. The day we reached Albuquerque was the day the CDC announced the cause of the illness: a new strain of a rare Asian virus called hanta that had previously been known to attack the renal system rather than the respiratory system. Acting on advice of traditional Navajo elders who observed that several outbreaks of severe illness earlier in the century had been associated with seasons of high rainfall, abundant pinon crops, and correspondingly high rodent populations (Schwarz 1995), investigators concluded that the virus was spread via the saliva, urine, and feces of deer mice. The deer mouse is a species not known to encroach on human habitation except occasionally in cold weather. The disease did not appear to be communicable among humans. The Centers for Disease Control, the Indian Health Service, the state of New Mexico, and tribal health agencies were all involved in spreading the word. The tribe issued guidelines for trapping and safely disposing of mice. Many Navajos heeded the health warnings and took precautions. The political situation deteriorated again, however, when one agency suggested that Navajos refrain from traditional ceremonies conducted in hogans with dirt floors. The implication that their sacred ceremonies were conducted in structures potentially made filthy by mouse waste was again deeply insulting. Hogans are used as dwellings or for ceremonies, and the earth inside them is carefully swept and regarded as very clean, indeed holy, by Navajos.

The predictable result of the episode was a degree of resentment and resistance. One Navajo woman, a sophisticated, bicultural person (i.e., equally conversant in Navajo and Anglo culture) who was active in tribal politics and health-care issues, drew the following parallel between the hanta virus outbreak and the serious flooding that was occurring at the same time that summer throughout the Mississippi Valley region. She indicated that the two events were linked—that given the proclivity of Mother Earth for keeping all nature in valance, it was not surprising that "whites" were having to undergo this hardship insofar as white people had slurred the Navajo with regard to the mystery illness. As evidence, she pointed out that due to the flooding, many whites along the Mississippi were being forced to haul their own fresh water, just as Navajos had been doing for many years in their arid homeland.

Even more telling was the understanding expressed by an elderly healer of the Native American Church (see below). He had been consulted by the relatives of a young couple who had been among the first to perish in the outbreak. They were concerned about spiritual danger, wondering what had caused the deaths and whether they should now be taking some ceremonial steps. His response was that it was incorrect to blame the mice, for they are harmless creatures with no apparent capacity to bear ill will to humans. In his estimation, the couple's death was caused by exposure to atmospheric contamination—poison in the air from some kind of government testing, or poison that had drifted over the ocean from a foreign sources (for example, Chernobyl or the Gulf War). The young people had succumbed because they had recently attended more than one funeral, thereby making themselves vulnerable by exposure to the dead in a way that is today quite common but is considered highly inappropriate in traditional Navajo practice. Traditionally only a few of the closest relatives take responsibility for a dead body and then only with careful ceremonial procedures that enable them to deal safely with the person's spirit and belongings. The healer's dual explanation is etiologically rational in invoking a factor of individual vulnerability combined with an agent to which others are also exposed but do not necessarily succumb. More impressive, however, is its cultural logic with respect to identity politics, whereby it combines lack of adherence to traditional practice by Navajos with the pathogenicity of the dominant society.

Stories of atmospheric pollution, occasionally linked to conspiracy theories, are relatively common on the reservation, which is to say they predate the hanta virus episode. Such stories are neither fictional nor delusional. Revelations over the past several years confirm that there have indeed been environmentally dangerous tests in the Southwest (see for example, the edition of ABC News' *Turning Point,* aired on February 2, 1994).

Government secrets only recently divulged in the media (to the surprise of most Americans) may well have been known from observation over decades by people living in the areas where such tests were carried out. Given the variety of ways in which Navajos (and others) see the environment and lifestyle of contemporary society as seriously out of balance, the authorities' implication of the humble mouse stimulated additional suspicion. It was common at the time to hear comments like, "My grandmother has mice around her house, and no one there has ever gotten sick." Some Navajos pointed out that people had lived side by side with mice for centuries, and only recently had it been reported that mice could be harmful—just as they had used the same water sources for years and only recently (for example, since the advent of uranium mining) had water contamination become a concern. In the rare instances in which healers acknowledged a possible role for mice, the mice tended to be regarded as messengers bearing a warning rather than as carriers of disease. In the even rarer instances in which mice were recognized as potential carriers of disease, they were more likely to be regarded as malevolent spirits in disguise than as vectors for a virus.

The dénouement of the episode occurred as more cases began appearing among non-Indians living beyond the reservation boundaries. As the affected area expanded, fewer cases were reported, and survival rates improved for those identified early. The media's appetite for the illness subsided, though occasional reports still appear, one being a brief *New York Times* article in October 1996 about a case in Utah. Perhaps the crowning irony—or crowning insult—of the hanta virus episode was reported by the *New York Times* in February 1999, as part of a story on mismanagement of funds held by the U.S. government on behalf of American Indians since the Nineteenth Century. At the document center in Albuquerque, New Mexico, records of the trust accounts were so poorly maintained that, according to government officials, they were contaminated with rodent feces that could contain the hanta virus (Egan 1999). With respect to the identity politics of healing, the outbreak had two effects on the ethnographic work we were doing. First, it largely precluded conducting interviews in the eastern area of the reservation where people had become most embittered about outside intrusion by the media. Second, it provided the occasion to include in interviews conducted in other areas a question about how the mystery illness, along with other "new" diseases like AIDS, fetal alcohol syndrome, and drug addiction, were being incorporated into Navajo understandings of health and healing.

### The Drought Apparitions

A second episode in spring 1996 also illustrates the identity politics of healing with respect to cultural representation. Two Navajo deities appeared to two elderly women in a remote area of the Navajo reservation. The apparition occurred in the context of a serious drought that lasted through the spring and into the fall of that year. Regarded as the worst drought since the 1850s, it caused considerable hardship for Navajos who were forced to sell part of their herds at a loss. On the reservation, the drought resulted in the largest livestock reduction since the government-enforced stock reduction of the 1930s. Aberle (1982) argued that the early popularity of the peyote religion among the Navajo was a religious response to conditions during the 1930s; similarly, I regard the 1996 apparitions as a religious response to the drought and its accompanying hardship. Of greatest importance, the deities left a message for the Navajo people. For reasons I will elaborate in a moment, I do not have the exact words as reported by one of the women to whom the deities appeared, but I did hear several interpretations of that message reported by Navajos in different regions of the reservation. Though these interpretations range in urgency and import, they bear a distinct family resemblance. The mildest interpretation was that the apparition was a warning that the drought was coming and that chanters (traditional Navajo ceremonial leaders or medicine men) should undertake the appropriate ceremonies to prevent or ameliorate the effects. Another was that the drought and other difficulties were occurring because the proper ritual offerings had not been made. The strongest was that the drought had occurred because Navajos were neglecting to learn about their own language and culture. Offerings should be made on the site of the apparition, and failure to heed the warning could lead to the end of the world.

These reports point to the relevance of the divine message to both healing and identity politics among contemporary Navajos. The requested offerings can be understood as healing rituals insofar as their intent is to remove obstacles to human existence and restore balance in natures and human affairs. The idea that this disorder is the responsibility of Navajos themselves for having forsaken their own identity is explicit in the strong form, while in the mild form it could be construed that the deities are simply doing the Navajo people a favor by instructing them in the ritual means for overcoming a difficult situation. The central theme is not a new revelation but reflects the sentiment of many Navajos concerned about cultural viability. It was well stated by a chanter I interviewed several years before the drought. For him, the central feature of chants, dances, and ceremonies is that they heal people. Not doing the ceremonies so often as in previous times, not knowing the older generation's teaching and planning, disharmonizing the ceremonies by secularizing them as "song and dance," or making fun of and fancifying them are all reasons "why we are easy targets for illnesses, tornadoes, lightning, things that harm us." In other words, he was suggesting that illness and natural disaster belong to the same category of events, that healing ceremonies address both, that both are exacerbated by failure to perform ceremonies, and that such failure is a consequence of weakened Navajo identity.

The critical feature of this episode for our present understanding is not in the message itself, however, but in the public response to it. In this respect, I must say, first, that the direct apparition of Navajo deities or Holy

People is rare[4] although they are pantheistically present throughout nature and human existence. Indeed, according to Navajo myth they terminated their immediate presence on earth long ago, departing with the following statement: "This day and this night alone you have seen Holy People. From this day on until the end of days you shall not see them again (in person), that is final!" (Wyman 1970:324–325). Second, as Aberle (1982) observed in his discussion of early Navajo resistance to the peyote religion, Navajo culture does not typically place high value on individual vision or mystical experience, and there is no clear tradition of publicly recognized visionaries.[5] Third, Navajo ceremonies are customarily organized around specific kin groups on a relatively small scale. While in Roman Catholic societies the apparition of the Virgin Mary or saints to gifted individual visionaries is widely publicized and leads to the establishment of permanent shrines (such as those at Lourdes or Guadalupe) as sites of pilgrimage, in Navajo society apparition to such individuals is not a typical mode in which traditional Holy People manifest themselves, and public pilgrimage is not a typical ritual practice among Navajos.

Thus it is all the more important that in this instance news of the apparition spread rapidly throughout Navajoland, and Navajos began arriving en masse—literally by the busload—to make offerings at the spot where the deities had appeared. A traditional diagnostician determined which of the Holy People in particular required offerings, and a variety of medicine men reportedly came to the site. One renowned and respected chanter performed a ceremony that included the appearance of masked dancers representing the deities who had come as messengers. A steady stream of pilgrims arrived, finally tapering off during the winter months. The president of the Navajo Nation granted time off to all tribal employees who wanted to make the trip, and he also made the pilgrimage. At the same time, however, he issued an appeal for Navajos not to talk about the sacred event, and for there to be no publicity about it. A few articles appeared in regional newspapers and then nothing. Not until December 31, 1996, did the tribe's own newspaper *The Navajo Times* carry the headline "1996 Top Story Is One That Never Ran." The article acknowledged the apparition but quoted the paper's publisher and managing editor as saying that the family of the two women had requested that no details be printed until they decided to give out the "correct story" of their experience and the Holy People's message, which at the time they had not yet done.[6]

The requests for circumspection by both the president and the family may or may not have been connected—that is, the president may or may not have been relaying the family's request to the Navajo public at large. The stated concern for accuracy of detail may be recognition of the stakes for the Navajo people in the transmission of a sacred message, or it may have been an attempt to control the message by requiring people to come to the site of the apparition to get the real story. In either case, the identity politics of this episode of collective environmental healing have important dimensions with respect to cultural representation both internal and external to Navajo society. Further, there is a sacred and a pragmatic aspect of each of these dimensions. Let me elaborate.

For Navajo people, sacred knowledge is powerful and potentially dangerous (*báhádzid*), and it must be treated with a respect that requires circumspection and even secrecy. Spreading knowledge too far could weaken its spiritual power, abuse its power, or turn its power destructively against the original knowledge holder. Widespread and uncontrolled dissemination of details of the apparition could be inherently dangerous in an analogous way. This understanding of potential danger dovetails with the pragmatic aspect of the politics of representation. Recall that the apparitions occurred within immediate memory of the media invasion surrounding the hanta virus episode. If a media circus is distasteful in itself, it is even more distasteful when it promotes misunderstanding and ridicule by outsiders who have no appreciation for Navajo traditions of spirituality. No less disruptive could be an invasion by well-meaning but unschooled New Age Indian "wannabes" who might have all manner of outrageous notions of what constitutes a proper offering to deities that are not theirs. From the traditional standpoint, that would be a dangerous situation indeed. In these respects the remarkably widely heeded call for circumspection was a notable political act of collective self-identification vis-à-vis the dominant non-Navajo society.

Internally, both sacred and pragmatic issues surrounded the interpretation of the apparitions, particularly with respect to their authenticity. Medicine men from some parts of the reservation objected that they had already been performing ceremonies and performing them correctly; therefore, they were skeptical of the apparent need for the deities to descend and deliver such a message. In addition, plans for a public ceremony at the site of the apparition for the benefit of the entire tribe seemed unorthodox to some elders who thought ceremonies for rain were more appropriately carried out in a localized and private manner by individual families. Yet others were concerned that traditional Navajos were (again reminiscent of the hanta virus episode) being unduly singled out and that the message of the Holy People was also relevant for Navajos of other religious persuasions and even for non-Navajos. In this view, all people, including Christians, need to return to their traditions. Pragmatically, some Navajos expressed concern that the legitimacy of the apparitions, or at least their positive impact, was being undermined by financial profit being made from the events. Making profit from a scared event is strictly distinguished from the fees paid to a chanter that establish respect for his ceremony and legitimacy before the Holy People. In this case, some people complained that the host family was "selling tacos" out there and that there was a medicine man (*hataalii*) asking for money from visitors. In contrast, a respected medicine man who did perform a ceremony at the site said that even with all the money being taken in, the fee he received was too small to pay his ceremonial helpers appropriately.

## Healing and the Negotiation among Traditions

The above discussion presumes a certain uniformity among Navajos with respect to both healing and identity. Within Navajo society, however, religious identity, multiple forms of healing, and interpersonal politics among Navajo make the picture considerably more complex. The critical ethnographic fact is the coexistence of three forms of spiritual healing in contemporary Navajo society: Traditional Navajo healing, Native American church healing, and Navajo Christian faith healing.[7] Traditional healing is practiced by the medicine man with his chant and sandpainting and the diagnostician who works by methods such as hand-trembling, crystal-gazing, or star-gazing. Native American Church healing is practiced by the road man, with his earthen altar, sacramental peyote, and Plains Indian-style sweat lodge. Finally, Christian faith healing is practiced by the independent Navajo Pentecostal preacher, with his revival meetings and laying on of hands, and by Catholic Charismatic prayer groups with their communal integration of Navajo and Roman Catholic practices.

All of these forms of healing are resources on the Navajo reservation, but only the one based on the traditional religion is indigenous to the Navajo people (Farella 1984; Reichard 1950; Witherspoon 1977). The Native American Church (NAC) is a pan-Indian movement that developed the sacramental use of peyote in its contemporary form beginning around the turn of the twentieth century with Plains Indian tribes. With its introduction to Navajoland in the 1930s, adherents faced legal pressures from their own tribal government, which decreed peyotism illegal in 1940 and did not move for tolerance until 1966 (Aberle 1982; La Barre 1975; Stewart and Aberle 1984). The introduction and influence of Christianity in many of its contemporary forms has been only sporadically documented for Navajo society (Bowden 1981; Hodge 1969; Sombrero 1996). Catholicism came largely with the influence of Franciscan missionaries, and Mormonism arrived with missionaries from Utah. Many of the major Protestant denominations are represented, but as is true among Christians in other societies, most ritual healing is carried out by adherents of various forms of Pentecostalism. These include branches of denominations like the Assemblies of God and participants in Charismatic prayer groups within Catholic parishes. Notably, however, they also include a number of independent congregations and networks of congregations that appear to be proto-denominations, all headed by indigenous Navajo pastors. They constitute an emergent and distinctly Navajo form of Christianity.

It is possible to outline a model of the relationship among the forms of healing associated with these three religious traditions with respect to what they have in common as aspects of Navajo culture and what distinguishes them as components of a cultural system of health care vis-à-vis one another. To summarize a longer argument (Csordas 1992), all three have as a common goal that the patient understand—Navajo healers typically say that a healer must talk to them so they understand. In contrast with a psychoanalytic emphasis on "insight" into the conflictual origins of the problem, this kind of understanding has more to do with a person's current place in the world, and is in accord with the often-observed preeminence of language and thought in Navajo culture (Farella 1984; Witherspoon 1977). However, each of the three Navajo healing forms approaches the goal of understanding in terms of a distinct philosophy and by means of a distinct therapeutic principle. Traditional Navajo healing is predicated on what might be called a philosophy of obstacles. Nothing happens without a reason, and the reason for misfortune is encountering an obstacle. The therapeutic principle of traditional healing is didactic, as the healer engages the patient in the therapeutic process using methods that guide thought toward the goal of understanding. In contrast to traditional healing's philosophy of obstacles, Native American Church (NAC) healing is predicated on a philosophy of self-esteem. Through the sacramental ingestion of peyote, patients achieve a profound personal connection with the sacred, and their voices and presence are valorized. The therapeutic principle in Native American Church healing is confessional, as patients pray, confess, or tell of their problems, are moved, and cry. Navajo Christian healing is characterized by a philosophy of moral identity, answering the question "Who am I?" in a way that among fundamentalists often includes the understanding that the person answering the question is not a traditionalist or peyotist. Finally, the therapeutic principle of Christian healing is conversional, with healing typically predicated on adopting Christian values and a Christian way of life.

Beyond these relatively abstract relations in principle, identity politics within Navajo society are played out in the interaction and negotiation among these three healing forms in everyday practice. The three tolerate varying degrees of eclecticism with respect to mixing forms, the most among peyotists and Roman Catholics, the least among fundamentalist Christians and conservative traditionalists. Fundamentalist Christians, including Protestant Pentecostals, typically require converts to burn ceremonial paraphernalia pertaining to Traditional or NAC practice—giving such objects away to unconverted relatives is not sufficient. On the other hand, conservative traditionalists regard Christianity and the Native American Church equally as foreign intrusions with no proper place in Navajo life. In practice, however, particularly in the pragmatic matter of trying to find the most effective type of healing in any episode of illness or distress, Navajos often have recourse to all three forms with little or no sense of contradiction.

The 1996 drought provides an initial example of this on-the-ground process of negotiation among religious forms writ large across the social field of therapeutic practice. It was reported that some Christian employees of the tribal government were none too pleased at the official leave granted to Traditional Navajos for making the pilgrimage to the site of the apparitions. A Traditional chanter who prayed at the site felt that one of the problems that needed to be ceremonially addressed was that the family had allowed a Native American Church meeting to be conducted on their land. On the other hand, a major public event was held in the tribal capital of Window Rock and broadcast across the reservation by the tribal radio station KTNN, during which a Traditional chanter, an NAC road man, and a Christian minister took turns offering prayers for the end of the drought.

In the remainder of this section, I will begin to elaborate on this complex situation through an examination of healers' understandings of one another's practices and worldviews. There is, to begin, a profound amount of cultural cross-pollination between Traditional and NAC among Navajos today. Among Traditional healers interviewed, a very small number dismissed peyotism outright. These were not always the eldest, contrary to the expectation that older Navajos would be most conservative. Indeed, some of the eldest have consented to be patients or lay participants in peyote meetings, and may have children who are active. Some appear to have integrated peyote into their pharmacopoeia, treating it simply as another among traditional healing herbs. One traditionalist who objected to peyotism appeared to do so on pragmatic grounds, and not without humor:

> That's a new fad. That is a new practice. They claim it as a Navajo ceremony, but you hear them sing happy birthday with their NAC songs. Us Navajo medicine men don't sing like that . . . . Now with the Native American Church, they use the peyote button and they sing songs. I don't understand the songs. But there's a lot of, still too much emotionalism in it. Also, the peyote button, it works as a painkiller. So to say it's real healing, there is two sides to it and doubtful.

Among NAC healers, multiple participation is also quite likely, especially as patients in Traditional ceremonies when illness strikes. Those who cite ignorance of traditional ways often appear to do so with the humility of an untutored layperson rather than with an attitude of aloofness or rejection. Today many of the most devout younger traditionalists appear to have been inspired by peyote to learn more about their own cultural roots. Some say that peyote is not an import from the Plains Indians, but was originally given also to the Navajo and lost, only now to return. Yet even among those who blend the two religions there are certain ways in which they are distinguished in practice: Certain traditional prayers should not be said from the peyote altar, one shouldn't participate in traditional and NAC ceremonies on the same day, some prefer to hold peyote meetings in a Navajo hogan rather than a Plains tipi, certain traditional herbs (as well as certain hospital medications) are known to be incompatible with peyote, and some Navajo road men insist on the legitimacy of holding funeral services (despite traditional constraints on proximity to the dead) on grounds that people should continue to feel connected to deceased relatives who will always be a part of them.

Among Christians, one striking example of multiple participation is that of the devout Catholic woman interviewed as a Christian healer who regularly attended Native American Church prayer meetings and who made use of our interview honorarium to pay for a Traditional blessing ceremony. On the other hand a conservative pastor made the following pronouncement: "With Christianity, our prayers float up to God spiritually. With NAC and traditional, the prayers go to the devil. In NAC, they have to go to Texas [where the peyote cactus is gathered] to get the medicine. In traditional, they get herbs from the mountains. We Christians don't have to go to the mountains or Mexico or Texas."

These varieties of conflict and cooperation are not only matters of principle and doctrinal positioning but have everyday implications for social interaction among individuals and especially within families. For example, three brothers participated in our study: One is a traditional chanter, the second a peyote road man, and the third a Christian pastor. The often-strained interplay of interpersonal relationships, religious commitments, and family loyalties among these men is evidence of the critical role of religion and ritual healing in the negotiation of contemporary Navajo identity.

## Healing and Personal Transformation

Examining the experience of individual patients in ritual healing calls attention to the little-addressed need to specify a theoretical connection between personal and collective identity. As a step in this direction, I have selected from among the patients interviewed in our work three persons, one treated in each healing form. For these three patients, the issue of identity is a life theme or locus of therapeutic attention. Each of these cases revisits the tensions animating the two levels of analysis I have treated thus far—the tension between Navajo and Euro-American values and ways of being-in-the-world, and the tension among the three healing forms and their implicit visions of what it means to be a Navajo today. Here on the level of individual experience, these are diffracted by the tensions between illness and well-being and between aimless existence and dignified self-worth.

### *Traditional Healing*

Sylvia is a 30-year-old woman in her third year of college at a small branch state university in one of the reservation border towns. She is very close to her family, especially her mother whom she admires for her strength, respect, comfort, and support. She regards herself as a well-rounded person confident in her traditional background in the face of non-Navajo friends, despite early problems with self-image because of being heavy-set in comparison to her "beautiful" sisters. She spoke explicitly about the relationship between self-identity and the traditional life philosophy summarized in the phrase "two walk in beauty":

> To me to walk in beauty would mean to know your whole self identity, to walk in harmony, you know with nature, your surroundings,

and even, you know, having your whole family, being aware of everything around you .... Walking in beauty will also be the person themselves. You know that uh, getting, knowing their traditional beliefs, their culture. Arising from that, one you know the whole background, that an be your backbone, to growth, knowing that's your self-identity, and from there, you won't get discouraged. You won't get disappointed. A lot of the negativity that one must feel won't be with your forever 'cause you'll know how to deal with it, once you know your self-identity. And that begins, I think that should begin at an early age. I believe that, you know, walking in beauty.

Her problem began with her father's death seven years previously when, as the eldest of six children, she assumed much of his role and responsibilities. Since then she has quite literally been carrying the weight of his death on her shoulders, with the onset of shoulder and arm pain on her left side. At the hospital, x-rays were inconclusive, and she developed a dependency on a pain medication that inflamed the lining of her stomach. Frequent treatment by chiropractors helped temporarily, but the pain always returned. She felt that there was a spiritual component to the problem and that this component could be addressed only through a traditional ceremony. This feeling may be due in part to the fact that her father was a strong believer in traditional religion, and, as in other aspects of life, she felt the need to carry on in his ways. Indeed, there was a significant emotional component to her distress. She reported that she wasn't herself (a common self-description among Navajos in illness or distress) and that she was lonely and unhappy. She said that she thought and dreamt about her father, had thoughts about the deaths of others, and experienced negative feelings: "Like there was something heavy weighted on me so much, on my shoulders, and I just couldn't take it anymore."

The most vivid dream she recounted was one shortly after her father's death. In this dream, he spoke to her lovingly, showing her where he was and who he was with, assuring her that he was fine and that he was watching over the family. Despite the positive nature of this particular dream, in Navajo culture dreams of the dead are invariably problematic. They require ritual treatment to determine the effects of the deceased spirit on the living. Sylvia herself acknowledged uncertainty about whether the dream was a good or bad thing, and her uneasiness was evident as she made it a point to report that every time she visited a medicine man he would ask about such dreams. Her first consultation was with a Traditional diagnostician who, by the technique of crystal-gazing, determined that Sylvia required an Evilway ceremony.[8] This diagnostician resolved the contradiction between a positive and negative interpretation of the dream with an elegant therapeutic move. He determined that her father's sudden death was due to witchcraft, the effects of which lingered in the family and were augmented by additional witchcraft performed since the death. He also determined that Sylvia and her second youngest brother, the two family members who were emotionally closest to their father, were most affected. It was thus possible to attribute any evil effects to the malicious intent of outsiders, removing blame from the father's spirit and preserving the positive emotional valence of his memory, while implicitly recognizing the two siblings as vulnerable to an internal psychological process identifiable in clinical terms as bereavement or grief reaction.

In the meantime, the diagnostician ritually extracted objects from Sylvia's shoulder. He claimed that these were the immediate source of her pain, whereupon it subsided. She had to return to the diagnostician on several occasions for this procedure while she delayed having the more elaborate ceremony. According to Sylvia, she delayed the ceremony because she did not know how to prepare or how to find the right chanter. Only after becoming involved in a significant new relationship with a man whose family, coincidentally, knew a chanter with the appropriate ceremonial knowledge to conduct her ceremony, did Sylvia become ready to leave behind her debilitating emotional attachment to her father's memory. She found the most compelling element in the ceremony the moment in which she accompanied the healer outside the hogan to confront the evil and to pray that it no longer affect her. Subsequently, during the final morning prayers, she likened her experience to that of an eagle, flying high in the sky. She said that she felt clean and as if her sense were heightened.

Sylvia stated quite explicitly that for her the ceremony was the beginning, not the culmination, of a healing process. Three months after the ceremony, she reported:

> For me personally, I'm a traditional person and when I know I have the prayer done, to me that means a lot. It motivate me, and it knows that I can … it tells me that I can do it. And that whatever obstacles that may lay ahead that may be hard, hard to do whatever, you know, hearing the prayer and having that protection I need. All that stuff I feel. I guess that's what a traditional person does … . It's given me a lot of courage and determination in saying that I won't finish. I guess that's it's a motivator for, you know, within me not just the English way but also the Navajo traditional way. It just makes you want more, to strive for more and, you know, knowing that wherever you go, you're always protected … . After I knew—once the ceremony or the main part that was done—what I was there for, and what it was supposed to have done for me, it did because, you know, they say that prayer is very powerful. And you know the stuff you hold like they have the arrowhead and that stuff. Those are very powerful. All that comes with, you know, stories and behind that there's a meaning for all that … I could feel it within me. I could feel a mixture of all that he was praying about. And I could feel it. You have to really understand, you know, why your ceremony was being conducted, you know, and the reasons for it and the purpose of what it's gonna do for you in the outcome. You know why they used those prayers. And also behind every prayer

and every song that's sung traditionally, the medicine man always tells you why that song originated and what its purpose is and what it served for. So knowing that, knowing after he tells you, you know, you think "Okay, this is how I'm gonna get you over this thought process." ... So I think everything has diminished ... I don't know exactly when it was, or you don't know exactly, "Okay after that song I'm healed." It takes usually time.

Thus, Traditional healing initiated a series of changes for Sylvia. She was able to make sense of her father's death without blaming him; to feel reassured of his well-being in the afterlife while letting go of his distressful afterlife presence in her life; to learn to experience positive rather than negative memories of her father; to question and finally reaffirm her own identity in terms of aspirations, past, family, and culture; to relieve her physical pain and negative thoughts; and to become closer to her boyfriend and his family.

### *Christian Healing*

Nancy is a woman of 47 with 3 children. At the time of this writing, she was twice married and twice divorced. Two years previously, she was diagnosed with breast cancer, which is currently in remission. In addition to suffering from the after effects of surgery and chemotherapy, she was troubled by abandonment issues; after her diagnosis, her second husband left her for another woman. She recounted a troubled early life of violent abuse by an aunt with whom she lived, by boarding school personnel, and by her father, followed by a first marriage to a violent alcoholic who left her a widow. Of her recent abandonment she commented, "This is like a recycle." Despite this history of difficulties, she demonstrated her resiliency in two ways. First, she pursued her second husband in court and won a significant settlement, and second, she reenrolled as a full-time college student.

Nancy encountered Christianity after being diagnosed with cancer and before her husband left her. A friend invited her to attend a small independent Navajo congregation. Nancy characterized the atmosphere at the church as peaceful and open and said she felt more at home there than with her own family where everyone was constantly arguing. Prayer meetings at the church include prayers for healing by the pastor. On one occasion, Nancy asked for prayers for her daughter's strength and health and for her own education in the context of upcoming midterm exams. Such healing prayers are available on a regular basis for participants in the Christian prayer meetings, services, and revivals. Nancy said the prayers gave her strength, determination, faith, peace, wisdom, renewed sociality, and help with the stress of pursuing her education. She added, however, that she did not understand the meaning of the songs sung in church until she had the experience of being saved. In her words, the overall effect has been:

to be more open and to be like, I feel it did a lot for me. It really, I feel more at peace, and I feel they can make me stronger like health-wise, and I can communicate better with people now than before because I was always locked up in bedroom, you know, studying, studying, studying. I'd be with my books, and I feel like I just came out into the world again, and I just went with them. And you know I love those prayer meetings. I like to go to those prayer meetings. I like to be with Christian people. They're more understanding. They help me a lot.

Critical to her understanding this overall effect in the case of Christian healing is that it is compounded of participation, healing prayer, and salvation. Moreover, particularly among such independent Navajo congregations (that is, those not affiliated with a major denomination), the Christian community with its distinct lifestyle is an insulated society within a society.

Nancy's experiences with the other two spiritual healing forms say much more about the politics of personal identity acted out in social relationships than about those healing forms per se: An exceedingly negative experience with the Native American Church, of which her unfaithful second husband was an adherent, and a highly positive one with Traditional healing, in which her father is a medicine man. She recalls her second husband threatening her that peyote would somehow "come after you and do something to you" if she did not listen to and respect it. She allowed him to practice and keep peyote at home, and she even allowed a road man acquaintance to live with them for a time and to hold a peyote meeting for her. She felt, however, that this road man was also "using peyote for sex and love," causing her husband to leave her by involving him in this abuse of peyote's spiritual power. She chased the corrupt road man out of the house, saying she would go the route of chemotherapy instead of submitting to his treatment; she was skeptical that a one-night ceremony could heal her in any case. She claimed she was told by a physician that peyote could make cancer grow, and she concluded that her husband and the road man had been conspiring to bring about her demise.

On the other hand, though in principle she feels that she could no longer participate in Traditional ceremonies because she has "dedicated myself to the Lord," shortly after her husband left she had a Blessingway ceremony performed by a chanter contracted by her father. In this case, pragmatism in the logic of therapeutic choice, in the form of accepting an alternative treatment that may not only prove to be efficacious but that will also please a family member, impinges on identity politics at the personal level. Although Nancy belongs to a classic independent Navajo fundamentalist congregation, the friend she asked for advice beforehand said to go ahead as long as it was a Blessingway and not an Evilway ceremony and as long as she said a prayer, "so you can get the color of the Lord's light." In effect, her friend advised her to give the Traditional ceremony a Christian benediction. Nancy reported the following effect of the ceremony:

[It] kind of cleared my mind, and I used to cry a lot, and I couldn't eat, and I couldn't even think. And it just kind of fulfill my spirit again.

It's like it just woke you up, like you were dead for a while and you just came back to life, you know. Kind of made me feel that way, and then the medicine man told me, "Don't think about your past, don't think about, don't think about your ex-husband. Don't think about what he's doing, what he's saying to you. Don't listen to the gossip. Don't listen to rumors." And he told me, "Just think about yourself." That's what the prayer was, just for myself, just to get my life back together . . . . And after that, that took a lot of pressure off my back.

Nevertheless, she declined the chanter's offer of further treatment because it would have cost additional money.

For Nancy, the experience of "being pulled three ways" was resolved by being saved. She expressed the need to "know where I stand," to "find myself," and "to know what's going on in my life." This knowledge emerged through an act of commitment to Christianity. She says, "And so when I got saved, I had to say my own prayer from my own heart, from my inner-self, just give everything, all my problems, everything back to the Lord, let him take care of it."

### Native American Church Healing

George is a 24-year-old student at a state university in one of the reservation border towns. His experience is diametrically opposed to that of Nancy in that his grandmother was an early and strong adherent of the NAC. He refers to her as a "pillar" and speaks of her "teachings" about "peyote and the importance of life." The family has long used the same road man—he ran meetings for George's grandmother and watched George and George's siblings and cousins grow up. This stability and the long-time friendship and respect between the family and the road man is important to George, and the role of road man as an anchor for social relationships stands in sharp contrast to the marginal and corrupt practitioner portrayed in the previous discussion of Nancy. George had a traditional upbringing centered around caring for sheep. His father was a heavy drinker who regularly abused his wife. He is very close with and concerned for his mother—for him the Navajo word *shima* refers equally to one's mother and to one's home, the earth where one was born. He cites the connection symbolized by the custom of burying a baby's umbilical cord where it is born.

After boarding school, high school, and a year of college (all on or near the reservation), George enlisted in the Marines, traveling widely in the Far East and Middle East. He participated in the Gulf War, having peyote ceremonies both before his departure and after his return for protection. A prevailing theme for him is comparing the Navajo worldview, religion, and lifestyle with those of the larger society, trying to come to terms with and integrate his experiences, to figure out how he wants to live and where he wants to fit in. His problems amount to a general malaise compounded by helping his mother care for an alcoholic brother's children, his

own lack of motivation concerning his goals of finishing school and joining the Peace Corps, problems in his relationship with his girlfriend, and
several physical complaints including back pain from an old injury and
respiratory problems that began during the Gulf War. He used to run
regularly (as is prescribed in traditional spiritual discipline) but says recently
he has stopped running because of "low self-esteem." Behind these issues,
there is a sense shared among family members that the influence of witchcraft perpetrated against his grandmother years ago has been passed on
through the whole family, causing them to fight among themselves.

George and his mother asked for help from their family road man during a brief ceremony he was conducting for some other close relatives. This
request initiated a process, allowing them to identify their problems explicitly. According to George, the simple effect was "just knowing you're getting helped. That's about it. I mean, I've attended it since I was small, so you
just know it's going to work, I guess. It's an idea. And, you relate your problems to other things, maybe see the source of your problem, why it's going
on, why you are blinded by it." Later, as a result of a full-scale, all-night
peyote meeting, he described a feeling of familiarity and self-knowledge
that brought him back to the basics and let him know who he is. He
strongly got the message that "it's all up to you" and was able to engage in
significant self-evaluation. In addition, he was able to "release a lot of emotional baggage," as well as to express some feelings and issues in prayer that
he knew his girlfriend, sitting next to him, could hear. Three months after
the ceremony, he reported doing better in school, having better motivation,
and having an improved family situation. He remained with his girlfriend,
and several months later she became pregnant.

George appears to be explicitly concerned with his own identity and
proud of his open-mindedness and experience. He is a young man who
wants to do everything, who is interested in everything, and who says, "My
whole life is an experiment." According to George, his religion teaches him
how to "carry yourself." He is concerned that "society is going to take us
all down," and he is concerned about "new Navajos" who "keep themselves
blind," are materialistic, and do not want their peers to know they speak the
Navajo language or that they participate in ceremonies. For him, peyote
itself is less a spirit or identity than a means of protection and a medicine
that allows clarity of thought, expression, and the ability to release emotions. In this case, healing was a way to move beyond a difficult transitional
period, one in which identity as a Navajo—a responsible adult bicultural
Navajo man with deep ties to family and significant aspirations in the contemporary world—was immediately at issue.

## Healing and Politics, the Politics of Healing,

## Healing Politics, Political Healing

How then to formulate this relation—how are identity politics being
played out through the practice of ritual healing among Navajos?

Answering this question requires taking a position on the series of conceptual religions I identified at the outset between religion and politics, tradition and modernity, individual and collective, microsocial and macrosocial. In elaborating such a position, I want to review some of the literature in which anthropologists have begun to discuss the way ritual healing, itself a form of cultural power, is relevant well beyond a specific problem, illness, or disorder. Arthur Kleinman (1980, 1986) pioneered this area through studies of healing and psychiatric disorder in the aftermath of the Chinese Cultural Revolution, showing that suffering must be understood in the context of both larger political realities and local moral worlds. Also working in China, Thomas Ots (1994) documented changes in bodily practice and emotional experience in a cathartic healing movement based on *qigong* (breathing therapy) in relation to the course of the prodemocracy movement that culminated in mass demonstrations at Tiananmen Square. Marina Roseman (1996) described a ceremony involving Sri Kelantan, spirit of the Malaysian state of Kelantan, showing how ritual action articulates interactions among Malay, Chinese, and indigenous Orang Asli. Drawing on material from the Newar of the Kathmandu Valley in Nepal, David Gellner (1994) took up the issue of the predominance of women in the role of medium by considering both traditional notions of gender roles and contemporary political changes.

In Africa, Jean Comaroff (1985) examined Zionist healing within the legacy of colonial repression as a "mode of repairing the tormented body, and through it, the oppressive social order itself. Thus the signs of physical disorder are simultaneously the signifiers of an aberrant world" (1985:9). Matthew Schoffeleers (1991) disagreed with Comaroff, arguing that political acquiescence, rather than resistance is a characteristic of churches in South Africa. In his view, Zionist churches are acquiescent because healing, the central component of their practice, individualizes and depoliticizes the cause of illness. Lesley Sharp (1990, 1993) understands possession by tromba spirits in northwest Madagascar in relation to psychological consequences of conflicting moral orders and anomie. In a setting where ethnic boundaries had been blurred by colonialism and polyculturalism, tromba possession articulates themes of ethnic and individual identity and resistance to capitalism. Sharp focused on female participants who became involved in a fictive kinship system requiring adherence to restrictions that provide a rationale for manipulating economic relations and thereby undermined processes of capitalist exploitation. This process provides them with work as healers, liberating them from ordinary agricultural labor.

In the Americas, Michael Taussig (1980a, 1980b, 1987) regarded shamanism in southwest Colombia as part of the context of colonial violence and its aftermath. This violence is intricately linked to conceptions of persons, self, and other, to constructive appropriation of the other, to various healing systems of the Indians, and to cultural understandings of Indians as mysterious, powerful, and dangerous. Shamanic practice and its hallucinatory possibilities thus transcend the meaning of healing as an attempt to ameliorate the distress of individuals, becoming a central figure in the cultural discourse of colonialism. Libbet Crandon (1989) examined mestizos' adoption of Aymara healing methods during rural Bolivia's transition from a colonial society to a class-based agricultural society. These methods serve to explain social group participation expectations and to integrate the person into a new sociopolitical position in the cosmological system. In Ecuador, S. A. Alchon (1991) showed that with the advent of the Spanish presence and the rise in mortality, indigenous conceptions about etiology changed, but healing practices changed little. A critical sense of cosmological balance based on the need to propitiate both Andean and Christian deities became increasingly difficult to maintain, and preservation of traditional healing practices became a form of political resistance. Ramirez de Jara and Pinzon Castano (1992) discussed how Sibundoy shamans in Colombia integrate indigenous thought structure and the challenges of national society with Colombia's diverse manifestations of popular culture. J. Waldram (1993) examined symbolic healing in Canadian penitentiaries, looking specifically at aboriginal offenders in cultural awareness programs. The programs provide new meaning to disrupted lives and help inmates resolve identity conflicts. Since many offenders are from different native groups, the establishment of a common cultural ground and mythic world is also an adaptation to a situation of increasing cultural plurality.

Integrating these few sources is only a small first step in fleshing out a problem area where discussion too easily defaults to a polarity of simplistic interpretations. Either ritual healing is a futile expression of frustration—the opiate of the masses interpretation—or ritual healing is a subtle form of political resistance—the postmodern liberation of the indigenous voice interpretation. I propose that the kind of analysis I have begun of the Navajo situation offers more nuanced opportunity to clarify relations among the conceptual parings that I have identified here. To make my position explicit, scholarship must move beyond defining the project as a study of ritual healing in the context of politics, or as the opening of a performative window onto larger political processes, and toward an understanding of the kind of

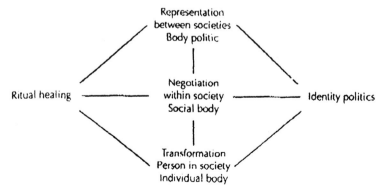

**Figure 5.1**   Relation between Ritual healing and identity politics

experiential transducers operative between the religious and political domains. Bodily experience may be a prime example of such a transducer, and moreover one that is relevant at all three levels of relation between healing and identity politics I have identified (see Figure 5.1).

Indeed, this notion corresponds with the kindred attempt by Scheper-Hughes and Lock (1987) to situate the "mindful body" of medical anthropology with respect to broader social issues. To elaborate briefly, cultural representations of the mystery illness and the drought apparitions contributed to the ongoing constitution of what Scheper-Hughes and Lock term the "body politic" of Navajo society in its vulnerable yet resistant confrontation with the dominant society. The ongoing negotiation among healing traditions is a process of constituting the "social body" by situating it within Navajo society as a subject of one tradition or as the node of intersection among traditions. The experience of personal transformation narrated by patients constitutes the "individual body" as a person with a contemporary Navajo identity in the politically charged space between tradition and postmodernity. In this formulation, each of the cultural processes I have singled out in the relation between ritual healing and identity politics (representation, negotiation, and transformation) is inherently political, and not only those occurring on the level of the body politic per se. In sum, reading Figure 5.1 horizontally, recall that it is no coincidence that the term *power* is essential for analyses of both religion and politics (Fogelson and Adams 1977); reading the figure vertically, recall the contribution of feminist theory's lesson that the personal is also political.

Again, in contemporary analyses of the relation between tradition and modernity, aside from an interest in fundamentalism of various stripes, there have been relatively few efforts to understand the place of religion in the contemporary world system or in the process of globalization (Beyer 1994; Csordas 1997; Friedman 1994; Ong 1996:745–747; Robertson and Chirico 1985; Schieffelin 1996; Wuthnow 1980). In this respect, it is critical both that the appeal to tradition is just as likely to be heard on a global as on a local scale (witness the recent proliferation of "fundamentalisms"), and that themes of modernity (and postmodernity) are evident not only in the global ecumene but equally in fourth-world enclaves such as Navajoland. For contemporary Navajos, tradition thrives in itself but also comes to be defined in relation to Christianity and the NAC, as well as in relation to modern technology, national politics, and global movements of indigenous peoples. The college-educated Navajo who declares in an e-mail message, "I am a traditional person," means something quite other than the stereotypical image of the old person adorned in turquoise or tending sheep in the desert. Especially in its religious aspect, tradition is more than a badge of ethnic identity; it is a mode of engaging the world.

Finally, just as power belongs to both the religious and political spheres, the concepts of self and identity belong to the psychological analysis of individuals and to the social analysis of collective processes. As Calhoun notes, in the historical context of democracy and the Protestant Reformation, "problems of individual and collective identity were joined, both because individual identity was shaped by what Foucault called new disciplines of power, and because the question was raised of what sort of individual identity qualified one to participate in the public discourses that shaped policy and influenced power" (1994:2).

I would argue that these issues need to be teased out by more explicit and frequent dialogue between psychocultural and sociopolitical approaches, by more attention to these issues between the locales where democracy and Protestantism have been cultural touchstones, and with respect to particular empirical problems such as that of ritual healing. In this way, they can continue to be made problematic as the point of intersection between, for example, Friedman's (1992) concern for the relation between the construction of identity and larger global processes and the concern raised by the study of ritual healing for the relation between construction of identity and smaller psychocultural processes.

One way to advance this problematic in future work might be to distinguish between a personal politics of collective identity, in which individual actors with clear commitments are struggling to assert shared identity, and a collective politics of personal identity, in which each among a group of actors with ambiguous commitments is struggling to attain individual identity. Such a distinction would certainly serve to set up comparisons by identifying relative emphases rather than absolute differences in the substance of identity politics. In contemporary North American society, for example, it would suggest a degree of commonality between a movement for gay rights to define a personal homosexual identity and a religious movement that places a priority on personal salvation or transformation. Likewise, it would suggest a commonality between a feminism that aims for a collective identity based on sisterhood and an ethnic movement intent on creating a community. With respect to the particular case of the Navajo, such a distinction would facilitate a specification of the sense in which individuals, though in identifiably Navajo ways, are struggling for personal identity, and the sense in which a personal commitment of the community is a contribution to collective identity. If, in practice, the distinction is difficult to tease out because both personal and collective sensibilities are highly relevant and because both center around the critical issue of being Navajo, it may, by the same token, serve as a useful tool for understanding the pairs religion and politics, individual and society, microsocial and macrosocial, tradition and modernity, as complementary aspects of the same phenomenon, sides of the same human coin.

CHAPTER SIX

_Talk to Them So They Understand_

This chapter has two goals in relation to Navajo healing, one ethnographic and one theoretical. The ethnographic goal is to present a thesis about the interrelation among traditional Navajo, Native American Church, and Navajo Christian healing in contemporary Navajo society. I will sketch the similarities and differences in principle among them, emphasizing that in practice many Navajos have recourse to all three in their search for healing. The theoretical goal is to advance an understanding of therapeutic process in terms of the model introduced in Chapter One in which the key elements are the patient's disposition, experience of the sacred, elaboration of alternatives, and actualization of change. In the process I will move toward further refining and elaborating the model itself with respect to the unique lessons to be learned from the Navajo setting. Both these goals can be approached in a very concrete way, by examining therapeutic process as undergone by individual Navajo patients who have shared their experience of suffering and healing with me and a team of Navajo and Euro-American researchers. First, let us look at the relation among the three healing traditions.

## Corn Pollen, Fireplace, and Cross

For Navajos who adhere to their people's traditional religion, the pollen of the corn plant that provides much of their nourishment is a powerful symbol of life, growth, and well being. Navajos often carry some of it in a small pouch, and it is used in many ceremonies. Traditional religion and healing are sometimes referred to as the "corn pollen way." This form of healing includes the practices of the medicine man with his chant and sandpainting, and of the diagnostician who works by methods such as hand trembling or crystal gazing. For those who follow the teachings of the Native American Church, a central symbol is the open fireplace around which participants gather for prayer, a source of warmth, energy, and inspiration sometimes thought of as a kind of nurturing grandparent. Native American Church healing is that of the road man with his crescent altar and sacramental peyote arranged in relation to the fire, and of the sweat lodge leader who often follows the Plains Indian ceremonial style rather than that practiced in the small earthen sweats of traditional Navajos. Finally, for Navajos who have converted to or grown up in one of the Christian denominations, the cross symbolizing the passion of Christ is the central symbol of power, healing, and salvation. Christian faith healing is that of the independent Navajo Pentecostal preacher with his revival meetings and laying on of hands, and of the Catholic Charismatic prayer group with its communal integration of Navajo and Roman Catholic practices. All of these forms are available as healing resources on the Navajo reservation. They appear to allow varying degrees of eclecticism and mutual borrowing of elements, the least among conservative traditionalists and fundamentalist Christians, the most among members of the Native American Church and Roman Catholics. Likewise, although it is not our central concern here, it is important to note that adherents of all three religious traditions seldom hesitate to use biomedical healing resources available in Indian Health Service facilities, and their understandings of illness often combine elements of Navajo and Euro-American culture (Chapter Ten; Csordas and Garrity 1992).

I would suggest that what makes all three of these forms distinctly Navajo, either by tradition or by adoption, is a shared criterion of success in healing: that the patient come to "understand." To be specific, when asked how their work helped people, Navajo healers do not refer to the spiritual effect of their ritual performances on patients, but typically say that one "must talk to them so they understand." This understanding is culturally distinct from the notion of insight into psychodynamic processes as the criterion of success found in Euro-American psychotherapy. It is instead a contextualizing of life experience in terms of a particular philosophy and by means of a particular therapeutic principle. Regardless of which of these three forms of healing they practice, contemporary healers are participants in Navajo culture, and articulate "understanding" as the criterion of therapeutic success. In fact, this is in accord with the often observed preeminence of language and thought in Navajo culture. As Gary Witherspoon (1977) has observed, "... the world was actually created or organized by means of language. The form of the world was first conceived in thought, and then this form was projected onto primordial unordered substance through the compulsive power of speech and song." Continuity among the healing forms with respect to these notions provides the conditions of possibility for their coexistence within the same cultural system of health care. Within that system, however, their approaches to the goal understanding are differentiated by distinct philosophies and therapeutic principles. I will briefly outline the typical healing practices of each tradition, and characterize each in terms of its typical philosophy and therapeutic principle.

### Traditional Navajo Healing

Traditional healing is predicated on what might be called a *philosophy of obstacles.* Nothing happens without a reason, and the reason for misfortune is encountering an obstacle. Obstacles are identified not only as causes of illness, but everywhere in life: growing sleepy at one's desk is an example. The problems we encounter are already there before we encounter them. As one chanter said, "It's like we walk into a room and among the problems, where we proceed to get stuck. We must separate them all like strands, make our way through them [the healer makes a gesture like parting bushes in front of himself]." This is an existential situation that has persisted from the time of creation. Even the food we eat is such a problem. It comes from alien, foreign, or enemy places, causing illnesses such as diabetes and cancer. Says the same chanter, "And what kind of food is alcohol? It makes you vomit and not think right, it kills people. Who makes such a thing that poisons and kills people?"

The therapeutic principle of traditional healing is *didactic*. There are multiple techniques available to the healer, and contrary to much of what is understood about ritual healing, they are not ritual manipulations that the patient observes as spectator. Instead, they are methods of engaging the patient in the therapeutic process, guiding thought toward the goal of understanding. This engagement of the patient in therapeutic process was suggested by a conversation with a traditional Navajo diagnostician. She related that she would instruct a troubled adolescent to "think of your parents and how they raised you," in order to cause him to review his life. She then turned to my field assistant and said, referring to me, "He's probably thinking about his parents right now."

Approaching ritual technique in this way throws an experiential light on practices such as having the patient sit on a sandpainting, or hold a sacred mountain pouch. The patient also becomes engaged in the therapeutic process by following the healer's enumeration of all the joints of her body, starting from toes and progressing to the head. To invoke a more specific example, the healer's use of stones gathered in the morning—black jet, mother of pearl, turquoise, red stone—has more than the symbolic significance of the dawn and the four cardinal directions with their cosmologically meaningful colors. For one traditional diagnostician I spoke with, the stones constitute a therapeutic mnemonic, wherein the patient incorporates in the first stone the intent to stop the urge to drink, in the second to better himself as a person, in the third to become whole again, thinking integrally of his home, his food, his water, and in the fourth to find a job for prosperity and to be able to get a car. The healer explains that the patient's problems are "packed in like a mud ball." She tells them healing is not an overnight thing, but will happen gradually as thoughts change.

Again, the common technique of having the patient repeat lines of the long chants after the medicine man has an engaging effect, especially, as one Navajo participant said, when the chant "speeds up into second gear" and one is carried along. As another noted, the proceedings are "ritualistic" only to one who does not understand the Navajo language. One who does understand experiences the contextualization of life experience within a cosmological and physical "home," the Navajo land and people. The chant is not magically efficacious, but existentially engaging. One Navajo commented that only those who cannot speak the Navajo language could ask how participants manage to stay awake all night long during a ceremony: those who understand are too absorbed in the beauty of what's going on even to think about sleep.

## Native American Church Healing

In contrast to Traditional healing's philosophy of obstacles, Native American Church healing is predicated on a *philosophy of self-esteem*. Through the sacramental ingestion of peyote, the patient achieves a profound personal connection with the sacred. His or her voice and presence are valorized by the opportunity to pray and speak in the ceremony. Finally, the contextualization of life experience extends to an embodied sense of participation in a fully animated world. As David Aberle (1982) has observed, this world is defined as a single moral community, with an important place in it for each individual.

Navajo peyotists define this contextualization in terms of the four elements, the five senses, and the network of family relationships. The four elements (earth, air, fire, water) make up the earth and are part of each person as well. In the words of one Navajo, "You are the same as the earth/universe, and it's unnatural to be alienated from it." He elaborated, saying that in addition to feeling the self as part of elemental nature, ritual practice fully engages the five senses: "The smell of cedar smoke goes directly into your brain and makes you feel good, so even someone merely saying they'll burn cedar for you picks you up. Hearing the words, you understand what's being said. Looking into the fire against a general background of darkness in the ceremonial tipi creates concentration, like a trance, focusing your thoughts. Tasting the bitter plant gives alertness/awareness to life—there's no use making it sweet. [You are] feeling the effects through your entire system, feeling the fire and smoke around your body." Finally, the family is mnemonically invoked by each ritual element in the peyote ceremony. The fire is one's grandparents, and each ritual object is kin or a relative. One is reminded that "I am a five-fingered individual, I am Navajo." The element of self-esteem is cultivated not so much in finding one's identity as in recognizing the value of the identity that one already has, and the place within the cosmos implied by that identity.

The therapeutic principle in Native American Church healing is *confessional*. People are in a circle. As one Navajo describes it, "Like Freud, the medicine man asks what went on in your early life, and your relatives are there to fill in the details." The patient prays, confesses or tells of his problems, is moved, and cries. Again, however, the confession in itself does not complete the therapeutic moment. From the healer's standpoint, talking to people and having them understand is the only cure. In addition, not all the therapeutic action takes place within the ceremonial context, as is evident in the following vignettes related by a peyote road man. A depressed widower for whom a prayer meeting was planned said if he had known the road man before, his wife would not have died. The road man's response was "Don't say that, everyone has a time to die, focus on your living children. We plan for the next day or week, but really we can die at any time." Again, a medicine man came to this road man for help to quit drinking. His response was "You never quit drinking, that's the wrong thing to say. Drink as much as you want of pure water when your mouth starts burning for alcohol—or coffee, juice, et cetera. You're dead when you 'quit drinking.'" As he tells it, the man looked down, smiled, and said, "You're right." He came back a year later and thanked the road man, and is reportedly now a good medicine man. He was spoken to, and he understood.

### Christian Healing

Pentecostal and Charismatic healing by and large have been neglected in studies of Navajo culture, but appear to conform to the cultural theme of understanding. They are characterized by a *philosophy of moral identity* specifically in the sense of answering the question "who am I?" Though this is notably less true among Catholic Navajos, part of the answer is typically that one is not a traditionalist or peyotist: Navajo fundamentalist Christianity requires adherents to destroy traditional ritual objects, and to regard peyote as a sinful drug or as Satanic. One Navajo woman, the wife of a Methodist Charismatic minister, explained that they oppose mixing traditional and Christian practices because it "confuses people." She told of attending church conferences in Nebraska and Oregon, following both of which participants staged intertribal drum dances. The Navajo delegation secluded themselves in their rooms and refused to attend any more such conferences. In their view, it was inappropriate and offensive for Indians who regarded themselves as Christians still to engage in such traditional practices.

The therapeutic principle of Christian healing is *conversional*. That is, healing is typically predicated on adopting Christian values and a Christian way of life. This is illustrated in the following account by a Christian woman whose neighbor came to her crying, asking her help. The neighbor was a "peyote eater" and had just returned from a peyote meeting in Utah. She had returned on Sunday and slept well that night, but Monday she felt movements all over her body and a voice threatening to kill her, her husband, and her children. Instead of going back to work she went back to bed, then later got up apparently in a kind of trance or sleepwalking, and broke everything in her house. Her daughter came in asking what was wrong. She put cold water on her face and woke up, coming to the narrator for help. The Christian woman prayed for a long time, read scripture to her, and got her kneeling by the couch praying. Following this episode, the neighbor and her husband both renounced long-time peyote use and became Christians. The narrator continued that after having "testified" or related this event to the congregation, a man who said he had problems with both alcohol and peyote and couldn't get out of either came forward during the "altar call." He said he really understood her words, and they really worked on him. Apparently he became free from the influence of both substances, equally regarded as dangerous and sinful drugs.

## Navajo People, Navajo Healing

In the presentation of Navajo patients' experience with ritual healing, I intend to convey a more experiential sense of the differences and similarities among the three forms of healing introduced in conceptual terms above. To do this, I have chosen to tell the story of three patients, one treated in each of the healing forms.[1] The three patients are roughly the same age, in their mid–sixties to early seventies, and to different degrees all three exhibit a mixture of psychological distress and physical symptoms. Unlike many anthropological accounts of healing, these emphasize the patients' understanding of their problems, their experience of the healing process, and the manner in which they integrate the results into their subsequent lives. Thus, I do not include detailed description of the ceremonial procedures carried out by healers on the patients' behalf. Nevertheless, it is important to note that there is a significant degree of variability in the ceremonial elaborateness of the three healing forms. Traditional Navajo healing, in addition to formal diagnostic or divinatory techniques, includes a substantial number of named chants that may last from abbreviated versions of several hours to those that last nine days and nights. Each chant is composed of a series of songs that often allow variation to suit the needs of a particular patient, as well as a number of important ceremonial objects and manipulations of these objects (such as wedding baskets, *jish* or medicine bundles, prayer sticks, corn pollen, water, sacred stones, buckskins), and carefully constructed sandpaintings with important mythical and cosmological meanings upon which patients are required to sit. Native American Church healing may include some formal diagnosis or divination, but its principal event is the peyote meeting that lasts a single night from dusk till dawn. The meeting is oriented around a central fireplace and altar upon which the road man's chief peyote button rests. It is composed of songs and drumming at which the participants take turns, confessional prayers and exhortations similarly at turns, the periodic passing of the peyote medicine among the participants, a water ceremony led by the road man's wife at dawn, and a final collective meal. Navajo Christian healing may take place privately in a session of intense prayer by the healer, but on many occasions it takes place in large public revivals. In such situations patients line up for prayer and laying on of hands that may last only a minute or two. Diagnosis and divination are inspired but rudimentary in detail if they occur at all, and often the healer knows little or nothing about a patient's problems. Despite these differences in degree of ritual elaboration, patients' experiences with healing in each of the three forms can be equally profound.

Following presentation of the experience of each patient, I will summarize its intelligibility in terms of the philosophy and therapeutic principle I have identified as characteristic of the healing form used. I will then analyze each patient's therapeutic process in terms of the rhetorical model outlined in Chapter One. To reiterate briefly, the four components of the model of therapeutic process include:

1. The *disposition* of supplicants, both in the psychological sense of their prevailing mood or tendency for engagement in ritual performance, and in the physical sense of how they are disposed vis-à-vis the social networks and symbolic resources of the religious community.
2. The *experience of the sacred,* taking into account the religious formulation of the human condition in relation to the divine, the repertoire of ritual elements that constitute legitimate manifestations of divine power, and variations in individual capacities for experience of the sacred that may influence the course of therapeutic process.
3. The *elaboration of alternatives* or negotiation of possibilities that exist within the "assumptive world" of the afflicted. Healing systems may formulate these alternatives in terms of a variety of metaphors, and may use ritual or pragmatic means that encourage either activity or passivity, but the possibilities must be perceived as real and realistic.
4. The *actualization of change,* including what counts as change as well as the degree to which that change is regarded as significant by participants. This may occur in an incremental and open-ended fashion, without definitive outcome.

### *Traditional Navajo Healing*

Jesse is a 69-year-old retired widower who lives in his own home, sparsely decorated with an old Chevrolet calendar and a velour Christ on the wall. His house is in close proximity to those of two of his sisters and their families, on a large piece of traditional family land where he was born and raised. Among the homes and outbuildings is one traditional Navajo house or *hogan,* which is used primarily for ceremonial purposes. The family also has a farm nearby, and his farm work is very important to Jesse. He is a jovial and talkative man, who laughs easily and often, and seems always to have a smile on his face. Though he does not appear particularly reverential or philosophical, he is earnest appearing and focused, an elder who still has vigorous strength. He speaks primarily in Navajo, and our interviews were in the Navajo language assisted by an interpreter.

Six of Jesse's eleven siblings are still alive. Family relations seem very important to him, and he said that "Our ceremonial tradition is to love each other." His childhood was "okay" without too much hardship, though his parents were strict and gave the children chores to do. He indicated obliquely that his father was fairly abusive and authoritarian, though his father ended up being his "drinking buddy" during his late teens and early twenties when they worked on the railroad together. Jesse went to day school briefly, but said that all they did was pick up trash and if they didn't pick up enough, they would be whipped. After leaving school he held many different jobs, including local and migrant farm work, making adobe bricks, railroad work, waterline construction, and a short stint of three to four months mining uranium. He reported having had twelve wives, though most of these relationships were short-term, and he has only four children by the two women to whom he had been officially married. His most recent wife passed away just two months before our interviews. Jesse seems to have cared for her deeply, but said that they did talk of divorce. He said she wanted to leave, partly because she believed there were an abundance of "skinwalkers" (sorcerers) around their home and that witches had planted bundles there as well.

As is typical of traditional Navajos, Jesse prays each morning at dawn with corn pollen, and his beliefs about the etiology of his various medical and social problems always reveal traditional causes. When asked how he prays, he replied, "Going to the hilltop, I usually recover with Navajo prayers. This is the corn pollen, the corn pollen boy, the corn pollen girl; and the earth, the dawn, sun, moon, twilight, darkness, and … whatever." He keeps a small clear plastic bag in his wallet, which contains corn pollen mixed with the gall bladder juice of either mountain lion, bear, or wolf. He said he carried this to prevent himself from being witched, particularly at social dances such as *yeibichei*. He mentioned that his grandparents often told him and his siblings coyote stories and other stories so that they would know good from bad behavior.

Jesse traces the root of his problems to a traditional Lightning Way ceremony in 1957, when he was around 30 years old. The ceremony was being given for a patient who was himself a Lightning Way singer. During a break in the ceremony, the *hogan* was struck by lightning, and a blue flame came down into the hogan through the smoke hole in the roof. Two of the eight people in the hogan were killed, including Jesse's father: "He was just asleep, wrapped in or curled in his blankets. There was another man laying over here … lying sideways, and he was thrown up and was ripped down the middle. My father was laying there, and the lightening went through him completely. There were two." The medicine man was laying down and was not injured. The patient was out of the *hogan* at the time and was uninjured also. Besides the two killed, another man was hit by lightning, but it only struck his leg and he survived. Jesse was thrown unconscious against the wall and went out cold, lying in the doorway. During the time he was out, he saw himself from above the *hogan,* out of his body.

Jesse and several other members of his family awoke with a jagged, lightning-shaped print across their chests. Two of his sisters who had not been attending the ceremony went into the *hogan* to help, thus also becoming exposed to the effects of lightning. Following the incident, and before being fully recovered, Jesse had sex with his wife. She died less than a year later. Jesse subsequently learned of his mistake and realized that he had caused her death. Jesse sees much of what has subsequently occurred in his life as being traceable to this event. Indeed, it was an enduring trauma for the entire family, all the more so because of the spiritual danger and ritual power associated in Navajo religion with lightning and ceremonies associated with lightning (see also Chapters Seven and Eight). Nevertheless he felt that his contact with lightning was a kind of initiation, and that having escaped injury he was "accepted by lightning," thus suggesting a kind of permanent affinity for this powerful force. In fact, the incident may have left him with powerfully divergent messages of his own spiritual strength combined with a sense of isolation or even alienation from other people in his life. Stated in other words, it may be describable as the combination of a post-traumatic stress response with the intensification of feelings associated with the violation of strong social conventions.

Jesse reported having had previous ceremonies done for the problem, but these were minor chants that he thought to have failed in part because "the chanters do not do a good, complete job .... Nowadays it's all done in haste, right after you make the payment." It is important that something as catastrophic as the lightning strike that killed Jesse's father had never been addressed with a major ceremony. Nevertheless, as is often the case, Jesse said that his symptoms didn't start until much later (15 to 20 years) because when he was young, he was "strong in body." In the Navajo view, these delayed effects could have been forestalled if the original chanter had taken steps to correct the situation, including a cooling rite, herbal rinsing rites for the people in the *hogan,* and construction of a separate shelter in which offerings and a blessing could have been made. These steps were never taken.

At the time he finally contracted a chanter to perform a Lightningway ceremony (traditionally to remove effects of exposure to lightning), Jesse reported a number of symptoms, including heartburn, aches in his bones and joints, blurred vision, lower back pain and leg pain, and dizziness. Plans were made also to integrate elements of another ceremony known as Evilway (traditionally to remove effects of exposure to the dead), which Jesse needed following the recent death of his second wife. "The wife that passed away," he told us, "I had to help her through her illness, for about nine years, thereabouts. Because of that, they tell me I'm affected by the Evil Way to a great extent." He said the main symptoms of this were a kind of nervousness, his "mental thinking" gone "astray." At one point he told us that he often has a lot of trouble when he's inside a large crowded place like a department store, where "My heart starts to race and causes me to breathe hard. I feel like I'm stumbling along the aisles, I'm afraid of running into something. You might black out and run head-on into something," he said. In addition, events in his family's life had recently become stressful, as some of his grandchildren had gotten in trouble and several family members had had health difficulties. Our clinical diagnostic interview suggested that he met criteria for dysthymic disorder, though since he appears to have been affected deeply by the wear, tear, worry, and workload of having to take care of his terminally ill wife, it is likely that this can be understood as due to the stress of bereavement. In addition, he likely would have met the psychiatric criteria for post-traumatic stress disorder in the period following the death of his father during the Lightningway ceremony years before. Incidentally, he reportedly had a history of cruelty to animals, alcohol use, and promiscuity, though in the absence of a criminal history this does not add up to an antisocial personality.

Following the ceremony, which took place over the course of a weekend and included two of his sisters as co-patients, Jesse slept the night in the *hogan* where it took place, beginning a period of four days in which a patient is supposed to remain still and reverent. He reported feeling that the ceremony had gone well. When we asked what had been the most important part, he said "the whole thing." He said he wasn't feeling or thinking about much during the ceremony, just focusing on the proceedings. He did mention being nervous throughout, however—"I felt like I was just a bundle of nerves," he said. "Maybe that was the workings of the tobacco and the mixing of the herbs working on me together," but mostly as they worked the tobacco and the herbs made "my whole being felt fine." Jesse also said he had a hard time following and repeating the chanter's prayers, a fundamental aspect of the traditional patient's participation. He indicated that he thought the ceremony would help him, but he also qualified this by saying it was best not to be too confident. He said that having faith in healing was very important, "They say you have to have complete faith, that is the only way you get healed, that's what they say. The Holy People are like that, you have to believe in them for them to help you. The same way with God." His attitude might be called one of faithful but cautious optimism. "You have to take caution here," he said. "You can't shout and say, I'm all well, and it's pretty good. You might be lying. Then you wonder if they were truthful. It's up to the Holy People." Like most traditional people, he felt that a cure does not come immediately, but rather, will come with time: "When you suffer from such an event, and you come down life's road, and you suffer all kinds of things from it. This suffering, in itself, is a teaching tool . . . . And now, finally, I had this Lightening Way ceremony performed and now, it seems like someone has said, 'Wait, Wait.' That's how it seems to me. In a few days, after four days, I'm sure everything will be fine. It is at that point that you become a real believer. Right now you do believe, too, and it has helped you, too." Once that healing occurs, one's belief is strengthened, and out of suffering comes understanding.

Following the Lightningway ceremony, Jesse reported that he was no longer feeling hot all the time and breaking out into a sweat spontaneously. By two weeks after the ceremony he told us that his vision had gotten better, and the difficulties he had been having while walking seemed to have disappeared, though he also mentioned the persistence of pain ("it feels like it has pins and needles") in his feet, especially in his heels. He said that working around his farm, which had been very difficult for him before the ceremony, had gotten easier for him. "It seems to be going in the healing direction for me, yes," he told us. At a follow-up interview three months later, however, things were much different, and he had had more difficulties. He indicated that his recent ceremony was somewhat invalidated by two occurrences. First, he had seen and killed a snake within a few days of the ceremony, significant both in ritual terms and with respect to his apparent ambivalence regarding such violations of convention. Second, he had a personal tragedy when a nephew who had passed out drunk on train tracks was run over by a train. This death figured importantly in Jesse's healing process because the chanter had warned him and his family against having anything to do with the dead. Unfortunately, after the accident the nephew's body was returned and his sisters hugged it in grief. Jesse believed that this in itself undid the effects of the ceremony, not just for his sisters, but for him as well. He specifically mentioned that his own pain had returned after this incident.

As a result of this incident the Lightningway chanter told Jesse that he should get rediagnosed and have another ceremony. Jesse said he was planning on having this chanter complete another full ceremony for him, "When I am financially able to." In addition, he said that he still needed a full Evilway ceremony following the death of his second wife:

> My late wife use to live here and be present here. Her presence is all over the place, her hand body oils are still on everything. She still has some of her personal possessions around in the house, there is one over there. Maybe that is what is affecting the house. Because of that, I won't move back in until the *hogan* has been properly cleaned. Then I will live in here again. Now, I'm being told to get well first, and sleep over at the other house for the time being. That is what my older sister tells me to do. She will clean this *hogan* out first, before I will move back in here. So that is a problem area, it causes me to have that problem, it seems that way. The Evil Way atmosphere is clouding up the air in here, it feels that way to me.

In conjunction with this concern, Jesse was worried that some of his wife's health difficulties may have been related to or caused by his own past disrespect for the values, attitudes, and rules of the Holy People. Finally, he was concerned that he had seen and killed a snake immediately following his recent ceremony. He thought the appearance of the snake meant that someone was trying to witch him or as he put it, that someone was "still" trying to witch him, intimating as well that witchcraft likely played a role in the death of his latest wife.

Jesse's experience shows quite clearly the sense in which Traditional healing incorporates a philosophy of obstacles in the sense outlined above. His narrative can be understood as the attempt to overcome a string of obstacles created by his own actions, attitudes, and ambivalence, the actions of others, the perceived inadequacy of healing ceremonies, or the force of events. The didactic principle is also evident in his sense of constant searching for the knowledge of what his problem really is and its ceremonial solution, as well as in his comment that suffering itself is a teaching tool.

Let us formulate the therapeutic process experienced by Jesse with respect to the rhetorical model outlined above. Jesse had a positive disposition towards healing, but his faith and optimism were temporized by his idea that too strong an expectation to be healed could be a set-up for failure and might incur disfavor with the Holy People (for example, the traditional Navajo deities). In addition, he expressed a theme of having many ceremonies but not fully being integrated by them, as well as a theme of ambivalence toward ritual conventions the violation of which could undermine ceremonial efficacy. In a number of instances in which ceremonies had not cured or even helped him, he was critical of the healers—either he regarded the traditional diagnosticians as frauds or felt that the chanters omitted certain key features of the rite, and their omission caused the cure to fail. The failure of these ceremonies quite likely caused him to have some caution about the imminence of a cure. Like many traditional people, Jesse felt that a cure could be a long time in coming, implicitly suggesting that the therapeutic process is a life-long process. From his comments, it does not appear that Jesse expects his ceremonies to end the troubles that stem from an event that happened over 40 years ago. The effects of the lightning strike implicated his entire family and will continue to affect them throughout their lives. Realistically, he is hoping for relief from of his symptoms. He feels that even if his condition improves in the short run, he will need to have additional ceremonies. Another salient fact with regards to Jesse's disposition toward healing is an apparent willingness to disregard taboos even as he maintains a belief in them. This may reflect an attitude that when one does something wrong, punishment may or may not result, an attitude that often leads people not to have ceremonies until something goes wrong. Finally, Jesse's disposition included the possibility of a partial cure. He had a plethora of symptoms, some of which he reported improved and others which remained the same. Nevertheless, by the time of our follow-up interview, other events had occurred that he saw as undermining his ceremony, events that did not concern him directly, but his sisters who had been co-patients in the ceremony. In his opinion, their contact with the dead was enough to undo his ceremony.

Our interviews suggest that Jesse has a vivid sense of the sacred and a life suffused with the sacred dimensions of traditional Navajo religion—the repercussions of broken taboos, ceremonials, prayer, witchcraft, and skin-walkers. His experience of the sacred in the ceremony itself also seems to have been strong. He said on several occasions that the ritual use of herbs and tobacco altered his mind and body in a positive way. His expressed fear that his inability to repeat the prayer properly might cause problems for him also suggests his belief that the Holy People are tangible, and because this ceremony was properly performed with all of the necessary elements, were present and would hear his prayers. A final episode that suggests a high degree of sacred experience in Jesse's case is his contact with a snake in the aftermath of the ceremony. He agreed that this was probably a bad sign, and cognizant of his own past with both lightning and snakes, must have seen it as meaningful. Nevertheless, it was not the snake but his sisters' contact with the dead to which he attributed the undoing of his ceremony. Moreover, he did not associate the snake with his own behavior, but read it as a witchcraft attempt by others.

With respect to the elaborations of alternatives, Jesse sees not a series of behavioral or emotional possibilities before him, but a variety of ceremonial possibilities, as well as a variety of possible outcomes ranging from success, to partial success, to failure. In addition to the Lightningway ceremony, he felt he still needed an Evilway because of the death of his wife and an Enemyway to remove the contamination incurred when he removed the remains of a dead white woman from a car he cleaned while working at an auto dealership. He seemed quite sure that he had needed all of the components that had been included in the Lightningway, such as the offering and the rinsing in addition to the sandpainting, since a diagnostician had told him that the reason past ceremonies failed was that some of these had been omitted. Jesse's story, and the particular sense it gives to the notion of an elaboration of alternatives, highlights a powerful theme in traditional Navajo therapeutic process, namely contamination via contact—with lightning, with a variety of animals, with the dead, with his dying second wife and sexually with his first wife, with the dead enemy, and with a witch or skinwalker.

Actualization of change is both tentative and incremental in Jesse's case. Despite his feeling that the ceremony went very well, he was initially reluctant to regard the healing as a success because such optimism could cause problems. Nevertheless, his interviews are punctuated with positive statements about its effects. These range from alleviation of pain and improvement of physical symptoms to an improvement in his outlook and the ability to get up earlier and get his farmwork done. Again, there are other symptoms, specifically some pain in his legs and feet, which he did not feel had disappeared and had improved marginally at best. Three months later things had changed somewhat because of the violent death of a nephew and his co-patients' subsequent contact with the corpse. It is difficult to regard this as an unrelated event, however, since it is of a kind that has direct bearing on therapeutic process insofar as that process merges with a person's overall life trajectory.

### Native American Church Healing

Marvin is a 64-year-old man who lives in a remote mountain area on the Navajo reservation. He and Nellie, his only wife of 40 years, along with two of his grown sons (three other grown children live elsewhere) live in a modest, old, two-room log house with Christian mementos on their living room wall and two easy chairs in which he and his wife appear to spend many comfortable hours. They seem to lead a fairly traditional Navajo lifestyle. Marvin is a neatly groomed, very alert, and vigorous appearing elder who, despite his absence of formal education, formed very coherent and insightful descriptions of his life circumstances (speaking in the Navajo language) and appeared to review in significant detail his medical history and his understanding of the importance of ceremonies in his healing. He is retired for health reasons from sawmill and railroad work after 16 years of employment and now ranches part-time. Although he began participating in the Native American Church in 1945, Marvin never became a road man (leader of peyote meetings), which is unusual for someone of his generation with such long participation. He attributes this to a bad peyote experience early in his participation when he decided to try to run a meeting while still too inexperienced to do so:

> I have done that before. I was single then, but I was not really into it. It was like I was not ready to take it then. But I prayed. I did it one time. Once when my uncle was not home I used his equipment and did the *nahagha* [peyote prayer meeting] alone. After midnight everyone became very affected by the medicine. I tried very hard to bring them out of it, but I was not successful. I prayed and prayed. The drummer fell over and I threw him in the truck and took off. It really scared me. We left them like that. I was scared. A man told me this, he said I am not mature enough, I don't have a steady home life, and I don't even have a wife, that's why this happened. One woman at home is what I didn't have. I guess it was true. I decided not to take it on. I know how to pray and do all of it but I left it alone.

The peyote medicine frightened him so that he subsequently could never "give himself completely to the medicine." This might also explain why he has continued to suffer from problems with drinking, in that road men say such surrender is absolutely necessary in order to get better, just as Christian ministers say that one must give one's life completely to Jesus. Today Marvin only takes peyote within the context of a prayer meeting under the guidance of an experienced road man.

At the time of his participation in our research project, Marvin was engaged in a process of healing on two levels. On an explicit, and perhaps more superficial level, he was having a ceremony in order to protect him during and ensure the success of upcoming surgery to remove a growth on his throat. On a deeper and more enduring level, he continued to seek healing (even if indirectly) for his chronic alcohol abuse problems; our clinical interviews suggested a long history of alcohol dependence and the possibility of major depressive illness. For Marvin, the two were experientially connected in that the crisis constituted by his physical illness provided an occasion for him to truly reflect on his life and his problems with alcohol. His treatment was conducted by a distinguished road man who is a close and long-time friend. Through many years of his drinking problems, the road man and his family have stood by Marvin and encouraged him to take up the NAC more seriously, regarding him at the same time as "a pioneer of NAC" and as "an alcoholic." In the present instance, prior to Marvin's admission to the hospital for surgery, the road man had conducted a "stargazing" diagnostic session and a traditional Shield/Protection Prayer as a "blessing that all goes well." Marvin's overall feeling was that NAC, traditional Navajo, and biomedical healing all worked together to help him.

Marvin acknowledged several factors he thought contributed to his physical illness, including witchcraft, heredity, pollution from the railroad work, stress and worries, diet, exposure to lightning and other traditional Navajo illnesses, and alcohol. The road man introduced another important diagnostic interpretation, namely that Marvin "has four personalities as a result of blood transfusions that he received after the accident where he was a pedestrian struck by a vehicle and seriously hurt many years ago ... there is the spirit of an Anglo man, a Hispanic woman, one other unknown person, and Marvin's own self all battling for control of him, and he requires long–term treatment." Marvin himself at first tended to hedge about discussing the accident and his drinking, instead preferring to relate the experience of his surgery and the possible causes of his tumor, but after awhile he began to discuss what appeared to be his main problem in life, drinking: "Well, it has been many years that this thing bothered me and with it came my attitude toward everything around me. I was drinking too. So the doctor said, no more drinking for you, if you want to continue to get better. So for me this means in order to live on this earth drinking has to be avoided. Never drink again. So I want to follow their instructions and because I want to live on this earth. About drinking, it is a terrible thing. Many a relative begged me to stop. They tried and I couldn't. Not until this happened to me. Look what it has done to me. That's what I learned."

In accord with traditional Navajo thinking on illness causation, the road man felt that Marvin's current health crisis was a cumulative result of past events in his life, such that now he was "worn out like an old blanket." In addition to the multiple spirits that he had previously diagnosed within Marvin, the road man had a vision of two coyotes that indicated to him the health problems were additionally caused by Marvin chasing these animals in his youth. He interpreted their appearance as omens of imminent death, and prayed accordingly. During the ceremony the road man also had a vision of the stars and heavens "breaking up" that seemed to indicate to him that the ceremony and prayers were working.

Marvin actually had two peyote ceremonies or *nahagha*. The first was prior to his operation, "for the purpose of going through this ordeal." Following the successful operation a second meeting was held "for appreciation" and thanks. Marvin reports that in preparation for this latter meeting "I took a lot of medicine. I started taking it during the day while I was herding sheep. If you counted them one by one there was a lot." In the heightened state of awareness induced by the peyote, Marvin experienced two profoundly moving moments. One was hearing the words spoken by the road man as the healer prepared a mixture of tobacco for ceremonial smoking, encouraging him to become yet more involved in the religion and become a road man himself:

When he was doing the tobacco. He spoke. He said you have accomplished what you wanted to do and it is here all done well. You have come to a point now where what you wished and hoped for can come to pass. Now you can do the things that I can do and am doing here for you he said, remember? In the future you can sit where I am sitting and where I stand and can do what I am doing. So it is up to me, but I still feel that it is not up to me, not yet. Way back in the beginning these two came over and prayed for me and set the course to follow in order to get better. One of those we took care of in this *nahagha*. It is up to God and it is his will that they came to me. He says I should take on this *nahagha* and do it for others. That is what he said then. In the future it is okay with me. It can come to pass.

The second profound moment for Marvin in the ceremony came when his son, who was also playing the important ceremonial role of "fire man," assisting the road man by tending the fire and the coals before the altar, took his turn to pray:

My child who took care of the fire spoke to God but he is speaking to me too. So my child is speaking of my past drinking and behavior. He asked God to provide him with a father whom he would be proud of and the father that he always wish to have. I understand what he means. I love him very much and I heard his words and know what he is asking for from me. I am very proud of him. I never ask him something that he didn't do for me. I value his words. He also has never ask me anything that I didn't do for him. He has come back to help me out that's it. All those who spoke they know and me have all come back to help me. He in particular does not like it when I drink. Whenever we had a *nahagha* he will speak up to this issue of drinking. Others have done that too. But now It is okay with me. I accept what they say, I value their concern. It does not make me upset. I know they are concern that if this what I had [the tumor] gets worse in the future it will be more difficult to heal. That's what they mean. That's why they make their thoughts known like that to me. That's it my grandchild.

Following the ceremony Marvin felt quite good and strong, able to work at such tasks as cleaning out the sheep corral. He reported being "happy for the *nahagha* ... I spend the whole day here thinking about it. I am sure the medicine is still in my body and doing it's work."

The philosophy of self-esteem intrinsic to Native American Church healing is evident in the overall therapeutic move to transform Marvin from a weak, self-doubting drinker into a confident practicing road man with the inner strength to help others. Although the narrative does not emphasize the confessional therapeutic principle in the sense of a cathartic moment for Marvin, it is present in at least two other senses. First is the importance to Marvin of his son's confession of his feelings about his father; second is the increasingly explicit confession or acknowledgment of the harm done by his drinking and the need to overcome it.

Let us summarize the four elements of therapeutic process for Marvin. He is a long time member of the NAC and hence quite familiar with the repertoire of ritual healing brought to bear in NAC ceremonies. He also is part of a close NAC network of dedicated members, and although he was ambivalent about adopting the role of road man himself, he appeared to have no reservations about involvement as participant or patient. As he said, invoking the most telling Navajo symbols of emotional security and intimacy, "I call the medicine my mother. It is like having been away for so long and then you come home and your mother is here with you." In spite of this overall positive disposition toward healing, he has still struggled with alcohol abuse for forty years. At the time he participated in our project, two levels of healing were being pursued simultaneously—the explicit problem of a life threatening growth on his throat and the implicit healing to resolve, or at least come to terms with, his alcohol abuse. The fear brought by his physical illness and the idea that his life could end because of the way he lived it were important in his disposition to be healed. The physical illness precipitated a crisis that served as an opportunity for Marvin to truly reflect on his life and his problems concerning alcohol. It is also of value to note that Marvin's immediate disposition in the moment of the ceremony was most likely enhanced by having consumed a large quantity of peyote before and during the ritual, while his long-term disposition was evidently enhanced as a result, as evidenced by his comment that "I don't think I will ever doubt again the power and the presence of this medicine and what it can do for me. I won't be changing my mind again."

Two moments immersed in an experience of the sacred for Marvin during his ceremony were the words of the road man encouraging him to dedicate himself more deeply to the medicine, and those of his son, who prayed that God give him a father that doesn't drink, one he can be proud of. Extemporaneous prayers in NAC ceremonies provide an indirect yet powerful vehicle for participants and family members to address one another while, under the influence of peyote in a sacred setting, they are more sensitive and receptive. It is worth reflecting on this form of sacred experience with respect to the sense in which peyote is characterized as having "hallucinogenic" properties. Although it was the case that the road man had sacred visionary experiences relevant to Marvin's diagnosis and the efficacy of the ceremony, Marvin's own experience of the sacred had to do with an intensified or amplified sense of the meaning of his life. The principal point is that, although peyote-induced visionary and other sensory phenomena occur and have their place, the ritual context defined by a focus on healing tends to give the experience of the sacred a form of immediacy and profundity, and a content of interpersonal and emotional meaning.

In the ceremony the road man elaborates a particularly salient alternative for Marvin, that of adopting the role of road man himself. Its salience lies in Marvin's status as a traditional elder and NAC pioneer who has been unable to successfully take up either role because of his problems and stigma with alcohol abuse. At first glance this appears to be an instance of an initiation pattern observed in many societies, in which the process of being healed requires the patient to become a healer. However, here it is quite specific to the patient's unique life circumstance as a member of a generational cohort in which healthy maturity often presumes adoption of the road man role. (This elaboration of an alternative life path is supported by elaboration of alternative explanations for his problem with drink.) The road man's diagnosis of "four personalities bothering him" partially relieves him from responsibility for his condition and places it outside on spiritual forces. The supplementary diagnosis of "bothering coyotes in the past" is both something for which he bears responsibility and an indication of a certain urgency in his situation. Both are elements drawn from the past that are less stigmatizing than being an "alcoholic" in the present, however. The healer then makes it clear to the patient that everything is different now from this point on and they have the opportunity to begin anew, emphasizing that it is the responsibility of the patient to take the initiative and do so. Finally, he had discussed with the road man another ceremonial alternative, the traditional Navajo Enemy Way, which still remained a factor because of his exposure to the blood of strangers during his long ago transfusion.

With respect to actualization of change, it appears that Marvin's interim goals for the operation to remove the growth from his throat was a success and that he is healing well from it. As to his more essential goal of healing from alcohol abuse, in our follow up interviews one year later it appeared that he had successfully remained sober throughout this period following his ceremony. His own words reveal the personal meaning of this achievement:

Well, I said that before in your presence. Before I became ill, I lived a carefree kind of life. I was careless with my life and probably did things I should be ashamed of. I cheated on life right along until this illness happened to me. I found out that my belief in God was shallow not deep. At my age and realizing this is enlightening. I discovered this. My words were shallow and meaningless before. I said I would do this and that and sometimes bragged about things. Well, I realized what that is now. It is like cheating on the truth. I want to straighten this out, I said that in your presence. I used to take the medicine and soon I would be drinking wine and whiskey again. I would pray and then forget what I prayed about. I was on the edge of the flow of life around me. It was like that before I got sick. Now I regret and wish I made a better thing out of my life. I should have done better with my life. Now that I had the *nahagha* done for me I want to be like the road man. Sure of my role right here and sure of my future. That is what I learned, my grandchild. And then all the things that were said by people around me at the *nahagha* was good. I heard what people had to say and I like what they said. I am very happy that people came to pray and offer their support. That is good.

This may be a significant actualization of change for him, though it cannot be said for certain without more concise information about his drinking patterns and the duration of previous periods of sobriety. Indeed, ministers and roadmen who have successfully stopped drinking appear to discuss it much more openly than Marvin was willing to do. In terms of his case defined within the therapeutic logic of the NAC, the ultimate actualization of change for Marvin would be if he finally "takes up the fireplace," to which his road man friend has been offering initiation for 40 years, and becomes a road man.

### Navajo Christian Healing

Eileen is a 70-year-old Catholic woman who has four living children (three girls and one boy) ranging in age from 30 to 51. At least two other children (as well as her first husband) died during the 1950s. Eileen was raised in a traditional *hogan* and was educated up to the eighth grade. She has fond memories of her childhood and says that she is the product of loving parents and a good upbringing. She was the middle child in her family. She remembers walking six miles to school every day, but warmly recalls that her father would take her by horseback when the weather was bad.

Eileen is a housewife who looks far younger than her age. She had bright eyes, a good sense of humor, and was articulate in reflecting on events of her life. She demonstrated an optimism and a sense of clarity about her life challenges and how she would face them. Eileen and her husband seem to have an exceptionally close relationship, truly enjoying each other's company despite occasional disagreements. They attend church every Sunday together and sometimes on Wednesdays. They also go to prayer meetings and 12-step meetings together. Eileen states that both she and her husband are particularly affectionate people and comments that he is stubborn and does not like to go to the doctor. She notes that he is a recovering alcoholic and that she sometimes goes with him to AA meetings. Eileen says that the things she wants most in life are peace and joy. Additionally, Eileen says she doesn't want her own children to go through the same hardships she experienced, such as raising children alone. She gains considerable joy from sewing and cooking. Eileen says her worst quality is putting herself down.

In her late twenties, Eileen was diagnosed with tuberculosis, contracted while working as a health aide on the reservation. On the day she was to be discharged from a 10-month stay in a sanitarium she was waiting for her husband and children to pick her up when she got word her husband and her two-year-old daughter had been killed in a car accident. She noted that she was given a couple of tranquilizer injections by the doctor before she was told of this news. After the funeral, she had nowhere to go with her two surviving young children. She recalls having felt troubled with her relationship with her husband prior to his death because of his lack of visits and because of her worries about his drinking problems. She notes too that this was only a year after the government had given Indians the right to vote and the right to buy alcohol. She reflects that "I didn't realize that God was still there for me. It took me a long time to believe, you blame yourself. There were times when I didn't want to talk about it but it's a part of healing. It's still not all over, it's like peeling the layers" (as she was saying this she was making motions as if she were peeling an onion). She denies having experienced disabling post-traumatic stress or mood/grief symptoms, on the grounds that she could not be because taking care of her children prevented her from being "dragged down" by the burden of her loss. Nevertheless, what she described as being numb and low for many years thereafter was doubtless her response to those stressors and traumas (residual PTSD symptoms). Soon after this time she met her present husband, already a Christian, and she explains that he really helped her to pull herself up. As Eileen puts it: "I think that sometimes, you know, he was there and my spiritual person was there all the time with me."

As a Catholic, Eileen was first exposed to Christian healing by participating in prayer meetings of the Charismatic Renewal movement (see Chapters One through Four):

> [When] I first went [to charismatic prayer meetings] we studied scriptures and the spiritual director would say to read specific scriptures so that the Holy Spirit would come, this was 20 years ago, I used to be so cold, I had no energy. There was something wrong and I couldn't put my finger on it. Men and women were split up at the meeting and the woman asked me about my life and she prayed with me and I felt warm and the warmth hasn't left me since. All I could say is my husband and daughter had died in a car accident and I was in a TB sanitarium and I was just coming out and feeling totally lost. I had no place to go. I took the kids and we scrounged around. I told this to this woman [at the prayer meeting, perhaps 15–20 years after the death of her husband] and the love just came back into me. I started reviving again. That's how I know there's a spirit stronger than me. I believe. The love and the fire and the light came back into me.

Eileen compares her first experience being prayed over as a Christian to plunging yourself into a pool of water and not knowing how to swim. It was sink or swim and, from Eileen's recounting, her faith enabled her to swim rather than sink: "You don't see anything, you don't hear anything, you just pray .... My whole body was always cold. I was always shivering cold, and they wouldn't get warm. And some other people prayed over me. And that's when my body was really rejuvenated. And from there to this day I am not cold [laughs] .... So that's when I started believing."

Eileen's faith has remained strong ever since, but it is also evident that this experience is overlaid on a preexisting spirituality:

> ... prayer had never been hard for me though. My father was very religious in a traditional way. He and my mother had a prayerful life and passed it on to me. Yesterday thoughts came back to me reminding me of when my son was a baby and he was sick and I had TB and was in the Sanitarium, the doctor thought my son wouldn't live but my

father stayed all night with my son, praying and singing and my father knew that he would be alright. And I heard him crying in the morning.... I believe that when I pray something will happen. I picked up the spirit of prayer from my father. You have to pray in the morning and the evening. You have to give blessing every morning and evening.

When asked how a priest might explain this, she stated:

Priests talk about Jesus as a healer. He doesn't encourage you to go to prayer meetings. Some people have a harder time understanding but I've learned a lot. Not only a priest can lay hands on sick people, I can do that too. I can encourage people. Some people you turn them off by talking about God, so I have learned to say "a greater power than ourselves," someone to pray to for strength. I myself used to get upset and didn't want to hear it when people would tell me that. I used to also think that this was a white people's group, a prayer group for white people's god and that the Indians would pray to their Indian God [she laughed as she said this].

At the end of the interview Eileen commented: "Since we've met, my husband and I, we've learned how to pray together. We pray over each other when we need help. We do that a lot. We do our own devotion every day. We'll pray with anyone. We'll pray with our grandchildren, our great-grandchildren and we pray when they're asleep." When the interviewer commented that that was covering four generations, she said, "It's very good to be in this situation with the spiritual life. In the Navajo way I'm not initiated so I can't pray in some ways and I have to pay, but with Catholics, we can go into any area and pray and it's free."

Despite her strong Catholic faith, Eileen is also tied to Navajo culture in many ways. For example, although she goes to church to renew spiritual strength she prays in the Navajo language. Eileen states that she has considered having a traditional ceremony for her failing eyesight. However, she decided against such an action because she does not have much have faith in today's younger lot of medicine men who "tend to drink." She remembers the old days when her grandfather, himself a medicine man, knew which ritual specialists to trust. Eileen also mentions that she would have used herbs as a form of treatment if she could have remembered the appropriate remedy used by her mother.

Indeed, Eileen's narrative provides some particularly interesting, and not atypical, attitudes of comparison between traditional healing and Christian healing. She explains, for example, that throughout her life she had always believed that Christianity was only for non-Navajos who have a particular conceptualization of "God." Her opinion was changed, however, through a discussion with a priest who "set her straight." Additionally, Eileen says that her identity as a Christian was further developed when she began attending a 12-step program where discussions centered around "a higher power" rather than "God" or "Jesus Christ." She explains: "That really made it easy for me. Believing in a higher power that was just like in traditional ways, you know." Thus Eileen overcame her initial skepticism of Christianity by grounding it within the context of Navajo religion.

Eileen also draws some important distinctions between the two traditions. Foremost among these distinctions is Eileen's perception that Christian healing involves a much more active role on the part of the patient than does traditional healing. She explains: "I used to think that only medicine men can help you and heal you. He's the only one. But there's a difference, I found out that he does all the praying. I don't have to do anything. That's the way. That's the traditional way. But this Christian way, I too myself can pray, I can talk to God myself and tell him my needs each day and begin to experience little healings or big healings in my life." These remarks are probably best interpreted as expressing a feeling that Christian healing affords her a more personal relationship with God. It is worth noting that proponents of traditional Navajo religion might reverse the characterization of their religion as requiring a less active role than Christianity, on the grounds that a traditional patient must actively repeat the prayers of the medicine man, while Christian patients passively receive divine power as mediated through the healer.

Similarly, Eileen also criticizes traditional healing on financial grounds, remarking that one does not have to pay for Christian healing. This seems to serve as an indicator for Eileen that Christian is more pure than traditional healing because it is all about love and sharing rather than money. In addition, the absence of a financial burden to the patient makes it more accessible than traditional healing. With Christian healing Eileen does not need to find someone outside her immediate sphere to pray over her. She can simply pray for herself or have her husband pray for her. She states, "Whenever I have a problem with my whole body or aching somewhere, I turn to my husband to ask him, 'this is what's wrong with me, can we pray?'" It seems particularly important to Eileen that she can turn to her husband for healing. The two are very comfortable and close with one another and the act of praying seems to simultaneously demonstrate and cultivate their sense of togetherness.

In sum, when asked if she ever had any unusual religious experiences, she said "Yes, knowing in yourself that you have the ability to understand both spiritual lives, both the Catholic and the Traditional, I am grateful that I have that. It is special." Thus her strong sense of Christian faith is characterized by a distinctive Navajo touch.

At the time of her participation in our project, Eileen's primary problem was deteriorating eyesight due to cataracts, in addition to which she reported problems with pain in her knee. She has received both biomedical treatment and healing prayer from her husband. The couple recently moved from another locality on the reservation, and live in a warm, welcoming, and well-organized home. Eileen seems to have a fairly broad network of friends and family, but reports feeling lonely since their move, feeling and being treated by others as an outsider. The only psychiatric diagnosis for which Eileen met criteria was social phobia. Interestingly, she did not feel impaired by these symptoms until she was about 50 years of age. "I have a question for you," she said, "If you've grown up as a girl and your peers make fun of you and your parents, it makes you withdrawn. Well, that's what happened. My older brothers and sisters weren't like me. I was the one who was always with Dad. He was almost blind and my mother had a hunchback. Kids are so cruel." Eileen's social phobia symptoms disappeared in the years following the death of her husband and child because she was too absorbed in just trying to survive. At 50, she became more involved in church activities (which is how she, 15 years previously, met her current husband). She also seems to be more comfortable with public speaking and has no trouble talking in a group if it's related to prayer.

Eileen describes her symptoms as follows: "[My eyes were] watery and always tired, even though I had glasses, sometimes they'd be itchy. But I really felt it had to do with, I needed [to get] another pair and to change the lens. I thought that was what was wrong and I went and they said 'No, we'll wait to get your lenses changed.' So we waited till [they discovered] way in the back of my eyes little blood vessels that were damaged and they were getting, getting deteriorated or something ... so I couldn't get my new glasses. And my driving was kind of getting bad [laughs]." Eileen says her eyes hurt when she is tired. She experiences soreness and aching on one side toward the corner of her eye. She cannot see very well, and her eyes hurt her when she is tired. Eileen expresses concern that if she loses her sight she will become dependent upon other people, especially her husband. She is clearly afraid of growing blind and also appears to fear going to the doctor. Although Eileen's eye problems do not seem to seriously restrict her life, they are an important source of concern.

Eileen mostly attributes her failing eyesight to aging. She states: "You start going down instead of getting better and you wonder why, you know, all of a sudden you realize that you can't do the things that you used to do. And then you wonder, you get kind of fearful, because you know, why can't I go on being strong and healthy, why all this problem? Why I can't hear too good, you know, this stuff." In addition to aging, Eileen also discusses several other factors that may have contributed to her condition. She notes, for example, that her father has been blind for a long time and that there may be a genetic component to her problems. She also says that she was involved in a car accident about seven years ago and hit her head on the windshield causing injury to her neck and shoulder. Eileen also briefly discusses stress as a potential factor in her condition and explains that she is a borderline diabetic, which could also contribute to her eye problems.

When asked if some personal loss or tragedy could have been the cause of her eye problems, Eileen responded with an equivocal reference to traditional Navajo ideas about the effect of exposure to the dead in contemporary-style funerals: "If I look at it in Traditional ways it might have. They tell you when you're that person, in the coffin, you're not supposed to stand there and [let] a tear drop in there, you know. But I'm always aware of that so I always tried

to keep away, but sometimes an emotion, at that moment you don't know—
but I really can't say that's what it is or anything." She made another reference to tears and grief when discussing the relation between her condition
and the difficult times she has faced throughout her life, particularly in the
1950s when she was hospitalized with TB and lost her husband and children.
She suggested that "...just being frustrated and just crying so much, griev-
ing so much, that might have damaged my eyes, I don't know."

Though she had initially intended only to have her eyeglass prescription
updated, on diagnosis of deteriorating vision she had been referred to an
off-reservation hospital where she had laser surgery on both eyes. Her treatment with Christian prayer, first by a well-known Navajo healer and then
by her husband, were intended to complement the biomedical treatment.
Indeed, before every doctor's appointment her husband would pray for her
not to feel any pain or experience any problems, and to help the doctor be
successful so that her eyes would heal. Eileen reports that she was particularly afraid of going blind and that the prayers helped build her confidence
by easing her fears. It appears that Eileen often turns to prayer in times of
need, particularly since she is sensitive to many medications and can't even
take aspirin for pain. As a result, she notes that "Most of the time when I'm
really having pains I just go to prayers. I just go to prayers."

The philosophy of moral identity outlined above as characteristic of
Christian healing is evident in several ways for Eileen. In contrast to
Traditional ways, she sees Christian healing as requiring more active participation and hence commitment by the patient and as less motivated by
financial considerations. I would suggest that there is also a moral sensibility at play in her highly personalized account of how healing prayer has
consolidated the marital bond with her husband. The conversional principle is not so pronounced as among many Pentecostal Navajos, particularly
given the Catholic propensity to synthesize Traditional and Christian elements, but it is worth recalling that Eileen and her husband met at a church
function and are both intensely involved in their parish upkeep.

Turning now to our understanding of the elements of therapeutic
process, Eileen's disposition is solidly rooted in Christian healing, built on
a preexisting groundedness in Traditional spirituality. She also had faith in
her husband's giftedness at healing prayer. Although Eileen's social network
does not seem particularly well developed, especially since her recent move,
she appears to have all the support she needs in her husband and the local
church community. Eileen's experience of the sacred can best be described
as a sense of peace. She reports that during the prayer she was really con-
centrating on how she was being helped. The primary feelings reported by
Eileen, as with many Pentecostal and Charismatic Christians both Navajo
and non-Navajo, were peace and love:

> I felt peace. Very peaceful 'cause I think when you worry about your-
> self.... I worry about myself, how I was going to ... how I was going
> to take care of myself and how I was going to do what I'm supposed

to, you know. And taking care of the house or how I'm going to get around. And it just worried me, but after he prayed over me I just felt confidence that I'm going to be all right. I don't need to worry about it and that peacefulness of Christ just came and is still with me. I think each time whether my husband or somebody else prays with me, I think you experience more of God's love coming into you and a peace that comes. Because on our own we think we know how to love but we don't. That's what I found out too. We think we have peace but there's always that turmoil in our minds but this is different. It's different. Your expectations on other people are not the same. You think of a person in a different way.

In this context, Eileen's experience of the sacred seems to revolve around the feeling that God had entered into her, providing her with a sense of peace or calm, and helped to restore her confidence that everything would be alright with respect to her eyes, specifically, as well as her life, more generally. In another respect, Eileen's experience of the sacred might also be conceptualized as extending into a broader process of spiritual growth as well as a self-perceived sacred relationship with her husband:

We learned that in our spiritual growth we can do that [pray] for each other. And we help each other that way. I don't have to go out and look for somebody to pray over me. And I had that confidence in him that he could do that for me ... now, when I see people that do need prayers and they don't [know it], there's no connection with their spiritual growth. They haven't experienced a spirituality part of their life. So it's kind of difficult for them. I know some other churches, they right away say we're going to do this for you, and this is how it's going to be. They just say well, you know, God would heal you. You know. But for me it was like I had to learn. And have their confidence. And have it deeply, praying for myself all the time.

The elaboration of alternatives for Eileen seems to be contingent on the abovementioned sense of confidence that enabled her to face her fear of going blind and becoming dependent upon others. In addition, her healing experience also offered a means for her to express and consolidate her bond with her husband, who served as her primary healer. Eileen elaborates in the following way: "I was just grateful, very grateful that he can help me that way. Praying with me. It's a beautiful thing that we can do that for each other, which most married couples do not have. And ... I've never saw my grandfather have anything done for my grandma." Finally, the primary actualization of change for Eileen seems to be precisely that she was less afraid and more at peace and confident before her doctor's appointments, specifically, and in her life, more generally. She had faith in the prayer, in her husband, and in God. In addition, during the 12-month follow-up she reported that her eyes had not gotten any worse.

Therapeutic Process and Life Trajectory: Understanding

Healing as Experience

The analyses of Charismatic healing in Chapters One through Four (see also Csordas 1994a) specified the incremental efficacy within therapeutic process in terms of discrete self-processes of language, emotion, imagination, and memory and showed how these self-processes are grounded in bodily experience. Our subsequent work with Navajo patients addresses the aim of contributing to a cross-cultural, comparative theory of therapeutic process by again placing the model in dialogue with empirical data. To be precise, we have set up a dialectic in which the model serves as a framework for understanding distinctly Navajo cultural elements of therapeutic process, while its application in a cultural setting distinct from that in which it was originally elaborated serves to further refine the model itself. So far in this chapter I have emphasized the second half of this dialectic, using the rhetorical model as a tool to interpret narrative accounts of individual cases. In this final section I summarize the ways in which these cases suggest refinements of the model itself.

The formulation of *disposition* based on the Charismatic study emphasized the patient's knowledge of a healing form, favorable inclination to participate, and social access to healing. This formulation emphasized a patient's orientation toward the healing system, but what is presupposed without elaboration is the presence of suffering, or at least of discomfort. Our Navajo work has brought home the necessity of including suffering as one of the determining factors of disposition, such that an account of disposition necessarily includes an account of suffering (see Kleinman, Das, and Lock 1997). Specifically, this includes analysis of the kind and degree of suffering that motivates a particular person to engage in a particular form of healing, the sense of urgency a patient and the patient's family bring to healing, and the nature of a person's personal and social capacity or incapacity to become engaged in the healing process. The dimension of this component that is of interest in the Navajo setting is what we have begun to call the "psychology of postponement." Even among patients well disposed in our original sense of orientation toward the healing system, prescribed ceremonies may be mounted only after considerable delay, and once scheduled may be put off any number of times and replaced with ceremonial stopgap measures understood as having only a temporary benefit. Our intuition is that this cannot be accounted for entirely in terms of the pragmatic difficulties and logistical burdens of assembling resources and social support, since a person may go for many years with the unacted-on intention of having a ceremony for a particular purpose. There is likely to be a motivational threshold that one must cross in order to actively seek change in one's circumstances or level of distress. Accordingly, in our ongoing analyses we are examining the psychology of postponement with respect to the relative salience of what could be called the ratio of the burden of suffering to the burden of change.

Already implicit in our understanding of the *experience of the sacred* is the idea that it is more than a means of eliminating resistance to change by overwhelming or bedazzling participants. As a result of this phase of the research program we can more specifically describe experience of the sacred as the medium in which the elaboration of alternatives takes place, a means by which those alternatives are vivified and legitimized for participants, a wellspring of reinvigoration and relief of demoralization (see Frank and Frank 1991), a source of motivation toward the actualization of change for persons in whatever state of initial disposition, and an ongoing validation of that change once it has begun. What this formulation still presupposes is that we look for the sacred in more or less dramatic form, a position ultimately beholden to William James's (1961) injunction that the best way to study religion is to focus on the most religious moments of the most religious man. In addition to the most obvious fact that experiences range from the inconsequentially mild to the most compellingly powerful, the Navajo work has allowed us to identify two related but distinct modulations of the sacred. First is that between implicit and explicit: Rather than being explicitly either mild or powerful, the sacred may remain implicit in the sense of not being overtly acknowledged as such. The most striking example is the Traditional healing patient whose response to the interviewers' interest in religious healing was that she had never really thought of the ceremonies as religious. Such responses require careful rethinking of the nature of the sacred and its role in healing. The second modulation is between what we can call transcendent and immanent experience of the sacred and has to do with the sensory quality of the experience. The former is numinous, supernatural, ethereal, and out of the ordinary, whereas the latter is intimate and embedded in the natural and/or the mundane environment. An example is the difference between revelatory imagery that appears transcendently "in the mind's eye" (as if on a screen with no determinate location before the person), and imagery that appears immanently embedded in or superimposed on perceptible objects (as in the appearance of significance in the shape of a cloud) (Csordas 2001).

We had conceptualized the *elaboration of alternatives* in terms of possibilities for the resolution of problems, distress, or illness that open up for patients by means of the healing process. It has been a staple of the study of medical pluralism that choices among healing resources are highly strategic and highly negotiated (Leslie 1980, Comaroff 1983, Brodwin 1996)— that, in effect, people shop around for the best solution to their problems. Likewise, it has been a staple of anthropology and transcultural psychiatry that healing alters the meaning of a problem (Bourguignon 1976) or changes a patient's assumptive world (Frank and Frank 1991)—that efficacy is constituted by this change in meaning. What these formulations, as well as our own, have presupposed is that there is a problem, the meaning of which is stable until it is transformed and thereby resolved by the rhetoric of healing. The Navajo work makes it clear that how to understand problems in the first place may be as much a component of the elaboration of alternatives as is how to resolve them. Prior to and independently of the transformation of meaning that constitutes efficacy, there may be a redefinition of the problem including more or less elaborate construals of causality. This is certainly the case when patients move among forms of healing such as Traditional, NAC, and Christian, but it can also be the case within systems. Successive Traditional ceremonies can redefine a patient's problem according to a number of causal schemas of contamination, transgression, or bewitchment; Christian healing prayer can address a problem in terms of sin, emotional woundedness, or demonic affliction. This reformulation allows us to go beyond documenting idioms of distress, semantic illness networks, and vocabularies of motives, to show how they work as expressive forms that define the experiential specificity of healing.

The principal feature of the model's account of *actualization of change* has been to shift the expectation that whatever efficacy can be found religious healing must be massive, sudden, and miraculous to a recognition that efficacy is often incremental in nature and a consequent focus on the significance of apparently minor or small-scale changes. Yet what this formulation did not yet take adequately into account is the scale of the problem in the first place, and from the standpoint of participants, how much change is needed or intended as a result of healing. The model still presupposes the goal of major transformation, regardless of whether the changes are incremental, with emphasis on a major healing event or series of events finally aimed at substantial change. Our Navajo work has required us to recognize that, in addition to transformation, ritual healing may be aimed at maintenance, in which case less change is intended, or at emergency intervention, in which case it may be intended to be palliative, to be a kind of triage to prevent deterioration, or to be a quick fix cure-all.

In addition to refinements in our understanding of each component of the rhetorical model, we can also refine our conceptualization of their interrelation. Although it may appear minor, presentation of the experience of the sacred after instead of before the elaboration of alternatives is intended to indicate that the sacred is not a precursor or precondition of elaboration, but fully integral to it. Likewise, we would now emphasize that the relation among the four components is best conceptualized not as linear or simply sequential, but cyclical such that the results in terms of change feed back into the disposition toward further participation. An important result of the earlier Charismatic work (Chapters One to Four, Csordas 1994a) was to have shown that an understanding of therapeutic process cannot be limited to a discrete healing event or ceremony, but must be traced beyond the ritual event of healing to a determination of how therapeutic process becomes integrated into the patient's life trajectory. In the Charismatic work this insight was linked to the religious understanding that "everyone needs healing," a formulation that explicitly linked healing with the overarching life process of spiritual growth. In the Navajo setting we observe in all three traditions multiple episodes of healing extending across the life trajectory of individuals. Indeed, it has become useful to introduce, between the notion of the discrete ritual event and the life trajectory, an intermediate concept of "healing career" (Garrity 1998) that spans multiple healing events, episodes of illness, and reformulations of already recognized problems. Finally, as we assess the contribution of the model by placing it in dialogue with the Navajo data, we are posing the critical question of whether it is primarily to be taken as descriptive in the sense of serving as a useful heuristic outline of experiential specificity in religious healing, or as evaluative in the sense of constituting an adequate empirical outline of criteria of efficacy in healing. In either case, bringing it to bear in future research on ritual healing traditions will perhaps lead to a more comprehensive and experientially relevant theory of therapeutic process.

# CHAPTER SEVEN

# The Sore that Does Not Heal

The problem of causal reasoning about illness is one of the enduring questions of anthropology, attracting perennial interest ever since the time of Tylor (Zempleni 1985). The Navajo ethnomedical system, one of the most extensively studied, is known to be particularly concerned with the determination and elimination of causes of illness. Two features of the Navajo literature are of relevance for the present argument. First, within the Navajo system anthropologists have identified etiological processes such as witchcraft, spiritual contagion, encounters with ghosts, and violation of taboo and, in practice, have classified Navajo healing ceremonies by the pathogenic agents they are intended to eliminate. However, not much attention has ever been paid to how these pathogenic agents are said to operate on or within people. Second, it is understood that Navajo ethnomedicine does not have a highly elaborated classification of diseases that can be matched with these general causal processes (Werner 1965); in principle, any cause can bring about any disease. Rarely has it been acknowledged that particular causes may be associated with particular symptoms or has a particular disorder been identified and analyzed (Levy, Neutra, and Parker 1987). Thus, the analysis of causal reasoning tends to stop with identifying causes and does not go on to a more complete account of cause and effect.[1]

In contrast to the Navajo medical system's focus on identification and removal of the *causes* of illness, the Anglo-American biomedical system is understood to be more concerned with the nature, classification, and removal of particular *diseases*. The disease in biomedicine is a quite specific entity that can be treated as a thing in itself, and even if it is the kind of disease that can be cured simply by removing its cause, the cause itself is a specific entity rather than a generalized process. It is easy from this point of view to draw the conclusion that our system is characterized by specificity, while systems such as that of the Navajo are nonspecific. The implication is either that biomedicine is thus superior by virtue of its precision or that we can easily understand nonspecific systems because they obviously heal through nonspecific mechanisms like the placebo effect or catharsis.

The question that remains unasked is whether medical systems other than biomedicine exhibit different kinds of specificity in their reasoning about cause and effect. This question is complicated by recent studies suggesting that systems of causes (etiologies) and systems of diseases (nosologies) are not as distinct as one might wish for analytic purposes. It has been reported among the African Evuzok (Guimera 1978) and the Iranians (Good and Good 1982) that disease classifications may simultaneously include categories that refer to cause and that describe symptom patterns. If we have in the past mistaken etiological categories for descriptive names of illnesses, we may also have made the error of missing the descriptive elements in what are thought to be purely etiological categories.

In this chapter I propose that the analysis of specificity of cause and effect be grounded in the concrete bodily processes said to be initiated by (often abstract) causal agents and said to characterize (equally abstract) diseases.[2] I focus on cancer, a disease for which the causes are uncertain and the manifestations are multiple, and examine how it has been incorporated into the medical reasoning of contemporary Navajos. Next I examine Navajo causal reasoning about cancer, based on how Navajo patients who have themselves experienced episodes of the disease construe the circumstances, and compare my findings with equivalent data from Anglo-American cancer patients. Focusing on lightning, the most frequent and most culturally distinct cause cited by Navajo patients, I then discuss the nature of the cause and effect relationship between lightning and cancer. Following a brief summary of the problem of specificity in these three areas—defining a disease category, causal attribution, and cause and effect—I conclude with a discussion of four pragmatic and methodological issues raised by the problem of causal reasoning about illness. These issues are defined in terms of conceptual distinctions between cause and symptom, between disease as entity or as process, between biomedical and traditional ethnomedical systems, and between body and mind.

## Cancer among Navajos and Anglo-Americans

Incidence rates of cancer among Native Americans remain typically lower than in the general population of the United States. From 1978 to 1981, incidence of malignant cases in all anatomical sites per 100,000 in the general U.S. population was 337.9. For the same period in the state of New Mexico (home to a substantial proportion of the Navajo population), incidence was a lower 285.2, but among Native Americans in New Mexico (including other groups as well as Navajos), incidence was only 164.2 (Horm et al. 1984).

In addition to this quantitative difference in total incidence, rates of different types of cancer vary between Navajos and the general population of the United States. Indian Health Service physicians typically mention relatively higher rates for Navajos of gastrointestinal cancers (stomach, colon, rectum), urological cancers (bladder and kidney), and cervical cancer, along with much lower rates of lung cancer (confer Kunitz and Levy 1981:353). The higher rates of some cancers have been attributed to dietary factors or the presence of carcinogenic trace elements in the environment; the lower rate of lung cancer (except among uranium miners) is attributed to the virtual absence of cigarette smoking among Navajos. The relative *prevalence* of different cancers among Navajos is depicted in Table 7.1, based on figures for the Navajo Reservation provided by the New Mexico Tumor Registry. Assuming a denominator of 150,000, the overall prevalence of cancer would be 0.63 percent among Navajos.[3]

According to figures cited by Kunitz (1983:67), hospital discharge rates for malignant neoplasms in 1972 were 23.3 per 10,000 for Navajos, while they

**Table 7.1**   Proportion of Cancer Diagnoses by Anatomical Site for Navajo Patients: June 1986

| Primary Cancer Site | Percentage | Number of Cases |
|---|---|---|
| Eye | 1 | 5 |
| Mouth | 1 | 9 |
| Throat/nose | 0 | 4 |
| Brain | 1 | 6 |
| Thyroid/pituitary | 5 | 50 |
| Other unspecified parts of nervous system | 0 | 2 |
| Stomach | 1 | 12 |
| Colon/rectum | 4 | 38 |
| Liver/pancreas | 1 | 8 |
| Gallbladder | 1 | 10 |
| Kidney | 2 | 20 |
| Urinary bladder | 0 | 4 |
| Bronchus/lung | 1 | 5 |
| Sinus/larynx | 0 | 3 |
| Blood | 2 | 15 |
| Lymph nodes | 1 | 7 |
| Bone/joints/cartilage | 1 | 9 |
| Connective, subcutaneous, and other soft tissue | 2 | 17 |
| Skin | 1 | 8 |
| Breast (female) | 6 | 55 |
| Cervix | 47 | 447 |
| Placenta/ovary/vagina/ vulva | 17 | 157 |
| Prostate/testicles | 5 | 48 |
| Special[a] | 1 | 5 |
| Unknown primary site | 1 | 5 |
| Total | 102[b] | 949 |

*Source*: Based on data provided by the New Mexico Tumor Registry for all IHS Service Units of the Navajo Reservation.
[a] Overlapping sites in nasopharynx (1 case), in pancreas (1 case), in urinary bladder (3 cases).
[b] Over 100 percent due to rounding off.

were 102.8 per 10,000 for the United States as a whole and 97.0 per 10,000 for the western United States. Reported mortality rates per 100,000 for Navajos were 24 to 27 from 1954 to 1956; 46 to 48 from 1965 to 1967; and 35 to 41 from 1973 to 1975. For the general U.S. population, the mortality rate in 1976 was 132.3 per 100,000; the percentage of total mortality from malignant neoplasms in the mid-1970s was only 5.8 percent for Navajos, while it was 21 percent for the general U.S. population (Kunitz 1983:65).

In spite of the lower overall cancer rates for Navajos as compared with the general United States population, the preceding discussion indicates a steady increase over the course of the century. This can be understood in terms of the theory of epidemiological transition, which suggests that "developing societies," among which Navajo society can in some respects be included, exhibit a shift from the prominence of infectious and parasitic diseases to chronic degenerative and man-made diseases (Broudy and May 1983). A somewhat simpler explanation is the survival of more older Navajos with chronic disabilities, survival brought about by the gradual improvement in medical care that has eliminated earlier causes of mortality such as tuberculosis (Kunitz and Levy 1981). For present purposes it will suffice to say that, although cancer is by no means among the leading causes of Navajo mortality, the gradual increase in cancer incidence has not gone unnoticed among Navajos and hence is an increasing source of concern.

The Navajo portion of this study was carried out among cancer patients who had utilized two Public Health Service hospitals on the Navajo Reservation, those at Fort Defiance and Tuba City (Figure 7.1). These two hospitals are located in distinct regions of Navajoland, and the patients they serve thus represent a range of internal diversity within Navajo society. The Fort Defiance cachement area is more densely populated than that of the

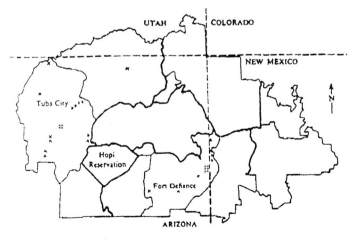

• = town or community
× = homesite

**Figure 7.1**   Approximate Homesites of Patients in 1988

Tuba City hospital and has more forest and grassland areas; its residents are more familiar with Anglo-American society. In contrast, the arid western area around Tuba City is more sparsely populated, and its inhabitants are more accustomed to a traditional Navajo life-style. The more traditional orientation of the west is borne out insofar as a greater proportion of Tuba City patients spoke the Navajo language, adhered to Navajo religion, and had less formal education (Table 7.2). In addition to gathering patient data, I also conducted interviews on more specialized traditional knowledge about cancer with four bicultural medicine men who also work as teachers in the public education system.[4]

The comparison group in Boston was drawn from patients in radiation therapy at the Massachusetts General Hospital.[5] The 55 patients were predominantly Euro-American; 10 were American blacks, and 2 were Haitian blacks. Demographics of the comparison groups (Table 7.2) indicate expected differences in level of education, with the Boston group considerably more educated, and in religious adherence, especially with regard to the number in the Boston group indicating no adherence. The distribution of cancer types (Table 7.3) conforms to the expected difference between Navajo and Anglo, with the former showing relatively more

**Table 7.2**   Summary of Patient Characteristics

| Demographic Data | Fort Defiance N = 12 | Tuba City N = 16 | Boston N = 55 |
|---|---|---|---|
| Male | 6 | 7 | 25 |
| Female | 6 | 9 | 30 |
| Age range | 19–86 | 27–82 | 19–78 |
| Language of interview | | | |
| Navajo | 4 | 12 | 0 |
| English | 8 | 4 | 55 |
| Marital Status | | | |
| Married | 6 | 11 | 32 |
| Widowed | 2 | 1 | 5 |
| Divorced/separated | 2 | 1 | 4 |
| Single | 2 | 3 | 14 |
| Education | | | |
| Graduate degree | 0 | 0 | 3 |
| College | 2 | 0 | 16 |
| High school | 5 | 3 | 20 |
| Less than high school | 2 | 3 | 9 |
| None | 3 | 10 | (missing) 7 |
| Religion | | | |
| Navajo | 3 | 8 | 0 |
| Native American Church | 1 | 4 | 0 |
| Catholic | 3 | 0 | 18 |
| Protestant | 4 | 4 | 13 |
| Mormon | 1 | 0 | 0 |
| Jewish | 0 | 0 | 6 |
| Greek Orthodox | 0 | 0 | 1 |
| None or missing | 0 | 0 | 17 |

**Table 7.3**  Summary of Cancer Types

| Type of Cancer | Fort Defiance N = 12 | Tuba City N = 16 | Boston N = 55 |
|---|---|---|---|
| Endometrial | 1 | 0 | 0 |
| Cervical | 1 | 1 | 0 |
| Ovarian | 1 | 0 | 0 |
| Breast | 3 | 4 | 19 |
| Testicular/prostate | 2 | 1 | 5 |
| Stomach | 2 | 0 | 2 |
| Colon/rectal | 0 | 2 | 6 |
| Liver/gallbladder/pancreatic | 0 | 0 | 3 |
| Kidney/bladder | 0 | 4 | 3 |
| Thyroid | 0 | 1 | 0 |
| Lymphoma | 1 | 2 | 1 |
| Brain/central nervous system | 1 | 0 | 2 |
| Leukemia | 0 | 1 | 0 |
| Lung | 0 | 0 | 8 |
| Bone/soft tissue | 0 | 0 | 4 |
| Unknown | 0 | 0 | 2 |

disease of the digestive and female reproductive tracts and an absence of lung cancer. This difference suggests that, in general, the comparison groups are diagnostically representative of their respective populations, in spite of the relatively small size of the groups by epidemiological standards.

## Navajo Conceptions of Cancer

It is not certain when the Anglo-American concept of "cancer" as a disease became widely known among Navajos. A survey of 4,826 Navajo admissions to Sage Memorial Hospital in the 1930s reported only three cases of cervical carcinoma in elderly women and one case of sarcoma, accounting for only 0.08 percent of hospital admissions (Salisbury 1937). Reichard (1950:97) cites two cases, probably from the 1940s, one of breast cancer and one of a man diagnosed with terminal cancer of the rectum who was healed by a traditional medicine man. In the 1960s a widely publicized outbreak of lung cancer occurred among Navajo uranium miners (Gottlieb and Husen 1982). This event very likely did much to disseminate the English term "cancer" among Navajos. In the 1970s personnel of the Navajo Area Indian Health Service, in collaboration with traditional medicine men, began a Cancer Control Project designed to increase cooperation between the two health systems. A major goal was to convince medicine men to refer cancer patients for simultaneous medical treatment rather than waiting to see if a traditional ceremony was effective.

Requisite to the validity of cross-cultural comparison is determination of whether an indigenous concept exists parallel to that of cancer as a discrete type of illness. To be sure, although oncologists technically regard each cancer as a separate disease, American popular culture recognizes cancer as a global entity. Aside from contact with cosmopolitan biomedicine, there is no immediate necessity for an indigenous nosological system to classify cancers that affect different parts of the body with different symptomatic manifestations as belonging to the same nosological category. In addition, it is relevant to recall Werner's (1965) observation that the Navajo language has never had a large list of named diseases, but rather a series of connotatively overlapping ways of referring to and describing sickness and pain. Thus, neither is there an immediate necessity that cancer be distinguished as a discrete disease entity in the first place.

The bicultural medicine men consulted placed the origin of cancer, along with other diseases, in the second mythic creation, the yellow world. One dimension of this origin is in sexual abuses committed by the yellow world's inhabitants, such as incest, homosexuality, and transsexuality, and in this way cancer is linked to the venereal diseases. A second dimension is the inhabitants' misguided attempt to control nature and their consequent misuse of natural forces such as radiation and electricity, and in this way Navajos understand why hospitals treat cancer with radiation and dangerous chemicals. At the same time, one medicine man speculated that the most common contemporary Navajo terms for cancer are probably of recent origin ("I don't know, I wasn't at the meeting where those words were decided on"), perhaps coined by people translating for doctors.

In fact, there are two principal Navajo language terms that denote cancer. Both *łóód doo nádziihii* (sore that does not heal) and *nááłdzid* (keeps on rotting) are in common use by patients and Navajo health care professionals alike. The Young and Morgan (1987) dictionary gives "rotten, gangrene, and cancer" as equivalent translations of the word "*nááłdzid*" and for cancer further specifies *nááłdzid k'ee'ąą nooséełii* (the rottenness that spreads as it grows). One medicine man stated that *nááłdzid* was the only correct Navajo term for cancer, while *łóód doo nádziihii* was a generalized term that could mean any kind of nonhealing sore. Another recognized both names but distinguished them as two types of disease.

In her medical lexicon, Austin (n.d.) includes the term *łóódtsoh* (big sore) as a translation of cancer, while Young and Morgan (1987) use the same term to denote smallpox. The term *łóód na'agháazhii* (sore that eats you inside) was cited by a Navajo health care professional and by one medicine man as referring to cancer. Young and Morgan (1987) again disagree, translating this term as ulcer. *Łóód doo yit'íinii* (sore you can't see) was also cited by a medicine man. In the term *nákid doo yit'íinii,* the word "*nákid*" refers to small worms or bugs of sexually transmitted diseases, which create sores and cause rotting regarded as related to cancer. However, none of these terms appears to be common in popular or current professional usage.

Whatever the correct relation among the terms, as a type of disease in the Navajo system, cancer has tended to become a composite etiological category, rather than a purely descriptive one (Good and Good 1982). Although Navajos recognize that cancer can occur in different parts of the body and may affect different parts in men and women, this is not precisely how they would understand the phrase "different cancers." Instead this phrase was described as a composite term in an etiological sense: cancer "caused by snakes, by tornados, or by [sexually transmitted] bugs [germs] all combined together is called *nááłdzid.*"

It is evident, however, that the Navajo terms conceptualize cancer as a sore more than as a growth or tumor. Indeed, another medicine man, speaking in English, indicated a similarity between cancer and boils. Neologisms for tumor exist only in the technical vocabulary prepared by Austin (*doo ákót'éégóó díníséehgo,* "mass") and the dictionary of Young and Morgan (*'atsį' bii' ni'ilts'id,* "compact mass within flesh"). This fundamental difference between Navajo and Anglo conceptions might be attributable to more than one source. An external sore is immediately apprehensible as a visible process. Likewise, rotting is a visible process quite familiar to people living in proximity to both domesticated and undomesticated animals, a process that furthermore could easily be extrapolated to the notion of decay as an internal, invisible process. Yet visibility and familiarity alone are inadequate to account for the difference between Navajo and Anglo conceptions, since many tumors can be palpated and since animal butchery could produce knowledge of pathological internal growths. I would suggest, instead, that negative, uncontrolled growth is a less culturally salient metaphor for Navajos than for ourselves. In Navajo thought, growth is inherently positive, whereas degeneration and decay are characteristically negative processes. The traditional Navajo conception of the life cycle is one of rising energy and achievement until age 50 and progressive decline and decay until death at age 100. To conceive of cancer as something that "keeps on rotting" is more consistent with such a view, while our own conceptualization of "unchecked growth" is consistent with our fear of nature (and society) out of control. Even the one Navajo patient who used the word "tumor," when questioned about her perception of how the disease worked in her body (pathophysiology), responded that it was probably "eating me inside."

The broader implication of this argument is for the role of metaphor in the relation between culture and illness. Not only can illnesses be used as metaphors of society and social process, as has been argued by Sontag (1978) and others, but the very features and processes that are attributed to illnesses and then projected onto social situations are themselves formulated in terms of dominant cultural metaphors (Lakoff and Johnson 1980). This is not the same as saying, for example, that our perception of tuberculosis is changed by its no longer being associated with hectic passion and creativity. We can still conceive of tuberculosis as a kind of "consumptive" process even if we no longer give the same connotation to consumption. Instead, if a disease is an apt metaphor for certain social processes, it is only because its pathophysiology has already been cast in metaphors generated in the process of social life, metaphors that may not suggest themselves in another society. Thus the metaphorical relation between cultures and illnesses must be understood as reciprocal.

To return to the more immediate question, however, we must determine whether the indigenous Navajo conception admits the possibility that cancer is curable or is invariably fatal. These questions are bound up with the issue, often raised by reservation health-care professionals, of whether Navajos tend toward "denial" of serious illness like cancer. In developing the interview, several Navajo consultants advised against direct reference to possible death and specific mention of the term "cancer," since to do so would appear to patients as an invocation of death and disease. In fact, few patients hesitated to name their illness when asked what it was, though only one patient referred explicitly to the imminence of death. Only one patient, who had consented to a hysterectomy in treatment of uterine cancer only after pain and bleeding became severe, exhibited a degree of overt denial, and even she acknowledged that her illness "would have become cancer" if she had not undergone surgery.

To us, the notion of denial implies above all an inadequate process of coping with impending death. Avoiding thoughts about and reference to death may appear rather different from a Navajo perspective. When asked how the illness affected their thoughts, it was common for Navajo patients to insist that they thought only of becoming well, with an overtone that to capitulate to the inevitability of death was a morally inappropriate stance. In a similar vein, one Navajo health educator expressed admiration for an uncle who had died of cancer precisely because the uncle "never gave up hope" up to the moment of his passing. This attitude suggests that it would in some sense be incorrect to acknowledge any disease as necessarily fatal, even if such fears are implicit.

The issue of curability is more complex. Of the four Navajo cancer patients who could not specify the name of their illness, one referred to it as *doo bi'déélníinii* (that which is not curable), while another emphasized that her illness could not have been "the sore that does not heal" (*łóód doo nádziihii*), since she was now cured. More indirect evidence comes from responses to the question of how traditional healing ceremonies and herbal remedies may have helped. Only two patients claimed to have been definitively cured, one by traditional herbs and one by peyote. Several others stated that the evidence of ceremonial efficacy was to be found not in their cure, but in the brute fact that they were still alive. Though by no means definitive, these statements allude at once both to a liberal criterion of efficacy and to recognition of the possibility of imminent death, while leaving open the question of curability.

In the specialized perspective of bicultural medicine men, the disease is curable. One medicine man who distinguished *nááłdzid* and *łóód doo nádziihii* as two types of disease stated that each has a distinct herbal cure. *Nááłdzid* is cured by *azee' hááłdzid*, literally "medicine for rotting," which itself is said to have a pungent smell like something that has spoiled. *Łóód doo nádziihii* is cured by his *yiyáąnii*, "that which eats or dries up pus." Several cancer patients did report having been treated with the latter remedy, although it appears to be used more broadly in treating infections and for patients who have undergone surgery. Adherents of the Native American Church claim that peyote can cure cancer, and narratives of such cures resemble Christian healing testimonies. Finally, one medicine man cited a traditional cure for cancer, known to Hopi, Zuni, Laguna, and Ute peoples but largely "forgotten" by Navajos, in which a dog is ceremonially killed and medicine prepared from its fat.

A more general statement comes from a medicine man informant of Adair, who cited three ad hoc categories of curability: (1) diseases, like tuberculosis, that the medicine men have given up on and left to white doctors—that is, intractable contagious diseases; (2) sickness caused by getting close to where lightning has struck, which medicine men can cure; and (3) illnesses, like snakebite poisoning, that both medicine men and white doctors can cure (Adair, Deutschle, and McDermott 1957). Combined with the statement reported above that cancer originates in part from the abuse of radiant energy by inhabitants of a previous mythic world, this statement prefigures what will become my principal ethnographic question, the role of lightning in traditional Navajo causal reasoning about cancer.

In sum, cancer is understood to have a mythic origin along with other diseases, although the terms that denote it are of contemporary origin. The Navajo concept of cancer is distinctive in that it defines the processual feature of the disease not in the idiom of growth but in the idiom of rotting, such that cancer is understood as part of a larger class of "sores that do not heal" and "keep on rotting." At the same time, relative to its use in biomedicine, "cancer" appears to have become transformed from a purely descriptive to a composite etiological category as it has been incorporated into the contemporary Navajo medical system. Although acknowledged to be sometimes fatal, cancer may also be cured. To go further toward understanding the cultural and existential meaning of cancer in the Navajo experience, however, we must examine Navajo causal reasoning about the disease.

### Causal Construal of Cancer

Traditional Navajo theories of disease etiology were summarized by Wyman and Kluckhohn (1938) under the concept of "infection," although a more appropriate contemporary concept appears to be "contamination." In Wyman and Kluckhohn's formulation (1938:13–15), disease could result from exposure to animals, natural phenomena, ceremonials, evil spirits, and enemies or aliens, with witchcraft as an additional source of illness. Kunitz and Levy (1981:356–360) take a somewhat different approach, distinguishing etiological processes from agents and including other processes in addition to infection. Thus illness may result from soul loss, intrusion into a person of alien objects or spirits, violation of ritual restrictions, and witchcraft. Any of these processes may occur by means of specific agents including dangerous animals, natural phenomena such as lightning, exposure to powerful ceremonials that are incorrectly performed or are conducted when a participant is in a weakened condition, and evil spirits or ghosts.

**Table 7.4**  Causal Construal of Cancer among Navajo Patients

| Cause | 1 | 2 | 3 | 4 | 5 | 6 | 7 | 8 | 9 | 10 | 11 | 12 | 13 | 14 | 15 | 16 | 17 | 18 | 19 | 20 | 21 | 22 | 23 | 24 | 25 | 26 | 27 | 28 | Cumulative total |
|---|---|---|---|---|---|---|---|---|---|---|---|---|---|---|---|---|---|---|---|---|---|---|---|---|---|---|---|---|---|
| | | | | | | | | | | | | | | | | Informant | | | | | | | | | | | | | |
| Injury | X | X | X | – | X | – | – | – | – | X | – | X | – | X | X | – | X | X | X | X | X | – | X | X | – | – | – | – | 15 |
| Lightning | X | X | X | – | – | X | – | – | X | – | – | – | X | – | – | – | – | – | X | – | – | – | X | X | X | – | – | X | 11 |
| Witchcraft | – | X | X | – | X | – | – | X | – | – | – | X | – | – | – | – | – | – | X | – | – | – | X | – | X | X | – | – | 9 |
| Exertion | – | X | – | – | – | – | – | X | – | X | – | X | – | – | – | – | – | – | – | – | X | – | – | – | X | – | X | – | 7 |
| Diet | – | – | X | – | – | X | – | – | – | X | – | X | – | – | – | – | – | – | – | – | – | – | X | – | – | – | – | – | 5 |
| Animal violation | – | – | – | – | X | X | – | – | – | – | X | X | – | – | X | – | – | – | – | – | – | – | – | – | – | – | – | – | 5 |
| Environment | – | – | X | – | X | – | – | – | – | – | X | – | – | – | – | – | – | – | X | – | – | – | – | – | – | – | – | – | 4 |
| Medication | X | – | X | – | – | – | – | – | – | – | – | – | – | – | – | – | – | – | – | X | – | – | – | – | – | – | – | – | 3 |
| Heredity | X | – | – | – | – | – | – | – | – | X | – | – | – | – | – | – | – | – | – | – | – | X | – | – | – | – | – | – | 3 |
| Alcohol | – | X | – | – | – | – | – | – | – | – | – | – | – | X | – | – | – | – | – | – | – | – | – | – | – | – | – | – | 2 |
| Stress | – | – | – | – | – | X | – | – | – | – | – | – | – | – | – | – | – | – | – | – | – | – | – | – | – | – | – | – | 1 |
| Illness | – | – | – | – | – | – | – | – | – | – | – | – | – | – | – | – | – | – | – | – | – | – | X | – | – | – | – | – | 1 |
| Old age | – | – | – | – | – | – | – | – | – | – | – | – | – | – | – | – | – | – | – | X | – | – | – | – | – | – | – | – | 1 |
| Ceremony | – | – | – | – | – | – | – | – | – | – | – | – | X | – | – | – | – | – | – | – | – | – | – | – | – | – | – | – | 1 |
| Don't know | – | – | – | X | – | – | X | – | – | – | – | – | – | – | – | X | – | – | – | – | – | – | – | – | – | – | – | – | 3 |
| Personal total of causes | 4 | 5 | 6 | 0 | 4 | 4 | 0 | 2 | 1 | 4 | 2 | 5 | 2 | 2 | 2 | 0 | 1 | 1 | 4 | 3 | 2 | 1 | 5 | 2 | 3 | 1 | 1 | 1 | |

Reichard's account (1950:80–82) includes the influence of completely buried monsters of the mythic age and the malevolence of undependable deities, and notes the role of human frailties such as ignorance of proper behavior, dangerously weakened conditions or states, and, especially, excess in any activity. Luckert (1975:151–162) proposes a typology of Navajo theories of disease and healing based less on etiology and more on a kind of pathophysiology grounded in an ethnopsychology of the person, including transformation and retransformation, fragmentation and reassemblage, submergence and reemergence, infection and catharsis, and separation and unification.

My findings on Navajo explanations of the causes of cancer (Table 7.4) must be understood against the background of this diversity of causes and effects elaborated within the traditional system. At the same time, the possible role of naturalistic or nonritual causes not included in ethnographic accounts must be entertained, as well as the interaction between Navajo etiologies and those of biomedicine and the popular medical culture of contemporary North America. Injury, the leading cause cited, creates an immediate problem in this respect, since Navajos traditionally distinguish between being "hurt" and "sick," and a distinct category of Navajo ceremonies (Lifeway) is directed toward injuries (Wyman and Kluckhohn 1938). Nonetheless, the idea that an injury can "turn into cancer" appears to be compatible with the awareness that such an injury may not heal properly—that is, could become a sore that does not heal or keeps on rotting. Lightning, to which I shall return below, witchcraft, and animal violation conform to the traditional pattern of infection by powerful and dangerous forces. Exertion is understandable as a cause of cancer in terms of both the traditional concept of vulnerability in a weakened state and the traditional understanding that old age and death are the result of a gradual wearing down and depletion. Diet, environment, and medication, on the other hand, are typically associated with contemporary conditions of change in traditional life, respectively referring to increasing consumption of junk foods and foods with additives, environmental pollution, and adverse side effects of biomedical treatment. Heredity is a special case here, since two of the three Navajo patients who cited it came from an extended family in which there was a documented presence of a rare genetically based colon cancer. Alcohol consumption, stress, illness, and old age were cited rarely, and the one case of ceremonial contamination was reported by the only medicine man among our patient informants, who stated that the onset of his lymphoma occurred shortly after he performed a ceremony for a woman with a sore throat.

These Navajo data are placed in cross-cultural perspective by the comparative data presented in Table 7.5. For the Anglo-American data it was possible to distinguish answers to the questions of what patients believed "caused" their disease and what other factors they thought were "related" to their disease, whereas linguistic and conceptual difficulties made such an analysis impossible for the Navajo data. Thus, for Anglo-American patients, the most frequently cited cause was heredity, while the most frequently cited

**Table 7.5** Comparison of Navajo and Anglo Causal Construal of Cancer

| Cause | Navajo N = 28 | Anglo "Related" N = 50 | Anglo "Caused" N = 49 |
|---|---|---|---|
| Injury | 15 | 5 | 4 |
| Lightning | 11 | 0 | 0 |
| Withcraft | 9 | 0 | 0 |
| Exertion | 7 | 0 | 0 |
| Diet | 5 | 3 | 2 |
| Animal violation | 5 | 0 | 0 |
| Environment | 4 | 1 | 1 |
| Medication | 3 | 5 | 0 |
| Heredity | 3 | 0 | 14 |
| Alcohol | 2 | 5 | 2 |
| Stress | 1 | 13 | 4 |
| Illness | 1 | 1 | 4 |
| Old age | 1 | 0 | 0 |
| Ceremony | 1 | 0 | 0 |
| Weight | 0 | 1 | 0 |
| Psychological distress | 0 | 2 | 0 |
| Life-style | 0 | 3 | 0 |
| Smoking | 0 | 5 | 3 |
| X rays | 0 | 2 | 1 |
| Contagion | 0 | 1 | 0 |
| Breast implant | 0 | 1 | 0 |
| Bad luck | 0 | 0 | 2 |
| Total | 68 | 48 | 37 |

*Note*: The Ns of 50 and 49 for Anglos exclude patients for whom data on causal construal were missing.

related factor was stress. When "causes" and "related factors" are collapsed into a single category, the ten leading elements of causal construal cited by Anglo-American patients were stress, heredity, injury, smoking, alcohol, diet, medication, illness, x-rays, and lifestyle, in that order. Only five of these leading elements also appeared among the ten causes of cancer most frequently cited by Navajos, and they appeared in a quite different order of priority. Caution must be taken in interpreting these differences, however, as is evident by contrasting our results with those of Linn, Linn, and Stein (1982) on etiological beliefs among Anglo-American cancer patients. In that study heredity and stress were both cited but ranked fourth and seventh, respectively. The three leading elements were smoking, God's will, and type of work (in contrast to the leading elements of stress, heredity, and injury in the present study), and only six of the ten leading elements were also cited by my Anglo-American consultants.

Nevertheless, it remains significant that Navajos cited lightning exposure, witchcraft, exertion, old age, animal violation, and ceremonial contamination as causes of cancer—causes that not only were absent from the Anglo-American data but also for the most part were prominent in the Navajo data. The principal overlap that requires interpretation in the causal construals of the two groups is injury. A tentative hypothesis is that, just as the Navajo attribution may be based on the cultural conception that an injury can turn into a sore that does not heal, so the Anglo American attribution may be based on the cultural conception that an injury can initiate an abnormal growth process, assuming an analogy between the "lump" caused by injury and a tumor. Aside from this, we can conclude that, despite over a century of assimilative pressure and despite the fact that all Navajo patients had received biomedical hospital treatment, the Navajo causal construal of cancer remains culturally distinct from that of Anglo-Americans. Given this general conclusion, we shall take a step further into the Navajo ethnotheory of disease etiology by examining the second leading causal element for Navajos, namely lightning.

## Lightning as a Cause of Cancer

The single ethnographic fact that poses a dilemma for the present inquiry was clearly stated by Wyman and Kluckhohn (1938:15): "In most cases one factor is thought of as being able to cause a variety of maladies, with one or two outstanding. Likewise, the same disease may result from one of various factors." It is evident from the data in Table 7.5 that neither Navajos nor Anglo-Americans identify a single causal element for cancer in the way one might in the classical model of contagious diseases in biomedicine, although the Navajo causal construal includes a greater diversity of elements. If most factors also may cause a variety of diseases, the question becomes whether lightning bears a specific causal relation to cancer, or whether it is equally a factor in other diseases. Most scholars of Navajo culture agree that lightning is indeed a very commonly cited cause of illness, so data suggesting a more specific relation between lightning and a particular disease must be evaluated very carefully.

A tentative step toward determining how frequently lightning is associated with other diseases at first appeared to disconfirm the specificity hypothesis. A physician colleague reported on ten traditional Navajo patients none of whom had cancer; fully five of them attributed their illness in some degree to lightning. However, two of those patients explicitly told the physician that they feared their problem might turn into cancer. These included a patient who had suffered a direct strike and someone shot by a witch with wood from a lightning-struck tree. A third was suffering from stomach ulcers, which are related to cancer as a member of the class of sores that do not heal.

These data are inconclusive but warrant pursuing the issue. Given that Navajo ceremonials are primarily directed toward removal of whatever etiological factors are determined to be active, indirect evidence can be martialed on the basis of what kinds of traditional healing ceremonies are used for cancer patients. In this respect, we must consider Jerrold Levy's observation

(1983:132) that "no Navajo disease is known by the symptoms it produces or by the part of the body it is thought to affect.... Nevertheless, certain groups of healing ceremonies appear to be associated with some symptoms and not others, while several other ceremonies appear to be good for a broad range of symptoms." This issue of generality in the efficacy of healing ceremonies is complicated by the observation that cancer is not a single disease but a class of diseases exhibiting a variety of symptom patterns. However, I have shown above that the concept of cancer has sufficient integrity within Navajo thought to be typically associated with a more or less discrete cause.

The role of lightning in conceptions of cancer causation is affirmed by patients' accounts of their use of traditional healing ceremonies in conjunction with biomedical treatment (Table 7.6). The standard ceremony used to remove adverse effects of lightning is the Shooting Chant (*na'at'oee*). In contrast, the conceptualization of cancer as a kind of sore (*łóód*) does not appear to prompt the use of those ceremonies described as especially suited for sores and boils, namely Eagleway, Eagle-Trappingway, and Beadway (Wyman and Kluckhohn 1938:29; Sandner 1979:45).

A more systematic test can be performed following the method used by Levy, Neutra, and Parker (1987) to establish a degree of specificity in the use of certain ceremonies for seizure disorders and depression. Having ethnographically determined a group of ceremonies that appeared to be associated with these two disorders, Levy compared the proportions of types of ceremonies used by a diagnosed group and a control group. For both groups of disorder, the results were statistically significant. What is important for the present work is that Levy's analysis distinguished disease-specific ceremonies from generalized or "broad-spectrum" ceremonies and that prominent among the latter was the Shootingway group. However, when a similar analysis is done comparing the group of Navajo cancer patients against Levy's control group, the result is that significantly more cancer patients have had Shootingway performed than have members of the control group.

**Table 7.6**   Traditional Treatments Used by Navajo Cancer Patients

| | | |
|---|---|---:|
| Major sings | | |
| | Shootingways | 12 |
| | Nonspecific | |
| | (Evilway, Enemyway, | |
| | Lifeway, Blessingway, | |
| | Enemy Lifeway, Windway) | 14 |
| | Mating of Reptiles | 2 |
| Other treatment | | |
| | Navajo herbs | |
| | (Lifeway, Pus-eater, | |
| | protection prayers) | 17 |
| | Peyote | 8 |
| | Sucking cure | 3 |

*N* = 28.

This analysis is shown in Table 7.7, juxtaposed to the comparable analyses from Levy's work.[6]

An additional element of specificity is added by the kind of Shootingway used. Navajo ceremonies are typically divided into male and female versions. Not all informants specified which had been performed over them, but when they did, it was always the male variant, except in one case. This exception was the only medicine man among the patients interviewed, and he prescribed the female version for himself because he had already had the male version years earlier. In my interviews it was more common for patients to refer to the ceremony specifically as ó'oos'ni'jí (Lightningway) or ił hodiitłiizhjí (Lightning-Struckway). Those who were able to specify that the Shootingway was of the male version (na'at'oee bik̨a'jí) tended also to be the ones to specify that the ceremony was of the Lightning-Struckway, describing the effects of a direct strike. In fact, Wyman and Kluckhohn (1938:23) distinguish subvarieties of the Male Shootingway (Upper Regions Side and Thunderstruck Side) that roughly conform to the distinction between ó'oos'ni'jí and ił hodiitłiizhjí. However, they indicate that the former is probably associated with flash lightning while the latter is associated with heavy lightning accompanied by thunder. The present data suggest

**Table 7.7**  Specificity of Navajo Sings in Treatment of Seizure, Depression, and Cancer

|  | Seizure Patients | Depressed and Control |
|---|---|---|
| Specific for seizures (Mountainway, Coyoteway Handtremblingway, Frenzy Withcraftway) | 29 | 1 |
| Nonspecific and not for seizures | 66 | 160 |
| Total | 95 | 161 |

Chi-square = 48.8, $p = < 0.0001$.
Source: Levy, Neutra, and Parker (1987:92).

|  | Depressed Patient | Seizure and Control |
|---|---|---|
| Evilways | 12 | 76 |
| All others | 8 | 160 |
| Total | 20 | 236 |

Chi-square = 6.46, $p < 0.05$.
Source: Levy, Neutra, and Parker (1987:93).

|  | Cancer Patients | Control |
|---|---|---|
| Shootingways | 12 | 24 |
| All others | 14 | 117 |
| Total | 26[a] | 141 |

Chi-square = 11.02, $p < 0.001$.
Source: For control group: Levy personal communication.
[a] Excludes Christian fundamentalists.

more that a distinction is made between indirect exposure (passing by a lightning-struck tree) and direct exposure (having contact with lightning itself) or perhaps that *ó'oos'ni'ji* is a more generalized term for any Shootingways directed toward lightning as an etiological factor.

In spite of the statistical support provided by data on ceremonial use, one may justifiably remain uneasy about the specificity hypothesis associating cancer and lightning. A final piece of evidence that supports the association was provided by a reservation physician in primary care. I spoke with this physician following a particularly heavy summer of lightning strikes, which had resulted in numerous patients coming for treatment to the Indian Health Service hospital and a consequent wave of prophylactic Shootingway ceremonies among hospital staff exposed to these patients. The physician stated definitively that, although she would not have noticed it if months earlier I had not mentioned my theory about lightning and cancer, she had observed that lightning-struck patients invariably expressed concern that their injuries could turn into cancer. Based on this and the above data, it can be asserted that lightning has more than a chance association with cancer among Navajos.

Given this ethnographic fact, it remains uncertain how old the association may be. Of the two cases of cancer cited by Reichard (1950:97), neither was attributed to lightning. These cases could be interpreted as counter to the present findings or could indicate a change since Reichard's time in the traditional understanding of cancer, a change perhaps related to the awareness that the "radiation" treatment often used for cancer bears some conceptual similarity to lightning. Indeed, one of my medicine man informants compared Navajo and biomedical cancer treatments by noting that, like the hospital doctors, "we Navajos have a radiation ceremony, too."

While this question cannot be definitively resolved, more can be said about the place of lightning in Navajo myth, daily life, and the experience of illness. Although patients consistently used the general everyday term for lightning (*ó'oos'ni*), lightning plays a prominent role in Navajo myth, where it is distinguished into varieties of zigzag (*'atsiniltł'ish*), forked (*hajilgish*), and flash or straight (*hatso'oolghał*). In myth lightning belongs to the class of inherently dangerous or evil things, used by the deities as a tool, weapon, or conveyance (Reichard 1950). In another respect, lightning is itself the manifestation of a class of deities or Holy People, the Lightning People.

However, lightning is not only a cosmological fact of life for the Navajo, but an ecological fact of life as well. It is an extremely common feature of the Southwest desert environment, so much so that at certain times of the year one can see several distinct thunderstorms moving across the expanse of sky at the same time. Navajo children learn the same caution about playing around lightning that urban Anglo-American children learn about playing near street traffic. The lesson is brought home by periodic fatalities from lightning strikes, which according to one reservation physician occur at the rate of at least one per year.

The pervasiveness of lightning is illustrated by the diverse circumstances of exposure cited by informants. One man recounted that lightning hit the electric line going to his house and knocked out the power four years prior to his illness, while another told how it hit the telephone line while he was talking, knocking the receiver out of his hand and deafening him as well as initiating his cancer. A woman cited an incident from childhood in which lightning struck an abandoned car in which she was playing with other children, burning some of them, while another recalled from boyhood that lightning struck many times around a wagon in which he was riding. One informant stated that there was a lot of lightning around his ranch, and another related a series of incidents in which he helped revive a lightning-struck cow, the family cornfield was struck by lightning, and he, as a thoughtless youth, counted the bones from lightning-struck sheep.

One man reported that a relative's death from a malignant brain tumor was traced to lightning striking a nearby tree while she was herding. Half her flock was killed, as everything around her turned blue and she inhaled the smell of smoke and burned hair and flesh She partially blacked out, experiencing numbness throughout her body along with hot and cold flashes, and the campfire was perceptually distorted to appear as if it were a little glowing dot. Over the next several years, she experienced regular headaches, eventually began passing out, and finally had a seizure and was taken to the hospital, where the cancer was diagnosed.

Yet lightning is more than a cosmological and an ecological fact of life; it is also a metaphorical fact of life, insofar as the category of Lightning extends beyond storm-caused Lightning to other forms of radiant energy. Thus one woman stated that the principal cause of her cancer was that she had picked up her children after they had been knocked out by touching a live electrical wire and only secondarily mentioned that Lightning had also struck a building in which she was attending a peyote meeting, following which she inhaled its smoke; when the cancer subsequently spread to her back, it felt like a Lightning shock. For another, the cancer was caused by radiation from a uranium mine, also understood as a form of Lightning. Yet another informant was a welder who assimilated his exposure to the flames and fumes of his torch ("the smell got into me") with the experience of having been exposed to natural Lightning in boyhood while herding sheep. One informant cited the breathing of fumes while working as a firefighter against a Lightning-caused forest fire. Evidence from a patient and a medicine man informant also suggests that exposure to the sun may be considered to be in the same broad category of Lightning.

The broader ethnographic conclusion that can be drawn from these data is that the Navajo category of Lightning is in fact metaphorically extended in two directions, cosmological and ecological. The literature on Navajo cosmology has long reported that lightning is mythically analogous to snakes, arrows, and other "shooting" phenomena. Indeed, among my informants there were three cases in which snakes were involved in the etiology of cancer and in two of these appropriate ceremonies were performed (see Table 7.6); this may implicate the lightning-snake analogy in the analysis of specificity among causes of cancer. In addition, the category of lightning is metaphorically extended to include such ecological factors as nuclear radiation, sunlight, electricity used for lighting, cooking on ranges or in microwave ovens, and television. One informant reported that one should eat homegrown meat rather than store-bought meat, not because the latter contains chemical preservatives as Anglo-Americans might fear, but because commercial livestock is sometimes subjected to electrical stimulation to enhance growth. Thus it appears that "Lightning" in its various aspects is understood by contemporary Navajos to be a principal form of environmental pollution.[7]

## Construing Specificity: Four Methodological Problems

Specificity is taken for granted as a goal of scientific research, and it is therefore expectable that we be concerned with identifying forms of specificity in the ethnomedical systems we study. Theorists in medical anthropology have recently developed a variety of analytic frameworks for making sense of the complexity of etiological reasoning encountered in the ethnological record. I have already noted the necessity of distinguishing between etiological and descriptive principles in systems of disease classification (Good and Good 1982). Kleinman (1980) has elaborated a framework for the analysis of specific illness episodes, placing etiological understandings in the context of understandings about the course of illness, pathophysiology, appropriate treatment, and expected outcome, and has emphasized the need to distinguish professional, folk, and popular etiologies. Young (1976) has identified four categories of information coded in explicitly etiological systems, including agencies (immediate causes, precipitating agents, and intermediating or predisposing agents), events or circumstances, instrumental or efficacious actions, and biophysical processes. Zempleni (1985) points out that illness etiologies must be understood in the context of how other kinds of misfortune occur as well; not only may an illness be the result of several interacting causes, but a particular cause may produce negative occurrences other than illness. He asks that etiological analysis make careful logical distinctions among instrumental (how) cause, efficient (who or what) cause, and ultimate (why) cause, between causes predicated on temporal sequence and those predicted on a conjunction of circumstances, and between causes determined a priori and a posteriori). Laplantine (1987) offers a series of analytic distinctions between causes that presume an ontological disease entity and those that refer to a relation between the afflicted and his surroundings, between causes of exogenous or endogenous provenance, between causes that operate by adding a noxious element or subtracting a vital one, and between the ultimately maleficent or beneficent effects of the causal agents.

These considerations go far beyond the kind of specificity dictated by the canons of biomedical science, which has to do with the specificity of diseases as discrete entities (Campbell 1976) and the doctrine of specific etiology that presumes one cause for one disease (Dubos 1959). Their theoretical importance lies in the determination of which of these multiple etiological categories are elaborated in particular ethnomedical systems and of the way these categories articulate with broader cultural goals, meanings, and priorities. If, for example, our analytic purpose were only to identify alternative loci of the specific in the Navajo system, we could do so, but this would only beg the broader questions about cross-cultural differences in reasoning about illness. To be sure, my data from bicultural medicine men indicate a measure of specificity between causes and symptoms: Killing a dog may cause "gas," attending a funeral may cause numbness, mistreating an aquatic animal may cause diabetes, exposure to lightning may cause heartburn. Likewise, there is a measure of specificity between certain classes of complaint and herbal/animal/mineral remedies: Hearing problems should be treated with a preparation from mountain sheep, vision problems with a preparation of the stinkbug, diarrhea with white clay, cancer with a medicine that "smells spoiled." The cultural logic behind these associations is more or less evident, as in the production of heartburn by the burning radiation of lightning or in the sympathetic connection between the rotting of cancer and its treatment by a spoiled-smelling medicine.

The specificity of a relation between cancer and lightning is of a different order, in terms of both defining the disease and identifying causal attributions. As I have noted, even in biomedicine, cancer is as much a group of diseases as a single disease entity, with the link among the diseases being the processual feature of aberrant, unchecked growth. This disease concept is taken up into a Navajo system unaccustomed to specifically named diseases, with a tendency toward etiological rather than descriptive disease concepts, and which interprets the unifying processual feature not as growth but as rotting.

The choice of the term causal construal to represent my empirical findings reflects the nature of the data as a repertoire of causal elements brought to bear by patients afflicted with a particular illness. This term does not distinguish which elements are regarded as specifically causal, as precipitating events, or as predisposing conditions. It also does not differentiate between which elements particular informants were certain about and which they speculated might be involved. Neither does it delineate possible differences between lay attributions and those learned by patients through consultation with specialist medicine men. Most of the Navajo patients generated a causal construal consisting of multiple elements totaling as many as six, with only seven citing a single causal element. In comparison to the Anglo-American data, Lightning can thus be understood as a cause of cancer specific to Navajo ethnomedicine, but by no means does Lightning conform to the biomedical doctrine of specific etiology. Furthermore, we have seen that Lightning itself is a category representing a greater range of phenomena than the meteorological.

Because of the multiplex nature of both cancer and Lightning as cultural categories, their causal connection must be understood by a less direct method than those outlined immediately above. One must define the semantic illness network (Good 1977), the system of relevant interrelated concepts within the cultural system. The principal conceptual link in our case is between the understanding of cancer as a *rotting* sore and the effects

of *radiation* as burning and eating a person's insides. The primary mechanism by which the disease enters a person is inhalation, which can include
the electrical fumes of a direct strike, smoke from a lightning-caused forest
fire, the rottenness of flesh from a lightning-struck animal, and, by extension, the stench of a rotting road-killed animal. Lightning is the prototypical form of radiation, but radiation is nothing other than a contemporary
interpretation of the traditionally broad category of *shooting* phenomena
albeit with less emphasis on traditional exemplars such as snakes and
arrows. The category includes electricity, and the medicine man's statement
that "our bodies are made of electrical impulses," whether or not it indicates an acculturated opinion, is in conformity with the traditional notion
that inordinate exposure to such impulses causes disease by disrupting the
harmonious electrical balance of life. Radiation also includes the energy o
the sun, and it is therefore relevant to the modern occurrence of cancer tha
the present world is said in Navajo tradition to be destined for destruction
by the sun. To have sexual intercourse when the sun is out is said to cause
damage to the sperm, and this may be related to the mythical origin o
cancer in the abuse of sexuality.

A comprehensive semantic illness network would account for the other
causes of cancer represented in my data, perhaps with the conclusion tha
there is no necessary relation among elements in either the cultural repertoire of causes or the causal construals made by individual afflicted persons
In concluding the present discussion, I can go only so far as to sketch a
series of methodological issues that would have to be taken into account in
such an analysis, issues that tie the analysis of etiological reasoning about
illness to broader anthropological concerns. These issues can be framed in
terms of four underlying conceptual dichotomies: (1) between cause and
symptom, (2) between disease as entity or process, (3) between biomedical
and traditional ethnomedical systems, and (4) between body and mind.

First is the dichotomy between *cause and symptom* in ethnomedical
systems, which is related to the distinction cited above between etiological
and descriptive systems of disease classification (Good and Good 1982)
This dichotomy is directly relevant to the pragmatics of clinical practice, in
that determination of cause in many medical systems has implications for
choice of treatment. As we have seen, the Navajo disease classification is
based primarily on etiology rather than on symptoms and syndromes
Navajos' relatively greater concern with causal factors is empirically evident
in comparison to Anglo-American informants in the present study, as the
mean number of responses in Navajo causal construals was 2.7, while that
for Anglo-Americans was 2.1. Moreover, only 3 out of 28 Navajos (11 percent
offered no response to questions about causality, while 10 out of 50 Anglo
American patients (20 percent) offered no response.[8]

The relative Navajo elaboration and Anglo-American poverty of causa
reasoning reflect more than a cultural divergence in the attention paid to
different aspects of the illness experience. The elderly Navajo who
complains that they "don't tell me what my illness is at the hospital" ma

mean not that the doctors failed to inform him of a tumor in his kidney, but they failed to inform him why he has it. In addition, physicians are faced with the fact that their Navajo patients are concerned about lightning as a sufficient cause of illness; even though asymptomatic, a Navajo may be considered sick following exposure to lightning. In general, biomedical professionals unfamiliar with ethnomedical causal construals may remain ignorant of patient fears that a particular course of illness will be determined by exposure to an indigenously defined cause.

A second conceptual distinction is that between disease as *entity or process*. In a discussion of ontological and relational understandings of disease, Laplantine (1987) shows that both formulations can be found in the cultural history of Western thought about disease. This issue is relevant to a critique of the methodology of biomedicine insofar as comparative (historical and cross-cultural) study throws into relief the role of etiological principles in our own ethnomedical system. Our dominant paradigm is predominantly ontological, defining a "Disease" as a discrete entity or biological "thing," here described by E. J. M. Campbell, "A Disease is first recognized syndromally—a constellation of clinical features. The Disease has a cause (infective, nutritional, genetic, immunological, etc); this cause produces characteristic structural changes which in turn cause characteristic functional disturbances which in turn produce the clinical manifestations" (Campbell 1976:50). This author, a prominent biomedical scholar, makes it clear that the kind of specificity required in our paradigm of the Disease leads to a confusing multiplication of conceptual entities that name the same global problem, but which name that problem from etiological, genetic, structural, biochemical, immunological, or prognostic points of view. At the same time, our paradigm ideally seeks to *reduce* syndromal, functional, and structural understandings to an underlying cause.

The very notion of a cause, however, takes on a particular ontological character because it is understood in relation to the Disease as a thing or entity rather than as a process or event. In the case of Navajo ethnomedicine, it is thus not sufficient to observe a shift in the concept of cancer from descriptive to etiological. As cancer is incorporated into the Navajo cultural pattern, it also becomes less an entity and more an event or process, with a consequent shift in what can be taken to constitute a cause. In broader purview, comparison of etiological systems with or without explicitly defined disease entities should take into account not only their recognition of different kinds of possible causes and of different possibilities for multiple interacting causes, but also the possibility of a different ontological status of the very notion of a cause.

Third, the importance of these problems should not lead one to suppose an indelible distinction between *biomedical and traditional* systems of causal reasoning. This issue bears directly on the ethnopsychology of cognition, in that causal reasoning reveals the structure of the mind as a capacity for generating propositions and seeking explanations about the world. My data on causal attributions for cancer lead me to conclude that making sense of the illness calls into play distinct *modes* of causal reasoning, but that these modes apply across Anglo-American and Navajo *systems* of ethnomedicine Injury, diet, and environmental exposures such as radiation are included ir the causal construals of both groups, though to different degrees and with varying rationales. It is a matter for empirical determination whether such co-occurring elements are indigenous or borrowed. Likewise, it must be determined whether elements from different cultural repertoires are seen to be compatible or incompatible, whether they can be assimilated to one another by metaphorical processes, and whether the interacting cultura repertoires, as wholes, occupy disjunct or integrated cognitive niches.

A striking example of this complexity comes from an interview with a women in her mid-thirties, a high school graduate with job experience in medical social services, who was in apparent remission from breast cancer In response to a question about traditional treatment and ceremonies, she discussed the causal influence of lightning at some length. Later, when asked specifically about what she thought had caused her illness, she replied thoughtfully that there were three possible factors, which she ranked in order of importance. First was the fact that her grandmother and an aunt had contracted cancer, therefore it could be inherited. Second was that she had once been on a regime of the drug Depoprovera, which she felt could have had a carcinogenic effect. Third, and somewhat skeptically, she recalled having been in an auto accident in which she bumped her breast against the steering wheel; she did not lend much credence to this cause but refrained from ruling it our altogether. I then reminded her that she had earlier mentioned a fourth cause, namely exposure to lightning. Appearing somewhat startled she said, "In that case, I'll make the lightning third and bumping against the steering wheel fourth."

Surprised at having the product of traditional causal reasoning juxtaposed to the hierarchical product of a more Anglo-American explanatory model this woman nevertheless quickly proceeded to integrate the two. The impli cation is that the Navajo and Anglo-American etiologies are cognitively distinct, but not cognitively incompatible. The problem remains why an explicit question about cause might elicit a response that excludes element of the traditional repertoire, unless there is a cultural disjunction in forms o reasoning about cause and effect relationships. As I have observed above in citing Adair's medicine man informant and as Ruth Benedict (1934) noted long ago about the Dobu, traditional therapeutic systems faced with new diseases may not evolve new therapeutic techniques to cope with them o may define them as outside traditional competence. Likewise, traditional etiological systems may or may not incorporate either new causal element or new causal rationales.

Finally, the suggestion that there are different modes of causal reasonin leads us to reconsider our methodological reliance on the distinctio between *body and mind,* or in the more precise terms of Evans-Pritchar (1937), between *sensible and mystical* causes. This issue bears on the existen tial rationality of culture, for as Lindenbaum (1979:56) has observed

"Beliefs about the etiology of disease are statements about the nature of existence, explanations of why things happen as they do." For most illness, the literature on Navajo ethnomedicine assumes a mystical cause, conceived predominantly as spiritual contagion or violation of taboo. My data suggest that, at least among contemporary Navajos afflicted with cancer, a physical cause (injury) ranks prominently alongside a spiritual cause (lightning). Much more significant, however, the data challenge the assumption that Lightning itself can be comprehended solely under the concept of spiritual contagion. This issue was framed in a discussion of Navajo disease etiology by Lamphere (1969:292):

> Activities involving dangerous animals or natural phenomena [are understood to] automatically arouse the supernatural's attack by weapons or anger, which in turn brings sickness. Until more fieldwork can be conducted on Navajo disease theory, it is only possible to suggest that, in some sense, the natural elements are fused with the supernatural The snake with which the Navajo might have contact and such *diyin diné'e'* [supernaturals] as ... the Snake People are, at some level, equated. It is impossible to determine if they are different forms of the supernatural, if the snake is a present-day natural manifestation of supernatural figures of the mythical past, or if they are 2 separate types of phenomena, one natural and the other supernatural, which share common characteristics. Whether one of these possibilities or yet another set of relationships best characterize Navajo beliefs regarding these matters cannot be concluded without more detailed data.[9]

It can be argued that, stated in this way, the problem is in part an artifact of the distinction between the natural and supernatural that was prominent in anthropology 20 years ago. This methodological distinction has three relevant characteristics. First, it is in essence predicated on distinctions between physical and spiritual, material and immaterial, tangible and intangible, or sensible and mystical, all of which presuppose what is typically called a "Cartesian" distinction between body and mind. Second, it presupposes that the supernatural is more truly "cultural" than the natural, in a manner roughly analogous to the way Kroeber theorized that the superorganic stood in relation to the organic. Finally, the Traditional approach focused almost exclusively on the abstract cultural definition of the causal agent, bypassing the question of how that cause produced its effect in terms of a cultural phenomenology.

Anthropology today is better prepared to attend to the physical in a definition of the sacred, bodily experience in an understanding of culture, and concrete ethnomedical practices in addition to beliefs. Examining the causal process associated with exposure to lightning exemplifies this methodological shift. Patients in the present study who mentioned lightning typically referred to a specific event in which lightning struck so near to them that they saw a bluish flash and inhaled the acrid electrical fumes.

To describe this experience they used the Navajo phrase *shił hodiitł iizh*, which can be translated as "I have been contaminated by Lightning." Based on the vividness of informants' statements and on the existence of a linguistic convention to describe the experience, the concrete sensory dimension of this exposure cannot be downplayed in favor of a concept of spiritual contagion. It is not only the fact of proximity that defines exposure to Lightning; at once a person's body is enveloped (external exposure) in blueness (visual modality) and incorporates by inhalation (internal exposure) the acrid haze (olfactory modality).

In emphasizing this embodied dimension of the experience, we can begin to resolve the question of whether Lightning is a natural or a supernatural phenomenon for Navajos. It is certainly natural in that it affects people in a physical, organic manner. At the same time, the enormity of the experience, its overwhelming "otherness," qualifies it as a quintessentially sacred phenomenon, culturally elaborated in myth and in Shootingway healing rituals designed to reverse its effects. In addition it is elaborated by temporal extension, in that (for example) going too close to a tree that has been struck by Lightning at an earlier time is thought to be equally as dangerous as having Lightning strike nearby; but I suggest that the physical encounter is primary to the cultural phenomenology of Lightning.

The importance of this discussion is not only that it answers Lamphere's call for more detailed data on Navajo disease etiology, but also that it represents a particular way of looking at (or looking for) the data within an anthropological paradigm of embodiment. In the present case the methodological shift is away from the problem of defining Lightning per se as a cultural phenomenon and toward the embodied human experience of lightning in cultural practice. While the goal of this approach is to collapse the distinction between mind and body in the name of a more comprehensive existential anthropology, it by no means seeks to preempt the biological importance of the body. Indeed, to argue thus in the present case would be to misrepresent Navajo thinking itself. The patient who assimilated the inhaled fumes of lightning with the inhaled fumes from his welding torch as intimately related causes of his brain tumor may have been engaged in biologically relevant speculation as wel! as syncretic cross-cultural reasoning. Likewise, the Navajo physician whose thoughtful response to my data was that the Navajo theory of Lightning as a cause of cancer may correctly intuit a process in which "oncogenes" are stimulated is taking seriously a potential biological consequence of Lightning exposure. Medical anthropologists who insist on the priority of determining the biological relevance of ethnomedical categories (Browner, de Montellano, and Rubel 1988) might feel obligated to follow this type of lead. Such work must be considered logically and methodologically secondary, however, to a determination of the human meaning of health-related phenomena through the use of empirical frameworks such as those summarized at the beginning of this section, through careful attention to methodological distinctions such as those discussed immediately above, and through the development of analytic perspectives such as that of embodiment.

# CHAPTER EIGHT

## *Words from the Holy People*

Embodiment, in the sense I am using it, is a methodological standpoint in which bodily experience is understood to be the existential ground of culture and self, and therefore a valuable starting point for their analysis. In this chapter I will focus on two issues that must be clarified in advancing a cultural phenomenology that begins with embodiment, or if you will, two issues that, unclarified, could become limbs in the embodiment of a straw man. One is the relation between embodiment and biology, and the other is the identification of this phenomenological starting point in preobjective or prereflective experience. I will present each in terms of a problematic quote.

From Martin Heidegger comes a statement of the first problem: "The human body is something essentially other than an animal organism. Nor is the error of biologism overcome by adjoining a soul to the human body, a mind to the soul, and the existential to the mind, and then louder than before singing the praises of the mind, only to let everything relapse into 'life-experience.' ... Just as little as the essence of man consists in being an animal organism can this insufficient definition of man's essence be overcome or offset by outfitting man with an immortal soul, the power of reason, or the character of a person" (1977a:204–205). Heidegger implies that the *ad hoc* tacking on of components to an essentially animal body exposes the inadequacy of distinctions among mind/body/soul/person, and betrays the existential character of the human body as essentially human. Leaving aside the problem of whether Heidegger essentializes the human body to the point of denying the Being of animals (Caputo 1991), does not the requirement to recast our understanding of the body in phenomenological terms thus mistakenly negate the important relationship between biology and culture? In response to this question, from the phenomenological standpoint our answer is that both "biology" and "culture" (or, more specific to the case discussed below, neuropathology and religion) are forms of objectification or representation. Thus a first goal is to suspend our reliance on both—or perhaps to suspend our description between them—in favor of an experiential understanding of being-in-the-world.

The nature of this preobjective being-in-the-world is our second issue, formulated as follows by Maurice Merleau-Ponty: "My body has its world, or understands its world, without having to make use of my 'symbolic' or 'objectifying' function" (1962:140–141). ". . . It is as false to place ourselves in society as an object among other objects, as it is to place society within ourselves as an object of thought, and in both cases the mistake lies in treating the social as an object. We must return to the social with which we are in contact by the mere fact of existing, and which we carry about inseparably with us before any objectification" (1962:362). To deny that the social is an object calls into question the status of "social facts," the existence of which is taken to have been established definitively by Durkheim. Does not the requirement that cultural analysis begin in preobjective experience thus mistakenly presume a presocial or precultural dimension of human existence? The answer lies in defining the sense in which we carry the social "inseparably with us before any objectification." Accordingly, the second goal of this chapter is to show how cultural meaning is intrinsic to embodied experience on the existential level of being-in-the-world.

The anthropologist addressing issues framed in this way can be distinguished from the philosopher by a simple criterion: The anthropologist is satisfied only by making the argument in terms of empirical data. The data for this chapter's exercise in "fieldwork in philosophy" (Bourdieu 1987) are drawn from the case of a young Navajo man afflicted with a cancer of the brain. I will introduce and contextualize the case in neither biological nor cultural, but in clinical terms. I will then attempt to thread the discussion of being-in-the-world between the two poles of objectification, showing how he brought to bear the symbolic resources of his culture to create meaning for a life plunged into profound existential crisis, and to formulate a life plan consistent with his experience of chronic neurological disease.

## Clinical Profile

The patient, whom I shall call Dan, was a participant in a larger study of illness experience among Navajo cancer patients (Chapter Seven), carried out with cooperation from Indian Health Service hospitals at Fort Defiance and Tuba City on the Navajo reservation in the southwestern United States. I was able to follow his progress for two years, beginning a year after the onset of his illness in 1985, and ending in 1988, about a year before he succumbed. Dan was 30 years old when I met him, an English-speaking former welder, with an education including two and a half years of college. He came from a relatively acculturated bicultural family; his mother was a schoolteacher and his father a ceremonial leader or "road man" in the Native American Church (peyotist), and one brother was attending college. Dan was divorced and, since the onset of his illness, was closely cared for by his own family.

His diagnosis was Grade II astrocytoma, a left temporal-parietal lobe brain tumor. After the tumor was removed, he received chemotherapy and radiation therapy, and was maintained on medications for control of tremors (legs, right hand, and head) and seizure-type neurological indications. Post-operatively, he experienced chronic headache, hypersensitivity of the operative incision, and occasional olfactory auras and paresthesias. His psychiatric profile was characterized by loneliness, pessimism, self-doubt, poor sleep, low energy, difficulty expressing his thoughts, rigid thought processes, blunted affect, disorganized ideation, rambling speech, preoccupation with mixed strategies for a plan of life, feeling depressed every day, and a sense of irretrievable loss over estrangement from his wife and children; his formal diagnosis included organic personality disturbance and mental deficits.

Dan's rehabilitative status was dominated by the loss and gradual recovery of linguistic ability, accompanied by frustration that "I can't say my thoughts," and that he recognized his relatives but "can't say their names." He reported that something will "hit my mind, but I can't say it," and that "I have a hard time trying to mention some words I want to say." His ability with spoken language appeared to return more completely and rapidly than that with written language, and he complained that although his English was steadily improving, the moderate amount of Navajo that he had been able to understand was completely lost. A Test of Non-Verbal Intelligence conducted a year post-operatively resulted in a score of 66, technically indicating persistence of a mild retardation but of ambiguous value based on motivational and cultural factors that could affect test performance.

Dan declined recommended psychotherapy and vocational rehabilitation. His post-operative session of mental testing had left him trembling and with headaches because of the exertion, and he apparently perceived such intervention as contradictory to the doctors' advice not to rush things in returning to normal activity.[1] He expressed some resistance to long trips to an off-reservation university medical center on the grounds that the doctors there "don't do anything" and that he preferred home visits from health-care personnel. In the meantime, Dan developed his own rehabilitative strategy for relearning vocabulary upon discovering the existence of word puzzle books, which consist of pages of letters from which one must identify words along vertical, horizontal, or diagonal lines. Completing these puzzles constituted Dan's main activity. Although working too hard on his puzzles sometimes caused him headaches, this activity appeared to provide a self-motivated form of linguistic therapy and cognitive rehabilitation.

By the spring of 1988 Dan's memory and ability to read had returned. Though he had been depressed and remained unsure of himself and his abilities, he claimed once again to be "trying." He had continued working his word puzzles, timing himself to record the increasing rapidity with which he could complete them, and comparing the increasingly neat manner in which he circled the words with his first puzzles, which he said looked like they were done "by a little kid." He acknowledged having "become a perfectionist in dressing and appearance" to such a degree that his sisters teased that the surgeon "must have put extra brain cells in." He had begun once again to talk about having a family, and to joke about "going to Gallup to find a chick." His physical condition at the time was stable.

## Cultural Phenomenology of Language and Inspiration

Dan considered himself to be an active person, and expressed frustration at not being able to work. He recounted commenting to his doctors that it was so difficult following their advice not to rush things that "maybe you guys going to have to tie me up ... [or] go to a special doctor again, what do they do, they put you to sleep, but they talk to you [hypnosis]." Dan's own solution, however, was to "follow the Navajo way," and learn to be "the kind of person that helps people"—a medicine man. This in fact became Dan's preferred strategy for reconstructing his life, focused around his existential struggle for language and expressivity. He indicated that this strategy originated in a direct encounter with the Navajo deities or Holy People, who inspired him with words of prayer:[2]

> Yeah. Yeah. See I think, see, I never use to have these help, this kind of help coming, but ... ah, my life is changing, but right now it is still kind of hurt me ... sometimes I have some prayers—when I was very small, learning, I prayed in front of my mom and dad or I talked to them—see, some of these [new] words I never knew, I never knew. My mom and my dad just said they never actually heard them [spoken by Navajos]. But then I told them and I said there's stuff I can hear. I said "Somebody waking me up wants me to listen," but then I really have pain in my head so I use to get up. But [when it first happened during Dan's treatment in an off-reservation hospital] we were staying in the motel and my dad was, my dad took care me so we been in the motel all the time, so sometime we go to sleep about 7 then I can't sleep, but when I do go to sleep right about 11 or 12 midnight I feeling [pause] my eyes just open like I'm not sleepy so I say, well, I might as well just turn the tv on, then I go turn the tv on, at that time [shortly post-surgery] too I couldn't understand what tv was, you know, what a tv was or the, the different show, I just look at them, I just sit there like this no laughing about it 'cause I forgot [what it meant]—I don't any more—and from there I just sit there and I said well I guess I can go back to sleep and end up laid back I tried to go—shoot!—just open up my eyes wide again, just like a talking come in, come in, come in and they keep me here for a good hour and a half. So then I get a headache if I don't do no, if I don't do no talking. I can feel it so my dad, I know that he's tired too and he has, I say "Dad please can you sit up." I said, "Some words coming to me I'd like to mention it to you. I want you to tell me if these are right are or if they are wrong. The word that I, that I have." So it's the Navajo way about

a long time ago and my dad says, "How can you know because you never even knew that these as you wake up these are put into your brain." And he says then I have to ask someone, I have to talk to either my mom or my dad and ask them each things that sometimes are—I change it that's when they get upset with me ... [confused passage] ... so I switched one of those words and they got caught [i.e., tripped up] my mom and my dad they got caught they could not think why this came out this way so they were trying to think about which one [of his words] was going straight, they were saying that this what they used to do a long time ago. So then as I made it, change it just a little myself I could feel it, just like you're going to throw up, yeah, and the pain real empty again you're just going "Doo! Doo! Doo!" [makes sounds describing a bodily/cognitive sensation] well, and I just sat there and then I just got a hearing that says, "Turn it back, turn it back." So then I just sat there and I, I said, "Mom, Dad," I said, "You have to hear it this way." I said I made a mistake. So then I talked it the way it came to me. Then they could answer it so they got caught on the spot.

This episode occurred relatively early in Dan's post-operative recovery, when his cognitive abilities were so impaired that he could not even comprehend television programs. It comprised a lengthy auditory experience, followed by a compulsion to talk that relieved his intense headache pain and left him with a "happy and a good feeling." For Dan this spiritual help received from the Holy People was different from the way he learned to pray as a child. The help does not come as an answer to prayer—it is the ability to pray itself that constitutes help for Dan, and his family concurred that indeed he was never before able to pray like he does now. He also reported that when he asked his father, himself a ritual leader, why he was speaking thus, his father responded "Someday you are going to be a very powerful person to help people." Some of his younger relatives, whom he had already begun to encourage with his new-found wisdom, have said "Uncle Dan, we kind of know that you are going to be a medicine man."

Nevertheless, his condition made the cultural validation of his experience, which was critical to Dan, somewhat problematic. During the struggle for fluency that characterized Dan's recovery, his parents and others had trouble understanding him. This difficulty was compounded of severe linguistic impairment and of the understanding that the utterance was a direct revelation of a new, contemporary synthesis of traditional Navajo philosophy in a young person who had never before known such things. Dan indicated that his own efforts to correct his speech were ineffective, and that only when he consented to "turn it back" and let his speech come out as it was inspired did they begin to understand.

His parents also told him that he must speak before the elders in the peyote meeting for their confirmation. He did this during a series of four peyote meetings held for his post-surgical recovery. These meetings last an entire night, during which participants ingest peyote, sing, and take turns uttering often lengthy, spontaneous, inspired prayers (Aberle 1982; La Barre 1969). In this setting each of the ritual officiants, the patients, and frequently other elders pray formally, and there are intervals in which quiet conversation is permitted. These intervals typically include encouragement and exhortation of the patient. Dan reported beginning his speech with the acknowledgment that what he was going to say might not be understandable to his elders because he was a younger person and his words would be "brand new," and receiving their permission to speak on the grounds that "if you want to help people [with] what you are saying, we have to listen to you." He said, "I made my heart and prayed to God, first, ask him, is this what I want for the rest of my life [to be a medicine man]?"

He also told them of his ambition to spend four years as a cowboy (that is, a professional bareback rider) prior to becoming a medicine man. Dan's anomalous desire to be a rodeo cowboy, which persisted throughout the time I knew him, is very likely the element that led the consulting psychiatrist to report "mixed strategies" in his thinking. In Dan's narrative it appears as a consistent part of his plan, though quite unrealistic given his physical condition. He describes the hoped-for cowboy career as lasting for a specific period of four years prior to becoming a medicine man, indicating by the use of the sacred number four that it would constitute a preparatory period. In addition to satisfying a personal wish, it would prove that he was physically competent, making money, and would perhaps help to establish a reputation that could be transferred to work as a medicine man.

Although Aberle (1982) points out that, in contrast to traditional Navajo ritual, it is virtually impossible to make a mistake in peyote prayer, from Dan's narrative it was evident that his speech in the peyote meeting provoked criticism. As he said, "Sometimes they get after you, like if you talk wrong or if you speak and it's wrong, or if your thinking is wrong. That's when they get after you, they tell you." Given his linguistic disability, his innovative or idiosyncratic message, and his proposal to become a medicine man via a rodeo cowboy career, the reaction to his speech was, not surprisingly, somewhat mixed. Dan acknowledged that some people accused him of being "wrong" and "off" in what he was saying, even interrupting him in violation of the ritual protocol of peyote meetings, to the point of making him cry. This reaction appears to have been mediated by the leader of the ceremony, however, who both accepted the legitimacy of what Dan said and acknowledged that the problem is "just the way you talk." Noting that Dan's father would be getting old and that the Navajo deities or Holy People appeared to be indicating that Dan could eventually replace him, the leader concluded, according to Dan's report, " 'Some day you are going to be a helpful person,' that's how they said it to me .... So from there I'll always be helpful."

Six months later, in another account of the same significant incident, Dan acknowledged making mistakes in what he was saying that upset people, and a certain arrogance in appearing to "force" his youthful inspired message on the others. He said that in trying to explain his thoughts he had to "change it back and then there's no hard feelings," but in his view, Dan eventually won over the elders. There appear to be three aspects to the validation of his claim.

First, in crediting his healing to spiritual help obtained in earlier peyote meetings, he claimed that the divine inspiration that was a consequence of the healing should be listened to by other peyotists. He reinforced this claim by stating that he could speak in this manner even though his "brain was cut out," and that others with similar problems do not recover their speech at all: The very fact that he could talk at all validated his words. Elsewhere in our interviews he stated that he never used to talk as he does now, evidently referring not only to the content of his speech but to the fact that before his illness he was rather taciturn.

Second, invoking a bodily criterion of validity, the other participants acknowledged that Dan would have become ill if his prayer had been incorrect. It is well known that peyote ingestion can cause severe vomiting. One interpretation is that the Holy People cause this suffering as a punishment for wrong thinking or speaking. Since Dan was unaffected by the peyote his words were finally understood as incurring divine approval.

Third, Dan argued that part of the reason for his words being misunderstood was that they were addressed to the contemporary situation and problems of the younger Navajos. While acknowledging that the older people can help troubled youth with peyote meetings, he emphasized that there are now many more people in the world and that things are different than they were 25 years ago. Old people know only the reservation, but younger ones have traveled and have even been to colleges, and so are being inspired with different kinds of prayer. According to Dan, the participants acknowledged that his words could help them better understand their own grandchildren. He claimed that some were moved to tears by his words, a reaction in conformity with the common occurrence of weeping in heartfelt response when one is moved by the sincerity of the speaker during peyote prayers (La Barre 1969:50; Aberle 1982:156).

The experience of language again played a remarkable and poignant role with respect to the theme that his message was intended for Navajo youth. Even though English was his primary language and by both his account and that of his mother his knowledge of Navajo was quite minimal, he attributed a great deal of meaning to the apparently permanent loss of linguistic facility in Navajo while having recovered the ability to speak in English:

> So I said "I guess that's what you call I'm learning to be a Navajo person, I guess," [pause] but then I said the reasons why is that maybe I am going to be the helper because there is thousands of people, I think, I think that's why I lost my Navajo talk, because there's thousands, thousands of people that went to school there [in] colleges, they don't understand right but then sometime they want help but then

they can't understand the Navajo so they get upset. They're new people, they just do the whole work for them in the Navajo way and they can't really stand up, they don't know what's being said so [they] kind of get upset about it so they don't know which way to go again.

He concluded that he lost his Navajo because the Holy People wanted him to address his message in English to the young people who wanted help from traditional Navajo ceremonies, but who became frustrated because they could not understand ceremonial proceedings. Said Dan, "maybe I can give them that."

That Dan attributed a great deal of his recovery to the divine help afforded by peyote is evident in the following passage from our second interview. The context of the passage is my query about his response to peyote during the meetings held for his benefit, a query intended to identify any experiential interaction between effects of the psychoactive substance and his neurological condition:

[The peyote] didn't really bother me. It just [pause] the way, what it was, it just brought me to where that I can use the mind to think. ... But if I didn't [pause] I would have nothing. See, that medicine is what [pause] they [the Holy People] give us, it is what we use. That's the ones that I eat it [pause] and it goes throughout my mind. It brings things to me [pause] for new types are the ones are sent [pause] and that's when it goes into me [pause] I start doing my talking. Before I never use to talk this way but now there's different word that all comes out [pause] especially when I have prayers [pause] I ask questions and then just before I go to sleep, sometimes I [pause] I have a prayer [pause] and I then I wonder that's why I always say "Heavenly Father, I wonder." I'm a wondering person [pause] as young as I am, for some reason I want to be a helper of people, young people my age, some of them may be older than me. So I want to be that type of person .... So then sometime I sleep [pause] and when I sleep here [in this house?] everything's close to me. A feeling comes into me, some big words, everything that I never thought about so when I get up in the morning I speak to my Mom and my Dad [pause] especially to my Dad [the peyote road man] more, and I ask him are those types of word, are they right or are they wrong? ... [I ask: Can you tell what some of those words are?] It's like [pause] I say, "Heavenly Father," I said, "At this time my life is going through a very hardship but the type of word maybe I have not spoken heavenly father maybe that's what holding me back. I want you to teach me heavenly father [pause] my life has [pause] no [sigh] happiness within me. But then from type [unclear] in me father, is the type that would is more specially the Navajo people [pause] to put them together, don't force them, easy, don't force them." See these [kinds of words] started comma out, but I never use to do, because people they sit [in peyote meeting] and they look [pause] they just start praying real low. Then they speak to me in the morning [at the end of the meeting] they say "You coming Dan," [pause] they say, "You never use to pray that way." I said, "That's what I've been going through and it makes me feel a lot happier a lot better … that my mind is actually coming."

The ability to think and speak "straight" are given great priority in Native American Church practice (Aberle 1982; La Barre 1969) and in general among Navajos (Witherspoon 1977), and Dan explicitly attributed his ability to do both to peyote as it "goes throughout my mind" and "brings things to me" so that "I start doing my talking." The brief sample prayer that Dan spoke for me also appears quite appropriate in the peyotist context. "Heavenly Father" is a common formula of address in Navajo peyote prayers (Aberle 1982:153).

Dan's notion that there is a "word maybe I have not spoken" that is "holding me back" is of particular interest. If it is true that in traditional Navajo religion there is a concern for absolute accuracy of utterance within a fixed liturgical canon, there is an equivalent concern in peyotism with the spontaneously creative utterance of the ritual *mot juste*. This takes on a particular significance in the case of a person afflicted by anomie, for whom rehabilitation and inspiration are synthesized, and the ability to utter exactly the right words signify both personal healing and the ability to help others. The salience of this synthesis is borne out in the very next statement, in which Dan prays that the Navajo people will be able to come to terms with the future gently and asks the divine "don't force them"—precisely the advice given to Dan by his physicians concerning rehabilitation.

Other statements in our interviews indicate that Dan's message was in conformity with traditional moral themes of Navajo peyotism, including concern for students away from the reservation, the importance of education in the contemporary world, relations with white people, and identification with the United States as well as with humanity as a whole (Aberle 1982:154, 156). At one point Dan gave a lengthy discourse about the traditional concept of the rising and falling life cycle as symbolized by the crescent-shaped peyote altar, lamenting how old people were neglected by contemporary youth and left in nursing homes. Another instance highlights global moral concerns:

this earth is very small and it's packing and packing throughout the world, and … to [avoid] having a war and have a big what do you call that, the Russian people … . For some reason it hit me and made me speed up [sigh] trying to learn and trying to start praying, praying the good way, to hold down [the war], so we don't want to go over there and start to fight again, have to fight. I said this is our land, white people that came over here … but since we are all mixed together, forcing us Indians to help, lot of them are going to get hurt, a lot of them aren't going to come home … . And then sure enough [pause] what happened with them .... They have a big blow out. In Russia. Poison. Some of it went all the way around. Already hit me—I already knew it. So then we had to ask the United States to send some people over there that are smart so that they can plug that thing up. That's what happened. So then they didn't want to argue any more they were just happy with each other ... I said that's just like the Navajo way, you know. They trying to, they are learning ahead of time, I guess that's what it is. They try to warn people ahead of time. And sometime they can't understand, you know. It gets too close. But then you always, you just mention, you can't just force them. You can just help them.

In this excerpt Dan recounts a prophetic experience that both encouraged his spiritual aspirations and impressed upon him the moral urgency of global concerns. The experience was of knowing about the Russian nuclear disaster at Chernobyl before it occurred. The broadening circle of danger that threatens the Navajos by virtue of their cooperation with the United States in any conflict with the (former) Soviets, and that likewise identifies American goodwill in offering friendly technical assistance to the Soviets with the "Navajo way" of harmony, is consonant with the "prayer circle" in which the prayers of Navajo peyotists generate "gradual spread of the blessings from the immediate group of those present to the whole world" (Aberle 1982:153).

This cosmological implication caps the existential analysis of language for Dan. Being a real Navajo person meant having the actuality of language as a mode of engagement in the world, having the project of becoming a medicine man was the rationale that grounded the return of language, and peyotist spirituality defined the moral horizon of his discourse as a global horizon. His struggle was not for *langue,* an abstract cognitive or representational ability, but for *parole,* the ability to produce coherent, socially and morally relevant utterance.

## Neurology and Cultural Phenomenology

Let us remind ourselves that despite Heidegger's rejection of the notion of the body as an "animal organism," the phenomenology of embodiment has engaged in theoretical and empirical dialogue with biology, specifically in Merleau-Ponty's (1962) analysis of patients with neurological lesions. To this point my exposition has been a hermeneutic of Dan's struggle for meaningful utterance, a hermeneutic that shows how he thematized and objectified his embodied experience of language into a life plan in conformity with cultural and religious meaning. We must now enter this analysis into dialogue with the considerable literature on experiential and behavioral consequences of temporal lobe lesions. Although this literature focuses almost entirely on patients with epilepsy rather than the relatively more rare brain tumor, the persistence of a secondary seizure disorder following resection of a left temporal-parietal astrocytoma warrants an examination of left temporal-lobe epilepsy for insight into several formal features of Dan's post-surgical experience.

Two broad groups of behavioral syndromes are of possible relevance. First are the so-called schizophrenia-like psychoses of epilepsy, which are episodic in character and clinically similar to atypical psychoses, and in which the majority of patients have either bilateral or left temporal-lobe involvement (Slater et al. 1963; Tucker et al. 1986). Second is the so-called interictal behavior syndrome of temporal-lobe epilepsy, which describes a complex of behaviors persisting through the period between explicit seizure or ictal activity and indicating enduring personality changes due to the illness (Waxman and Geschwind 1975). A critical feature observed in notable proportions of patients in both categories bears the clinical label "hyperreligiosity." The first documented sample of schizophrenia-like psychoses included 26 (of 69) patients who exhibited such religiosity, of whom only 8 were religious prior to the onset of their illness, and of whom 6 experienced profound and sometimes repeated episodes of religious conversion (Dewhurst and Beard 1970). In a more recent study of patients originally admitted for behavior disturbance and secondarily diagnosed with seizure disorders, 5 of 20 patients exhibited hyperreligiosity (Tucker et al. 1986). In the series of cases cited to define the interictal behavior syndrome, 6 of 9 patients exhibited some degree of religiosity (Waxman and Geschwind 1974, 1975).

It quite reasonably has been hypothesized that in spite of the overlap with respect to features such as hyperreligiosity, there is a degree of distinction among classic ictal phenomena proper, the episodic clusters of affective features that define the epileptic psychoses, and the nonremitting interictal behavior syndromes (Tucker et al. 1986). For Dan, it appears that actual seizure activity was largely controlled by medication, and there is no evidence justifying the label of epileptic psychosis. However, the following features of the interictal behavior syndrome pointed out by Stephen Waxman and Norman Geschwind suggest its relevance in Dan's case: "Deeply held ethical convictions and a profound sense of right and wrong ... interest in global issues such as national or international politics ... striking preoccupation with detail, especially as concerns moral or ethical issues ... preoccupation with detail and clarity [of thinking] and a profound sense of righteousness ... speech often appears circumstantial because of the tendency of these patients to digress along secondary and even tertiary themes ... deepening of emotional response in the presence of relatively preserved intellectual function" (Waxman and Geschwind 1975:1584).

A second set of authors describes these changes not as a behavioral syndrome but as a personality syndrome that includes hypermoralism, religious ideas, an unusually reflective cognitive style, tendency to label emotionally evocative stimuli in atypical ways, circumstantiality, humorlessness, difficulty producing the names of objects, and an obsessive, sober, and ponderous style probably related to an effort to compensate and make sense of a world rendered exotic by language dysfunction and confusion about temporal ordering of cause and effect relationships (Brandt, Seidman, and Kohl 1984). These are all features that we have encountered in some form in Dan's case, and their relevance is supported by the observation that they occur primarily in patients with left temporal-lobe epilepsy, suggesting their origin in disruption of left temporal-lobe mechanisms (Brandt, Seidman, and Kohl 1984).[3]

We must give special attention to disturbances of language in temporal-lobe epilepsy, since language plays a critical role both in Dan's frustration over his anomie and in his divinely inspired ability or compulsion to pray. Two features of the interictal behavior syndrome are relevant: hypergraphia or the tendency to write compulsively and often repetitively, and verbosity or loquaciousness, the tendency to be overly talkative, rambling, and circumstantial in speech. Perhaps due to a combination of illness-related cognitive deficit and the persistent trembling of a right-handed person with a left-hemisphere lesion, Dan claimed to be virtually unable to write at all, and unless one considers his dedication to word puzzles as a form of hypergraphia, this feature must be judged to be absent. On the other hand, verbosity is evident in both his interview transcripts and his ability to generate lengthy peyote prayers. Accordingly, it is consistent with our argument that while hypergraphia is typically a characteristic of right temporal-lobe epilepsy (Roberts et al. 1982), verbosity is associated with left temporal-lobe epilepsy (Mayeux et al. 1980; Hoeppner et al. 1987).

An important hypothesis for our purposes stems from the observation that left temporal-lobe epileptics perform worse than other epileptics on confrontation naming tests and related dimensions of verbal functioning (Mayeux et al. 1980). The researchers suggest a location of the lesion in the inferior left temporal-lobe, and propose that the integration of sensory data requisite to naming occurs in the parietal lobe. This would affirm the importance of the temporal–parietal site of Dan's tumor in his acknowledged anomie. However, the critical point of their argument is that loquaciousness is quite likely an adaptive strategy compensating for a patient's inability to find the right word, and neither a direct effect of the lesion nor an epileptic "personality trait" as is commonly assumed. Janis Jenkins (1991) has observed the cultural predisposition in American psychology and psychiatry to explain behavior by personality attribution, a predisposition which, in this case, might obscure the importance of intentionality and reconstitutive self process. I would suggest that this hypothesis is supported by a review of Dan's interviews, where much of the apparent ponderousness may be attributable to the frequent false starts and rephrasings that indicate a search for the appropriate word.

Dan's desire to become a medicine man is closely bound up with his search for the "word maybe I have not spoken heavenly father maybe that's what holding me back" and his goal of teaching and helping as a medicine man through the utterance of lengthy peyote prayers. Compare this personal scenario with the case described by Waxman and Geschwind of a 34-year-old white man who had undergone several craniotomies for a left temporal-parietal abscess, precisely the location of Dan's lesion: "Three years after surgery ... his hospital records noted that 'he now expresses interest in religion and possibly in becoming a minister, and hopes to increase his "use of good words" towards that end ....'" He subsequently at his own suggestion delivered several sermons at his church. He spontaneously delivered typed copies of the texts of these sermons to his physician. The sermons concerned highly moral issues, which were dealt with in "highly circumstantial and meticulous detail" (Waxman and Geschwind 1974:633). Dan's anomia and this patient's aphasia—or, more directly to our point, the shared features of their recoveries—could certainly be objectified in neurological terms, but the inability to find a word bears a preobjective existential significance that also becomes objectified or thematized primarily in religious terms. Stated somewhat differently, the remarkable similarity of these cases is very likely due to the similar site of the neurological lesion, but while the lesion comfortably accounts for the linguistic disorder, it does not adequately account for elaboration of the moral significance of finding the right word. This leads directly back to bodily existence and "the social that we carry around with us prior to any objectification." In Dan's case we can discern this existential ground in two ways: with respect to bodily schemes that organize interpretations of etiology, prodromal experience, and post-surgical sequelae; and with respect to the existential status of language itself.

Dan's understanding of how his cancer occurred followed the traditional Navajo etiological schema[4] of contamination by lightning (See Chapter Seven). Two points must be made in order to understand the preobjective cultural reality of this attribution. First, for Navajos lightning is not a simple natural phenomenon, but includes all kinds of radiant energy such as radioactivity and emanations from microwave ovens. Dan cited a boyhood exposure to natural lightning while herding sheep as the origin of his tumor, but also cited the adult exposure to the flames and fumes of his welding torch as a secondary exposure to lightning that precipitated the onset of his current illness. Second, lightning is not an abstract spiritual phenomenon that acts by mystical contagion devoid of cultural phenomenology. The prototype of lightning contamination is direct physical encounter in which lightning strikes so near that the person sees a bluish flash and inhales acrid electrical fumes—*ił hodiiłłiizh*—(compare Young and Morgan 1987:453). The preobjective grounding of this contamination schema in bodily experience accounts for the assimilation of natural lightning and the lightning produced by the welder's torch.

The schema that preobjectively ordered Dan's experience of the prodromal period was that of the ritual importance of "fourness" among Navajos, as in four cardinal directions, four sacred mountains, the performance of ceremonies in groups of four, and so on. Highly salient in Dan's personal synthesis are three discrete events that occurred within several months of each other and culminated in his hospitalization and diagnosis. The first occurred on a trip with some co-workers to a large off-reservation city, when he fell unconscious from the second-story balcony of a motel. The second occurred while he was alone at home with his father, when Dan started shaking uncontrollably while sitting on the couch preparing to read a document that his father could not understand. The final episode occurred while Dan was at lunch with co-workers, when he collapsed in a restaurant and was taken to the hospital. For him the possibility of a life-culminating fourth seizure was deeply implicated in his daily life and in his plan to become a medicine man. To parphrase Merleau-Ponty, the social that Dan carried about with him as a feature of his embodied existence was the profound polysemy of the sacred number "four."

Dan did in fact experience something close to this fourth seizure in the post-surgical period of struggle to regain language and expressivity. He was sitting in his room listening to the radio when it began to emit the high-pitched test tone of the Emergency Broadcast System. Dan reported, "I had a feeling that something went through me, I was just in here going like this [shaking]." The cultural basis for this preobjective experience is the schema of a foreign object shot through one's body by a witch. This schema was explicitly elaborated by a medicine man who determined that there was a sharp fragment of bone inserted or shot into Dan's brain by a witch, and who removed it by appropriate ritual means. Dan felt that the bone had been inserted into his head so that he would "not to be ever able to have that kind of thought, not to be able to speak again."

The existential status of language for Navajos was captured by Reichard (1944), who called attention to their self-consciousness with respect to words and word combinations, and subtitled her description of Navajo prayer "The Compulsive Word." Reichard's phrase has a double meaning in that prayer compels the deities to respond, and that efficacy requires a compulsive attention to correctness of utterance. Dan's case adds a third sense for it is he who is compelled to speak by the Holy People. The implication is that because they inspired the words the deities will be more readily compelled by them, but with the paradox that the speaker must struggle to give them coherent and lucid form. This layering of compulsion upon compulsion conforms to a schema that undergirds all of Navajo thought about healing: "According to Navajo belief, that which harms a person is the only thing that can undo the harm. The evil is therefore invoked and brought under control by ritualistic compulsion. Because control has been exerted and the evil has yielded to compulsion it has become good for the person in whose behalf it has been compelled. In this way evil becomes good, but the change is calculated on the basis of specific results" (Reichard 1944:5). For Dan the evil of illness is transformed into the medium by which he can become a "helping person," that is, language that is at once compulsive and compelling.

Dan's experience cannot be understood outside the intimate and efficacious performative nexus among knowledge, thought, and speech. Reichard notes that unspoken thought and spoken word share the same compulsive potential (1944:9–19), and Witherspoon (1977) has shown that for Navajos, thought is the "inner form" of language, and knowledge the "inner form" of thought. All three are powerful, in that all knowledge is inherently the power to transform or restore, all thoughts have the form of self-fulfilling prophecies, and all utterance exerts a constitutive influence on the shape of reality: "This world was transformed from knowledge, organized in thought, patterned in language, and realized in speech. . . . In the Navajo view of the world, language is not a mirror of reality; reality is a mirror of language" (1977:34). In this light, the gift of speech, experienced in the modality of divine "otherness" as a spontaneous gift from the Holy People, is coterminous with the intuition of spiritual knowledge that should be used to help others. Also coterminous are Dan's struggles for a postoperative return to fluency and for the ability to utter a spiritually relevant message. Because of the existential status accorded to speech, the fact *that* Dan was trying to speak bore implications for *what* he was trying to speak.

With this glimpse at the habitus within which Dan's struggle for language took place we can return to an issue raised by a clinical study, this one of non-Navajo verbose epileptic patients with left foci and complex partial seizures. This group of patients was relatively older than others studied by the same researchers, and showed no notable anomie and no poorer performance on verbal tasks than other epileptic patients. However, their performance on a story elicitation task was characterized by "nonessential and at times peculiar details" (Hoeppner et al. 1987). Of interest to us are the three alternative hypotheses offered by the researchers to account for this verbosity: Patients are either driven internally to talk, notice things not noticed by others, or give particular significance to things others see as irrelevant. For the authors these hypotheses indicate a relation between verbosity and hallucination. In particular, they suggest that there may be a continuum between noticing or commenting on details trivial to others and hallucination. Reasoning from this idea, verbosity may be a "subclinical manifestation of a complex phenomenon, involving speech, perception, and affect, which in more severe form appears as psychopathology" (Hoeppner et al. 1987:40).

Dan explicitly acknowledged being internally driven to talk, with refusal to do so resulting in a headache. We can infer from his narrative that at first his experience of hearing the Holy People was independent of the words which he then repeated in order to verify their validity. It further appears that over time the experience became a direct non-auditory inspiration to talk. From the initial account of hearing words shortly after his surgery, his description of the inspiration changes to having the words given to him or put into him, and eventually to simply having the peyote spirit enter him and that's when "I start doing my talking." Thus beginning with what are probably best described as auditory hallucinations, which are clinically more common than visual ones in temporal-lobe epilepsy, it appears that there was a phenomenological fusion of what Dan heard with what Dan said. For Dan, then, the adoption of the goal to pray as a medicine man served as a self-orientation that allowed for the domestication of auditory hallucination into intentional (if inspired) utterance, for a devout attention to details ignored by others, and for a search for the deeper significance of the apparently irrelevant. Here we can go no further, for we are at the very frontier of the cultural phenomenology of language.

The question of hallucination leads directly, however, to the neurological and psychotropic effects of peyote ingestion. One might assume that the development of "verbose" prayer in peyote meetings was stimulated by a drug whose neurological effects parallel those of temporal-lobe anomalies, but this would be a reductionistic slight to firmly established canons of American Indian oratory that have no discernible link to specific brain structures. Nevertheless La Barre (1969:139–143), in his careful study of the peyote cult, lists effects of the cactus including talkativeness, rambling, tremulousness, rapid flow of ideas, shifts in attention with maintenance of clarity, incoordination of written language. Auditory hallucinations stimulated by peyote primarily modify the quality of sounds, although La Barre would perhaps accept the possibility of voice audition such as Dan experienced. Although peyote is known to produce convulsions in animals (one of its constituent alkaloids has properties similar to strychnine), it is not known to produce seizures in humans, and localization of its action remains unspecified. The important phenomenological work of Kluver (1966) suggests the involvement of the vestibular system and that the effects of parieto-occipital lesions exhibit some formal similarities with those of peyote, but the emphasis of his study was on visual effects, none of which play a role in the case under consideration.

The single extant study on mescaline administered to epileptic patients showed that the principal effect on 8 of 12 was drowsiness, lethargy, and apathy or sleep, concordant with the somewhat narcotic properties of that constituent alkaloid of peyote when taken by itself (Denber 1955). A more significant finding of this study was that, although there was no correlation between brain wave (EEG) and clinical results, there was a decrease or disappearance of delta waves, and a disappearance of spike-wave patterns and slow waves lasting for hours. The author observes that problem-solving activities also cause a disappearance of spike-wave patterns, but only for several minutes. Here we might suggest that as a form of problem-solving, Dan's word puzzle activity may have had some effect in suppressing seizure activity, conceivably reinforcing the effects of peyote. In addition, to the extent that inspired prayer became part of Dan's intentional synthesis of a life project it could be understood both as a kind of moral problem-solving activity, and as an action that reinvoked the peyote experience in the manner typically referred to as a psychedelic "flashback." In this light, the boundaries between the effects of neurological lesion and cultural meaning dissolve, revealing in the space between a glimpse of the preobjective indeterminacy of lived, embodied experience.[5]

Finally, brief mention is made of peyote in Levy et al.'s (1987) study of seizure disorders among the Navajo. Interestingly, one patient was dropped from the study because her family was peyotist and had not diagnosed her, implying either that Navajo peyotists do not recognize epilepsy as a disorder or regard it as a problem to be treated exclusively with peyote medicine. In fact, the authors noted that "some participants in peyote meetings would hold one arm to keep it from shaking. This was taken as a sign that some malign influence was at work and that the individual in question needed special prayers" (Levy et al. 1987:104).[6] I do not know whether Dan's arm trembled during his peyote meetings, but the general persistence of such tremors in his right arm may be closely connected to the medicine man's diagnosis of malign influence in the form of witchcraft.

## Meaning and Lesion in Bodily Existence

The connection between epilepsy and religion has long been a *bête noir* for anthropology. For decades it was assumed that shamans and religious healers were either epileptic or schizophrenic, and the struggle to normalize them in the literature was a struggle of cultural relativism against biological reductionism, against the pathologization of religious experience, and on behalf of sensibility and meaning where there appeared to be only bizarre and irrational behavior. Anthropology fought against the logical error that the religiosity of the epileptic necessarily implied epilepsy in the religious. Nevertheless, as late as 1970, Kenneth Dewhurst and A. W. Beard wrote an article describing the cases of several temporal-lobe epileptics who had undergone religious conversion, and juxtaposed them to a series of "possible epileptic mystics" from the Western tradition, including, among others, St. Paul, Theresa of Avila, Theresa of Lisieux, and Joseph Smith. One might suspect the presence of a rhetorical agenda in a study of religion that begins with the religiosity of explicit medical patients. Such an agenda is indeed evident in the following passage: "Leuba's (1896) so-called psychological analysis of the mental state immediately preceding conversion includes such theological categories as a 'sense of sin and self-surrender,' there is no mention of temporal lobe epilepsy" (Dewhurst and Beard 1970:504). The authors allow no analytic space between the theological and neurological, despite William James's (1961 [1902]) extensive use of Leuba's work that certainly acknowledged it as "psychological." Dewhurst and Beard (1970) clearly disdain the study of a "mental state" in favor of determination of a "neurological state." Yet even they acknowledge that the eminent neurologist Hughlings Jackson did not attribute religious conversion directly to epileptic discharge, but argued that it was facilitated by the alteration in level of consciousness, the increased excitability of lower nervous centers, and religious background.

A more recent series of papers by Persinger (1983, 1984a, 1984b; Persinger and Vaillant 1985) attempt to validate a more subtle hypothesis about the relation between religion and temporal-lobe function. He argues that "religious and mystical experiences are *normal* consequences of spontaneous biogenic stimulation of temporal lobe structures"

(Persinger 1983:1255, emphasis in original). Thus religious, paranormal, and ictal phenomena are distinct, except that they all originate in electrical activity of the temporal lobe. The hypothesis is predicated on the post-stimulation instability of temporal-lobe structures, susceptibility to transient vasospasms, and the tendency of cellular membranes to fuse. The latter contributes to unusual mixtures of cell ensembles especially sensitive to a variety of factors including tumorigenesis. Persinger proposes a construct of Temporal Lobe Transient (TLT), defined as learned electrical microseizures provoked by precipitating stimuli and followed by anxiety reduction, and his work attempts to demonstrate that persons who report religious and paranormal experiences also report signs of temporal-lobe activity. Whether or not one is convinced by these results, it is important to note that Persinger equivocates between regarding religious experience as normal and as an evolutionary liability. Even more salient is his equivocation between neurological reductionism (biogenic origin) and statements that people are predisposed to TLTs by cultural practices or "racial-cultural" interactions, and that "metaphorical language is the most profuse precursor to TLTs" (Persinger 1983:1260).

What is absent from such accounts is the analysis of the embodied, speaking person taking up an existential position in the world. Without this we risk a battle of causal arrows flying in both directions from neurology and culture, with no analytic space between. By the same token, eliminating the preoccupation with causality is part of phenomenology's radicality in opening up the space of being-in-the-world to existential analysis. Following this line, our argument in the case of Dan suggests that it may make sense to consider verbosity not as a personality trait grounded in neuroanatomical change but as an adaptive strategy that spontaneously emerges from a preobjective bodily synthesis. Likewise, it suggests that the patient's search for words is thematized as religious not because religious experience is reducible to a neurological discharge but because it is a strategy of the self in need of a powerful idiom for orientation in the world.

Here we rejoin Heidegger and Merleau-Ponty; one condemning the error of biologism that consists in merely adding a mind or soul to the human body considered as an animal organism, the other condemning the error of treating the social as an object instead of recognizing that our bodies carry the social about inseparably with us before any objectification. The errors are symmetrical: In one instance, biology is treated as objective (the biological substrate), in the other the social is treated as objective (the domain of social facts), and in both instances the body is diminished and the preobjective bodily synthesis is missed. For biologism the body is the mute, objective biological substrate upon which meaning is superimposed. For sociologism the body is a blank slate upon which meaning is inscribed, a physical token to be moved about in a pre-structured symbolic environment, or the raw material from which natural symbols can be generated for social discourse.

Approaching the same problem from feminist theory, Haraway (1991) argues that the body is not an objective entity because biology itself is situationally determined: "the 'body' is an agent, not a resource. Difference is theorized *biologically* as situational, not intrinsic, at every level from gene to foraging pattern, thereby fundamentally changing the biological politics of the body" (Haraway 1991:200). With the recognition that difference is not merely a cultural overlay on a biological substrate, our argument goes beyond the pedestrian assertion that culture and biology mutually determine the experience of illness, and toward a description of the phenomenological ground of both biology and culture. The struggle for correct expressive utterance has a global and corporal existential significance, whether the ones who struggle are Indian or Anglo, and whether or not they are neurologically afflicted.[7] Only late in the process of analytic objectification can we say that the features of an interictal behavioral syndrome are formally dependent on the nature of the neurological lesion, and that the options to become a medicine man or minister, to speak in the genres of peyote prayer or sermon, and to search for the correct word that will bring healing for oneself and help to humankind are formally dependent on culture. The problems posed by Heidegger and Merleau-Ponty thus point us not toward the end but toward the starting point of analysis. That starting point is a cultural phenomenology of embodied experience that allows us to question the difference between biology and culture, thereby transforming our understanding of both.

# PART III

## Modulations of Embodiment

# CHAPTER NINE

# Somatic Modes of Attention

Embodiment as a paradigm or methodological orientation requires that the body be understood as the existential ground of culture—not as an object that is "good to think," but as a subject that is "necessary to be." To argue by analogy, a phenomenological paradigm of embodiment can be offered as an equivalent, and complement, to the semiotic paradigm of culture as text. Much as Roland Barthes (1986) draws a distinction between the work and the text, a distinction can be drawn between the body and embodiment. For Barthes, the work is a fragment of substance, the material object that occupies the space of a bookstore or a library shelf. The text, in contrast, is an indeterminate methodological field that exists only when caught up in a discourse, and that is experienced only as activity and production (1986:57–68). In parallel fashion, the body is a biological, material entity, while embodiment can be understood as an indeterminate methodological field defined by perceptual experience and the mode of presence and engagement in the world. As applied to anthropology, the model of the text means that cultures can be understood, for purposes of internal and comparative analysis, to have properties similar to texts (Ricoeur 1979). In contrast, the paradigm of embodiment means not that cultures have the same structure as bodily experience, but that embodied experience is the starting point for analyzing human participation in a cultural world.

To best understand the theoretical origin of this problematic, it is useful to distinguish between what has come to be called the anthropology of the body and a strand of phenomenology explicitly concerned with embodiment. Although glimpses of the body have appeared regularly throughout the history of ethnography (for example, Leenhardt 1979), an anthropology of the body was inaugurated by Douglas (1973) and elaborated in the collections by Benthall and Polhemus (1975) and Blacking (1977). The historical work of Foucault (1973, 1977) provided new impetus, evident in the works of Scheper-Hughes and Lock (1987), Martin (1987), and likeminded sociologist Bryan Turner (1984). The work of Bourdieu (1977, 1984) shifted an earlier focus on the body as the source of symbolism or means of expression to an awareness of the body as the locus of social practice. This is powerfully evident in Comaroff's (1985) work, which exhibits a theoretical movement from the social body of representation to the socially informed body of practice, while still emphasizing the traditional focus on body symbolism.

Meanwhile, an opening for phenomenology in anthropological theory has come with the possibility of articulating a concept of experience around the edges of the monolithic textualist and representationalist paradigm dominated by Levi-Strauss, Derrida, and Foucault. Geertz's (1973) concern with culture as text was complemented by interest in the phenomenology of Alfred Schutz, and with the distinction between experience-near and experience-far concepts. It has finally become legitimate for Wikan (1991) to tackle the problem of an experience-near anthropology, for Turner and Bruner (1986) to espouse an "anthropology of experience," and for Joan and Arthur Kleinman (1991) to declare an "ethnography of experience," approaches that are more or less explicitly phenomenological. Among such approaches, a few scholars—influenced especially by Merleau-Ponty (1962, 1964) and occasionally by thinkers such as Marcel, Scheler, Straus, and Schilder—have highlighted a phenomenology of the body that recognizes embodiment as the existential condition in which culture and self are grounded (Corin 1990; Csordas 1990; Devisch and Gailly 1985; Frank 1986; Jackson 1989; Munn 1986; Ots 1991, 1994; Pandolfi 1990). They tend to take the "lived body" as a methodological starting point rather than consider the body as an object of study.

From the second of these two perspectives, the contrast between embodiment and textuality comes into focus across the various topics examined by an anthropology of the body. For example, the influential synthesis by Scheper-Hughes and Lock (1987) clearly lays out the analytical terrain claimed by an anthropology of the body. These authors rework Douglas's (1973) "two bodies" into three—the individual body, the social body, and the body politic. They understand these bodies as interrelated analytic domains mediated by emotion. To pose the problem of the body in terms of the relation between embodiment and textuality invites us to review this field with an eye to the corresponding methodological tension between phenomenological and semiotic approaches. This methodological tension traverses all three bodies sketched by Scheper-Hughes and Lock. That is, each of the three can be understood either from the semiotic/textual standpoint of the body as representation or from the phenomenological/embodiment standpoint of the body as being-in-the-world.

However, the contemporary anthropological and interdisciplinary literature remains unbalanced in this respect. A strong representationalist bias is evident most notably in the predominance of Foucauldian textual metaphors, such as that social reality is "inscribed in the body," and that our analyses are forms of "reading the body." Even Jackson's (1989) predominantly phenomenological formulation is cast in terms of the body as a function of knowledge and thought, two terms with strong representationalist connotation.Yet Jackson was perhaps the first to point out the shortcomings of representationalism in the anthropology of the body, arguing that the "subjugation of the bodily to the semantic is empirically untenable" (1989:122). I would endorse the critique that meaning cannot be reduced to a sign, a strategy that reinforces a Cartesian pre-eminence of mind over a body understood as "inert, passive, and static" (1989:124). This critique should not be construed as negating the study of signs with respect to the body, but as making a place for a complementary appreciation of embodiment and being-in-the-world alongside textuality and representation. That these are complementary and not mutually exclusive standpoints is demonstrated in the rapprochement between semiotics and phenomenology in several recent works on the body (Csordas 1994a and 1994b; Good 1992; Hanks 1990; Munn 1986; Ots 1991). Nevertheless, because for anthropology embodiment is not yet developed enough to be truly complementary to an already mature textuality (Hanks 1989), this article has the limited aim of taking a measured step toward filling out embodiment as a methodological field.

Reconsidering the work of Merleau-Ponty (1962, 1964b) and Bourdieu (1977, 1984) suggests bringing into the foreground the notions of perception and practice. Briefly, whereas studies of perception in anthropology and psychology are, in effect, studies of perceptual categories and classifications, Merleau-Ponty focused on the constitution of perceptual objects. For Merleau-Ponty, perception began in the body and, through reflective thinking, ends in objects. On the level of perception there is not yet a subject–object distinction—we are simply in the world. Merleau-Ponty proposed that analysis begin with the preobjective act of perception rather than with already constituted objects. He recognized that perception was always embedded in a cultural world, such that the preobjective in no way implies a "pre-cultural." At the same time, he acknowledged that his own work did not elaborate the steps between perception and explicit cultural and historical analysis (Merleau-Ponty 1964:25).

Precisely at this point where Merleau-Ponty left off, it is valuable to reintroduce Bourdieu's (1977, 1984) emphasis on the socially informed body as the ground of collective life. Bourdieu's concern with the body, worked out in the empirical domain of *practice,* is parallel and compatible with Merleau-Ponty's analysis in the domain of *perception.* To conjoin Bourdieu's understanding of the "habitus" as an unself-conscious orchestration of practices with Merleau-Ponty's notion of the "preobjective" suggests that embodiment need not be restricted to the personal or dyadic micro-analysis customarily associated with phenomenology, but is relevant as well to social collectivities.

Defining the dialectic between perceptual consciousness and collective practice is one way to elaborate embodiment as a methodological field (see Chapter Seven). It is within this dialectic that we move from the understanding of perception as a bodily process to a notion of somatic modes of attention that can be identified in a variety of cultural practices. Our elaboration of this construct will provide the grounds for a reflection on the essential ambiguity of our own analytic concepts as well as on the conceptual status of "indeterminacy" in the paradigm of embodiment and in contemporary ethnography.

## A Working Definition

Alfred Schutz, the premier methodologist of phenomenological social science, understood attention to lie in the "full alertness and the sharpness of apperception connected with consciously turning toward an object, combined with further considerations and anticipations of its characteristics and uses" (1970:316). Merleau-Ponty goes further, pointing out that attention actually brings the object into being for perceptual consciousness: "To pay attention is not merely further to elucidate pre-existing data, it is to bring about a new articulation of them by taking them as *figures*. They are performed only as *horizon,* they constitute in reality new regions in the total world....Thus attention is neither an association of images, nor the return to itself of thought already in control of its objects, but the active constitution of a new object which makes explicit and articulate what was until then presented as no more than an indeterminate horizon" (1962:30). What is the role of attention in the constitution of subjectivity and intersubjectivity as bodily phenomena? If, as Schutz says, attention is a conscious turning toward an object, this "turning toward" would seem to imply more bodily and multisensory engagement than we usually allow for in psychological definitions of attention. If, as Merleau-Ponty says, attention constitutes objects out of an indeterminate horizon, the experience of our own bodies and those of others must lie somewhere along that horizon. I suggest that where it lies is precisely at the existentially ambiguous point at which the act of constitution and the object that is constituted meet—the phenomenological "horizon" itself. If that is so, then processes in which we attend to and objectify our bodies should hold a particular interest. These are the processes to which we allude with the term *somatic modes of attention.* Somatic modes of attention are culturally elaborated ways of attending to and with one's body in surroundings that include the embodied presence of others.

Because attention implies both sensory engagement and an object, we must emphasize that our working definition refers both to attending "with" and attending "to" the body. To a certain extent it *must* be both. To attend to a bodily sensation is not to attend to the body as an isolated object, but to attend to the body's situation in the world. The sensation engages something in the world because the body is "always already in the world." Attention *to* a bodily sensation can thus become a mode of attending to the intersubjective milieu that give rise to that sensation. Thus, one is paying attention with one's body. Attending with one's eyes is really part of this same phenomenon, but we less often conceptualize visual attention as a "turning toward" than as a disembodied, beam-like "gaze." We tend to think of it as a cognitive function rather than as a bodily engagement. A notion of somatic mode of attention broadens the field in which we can look for phenomena of perception and attention, and suggests that attending to one's body can tell us something about the world and others who surround us.

Because we are not isolated subjectivities trapped within our bodies, but share an intersubjective milieu with others, we must also specify that a somatic mode of attention means not only attention to and with one's own body, but includes attention to the bodies of others. Our concern is the cultural elaboration of sensory engagement, not preoccupation with one's own body as an isolated phenomenon. Thus, we must include, for example, the cultural elaboration of an erotic sensibility that accompanies attention to attractiveness and the elaboration of interactive, moral, and aesthetic sensibilities surrounding attention to "fatness." These examples of attention to the *bodily form* of others also include attending with one's own body—there is certainly a visceral element of erotic attention, and there can be a visceral component to attending to other aspects of others' bodily forms. Attending to others' *bodily movements* is even more clear-cut in cases of dancing, making love, playing team sports, and in the uncanny sense of a presence over one's shoulder. In all of these, there is a somatic mode of attention to the position and movement of others' bodies.

It is a truism that, although our bodies are always present, we do not always attend to and with them. Let me reiterate, however, that the construct I am trying to elucidate includes culturally elaborated attention to and with the body in the immediacy of an intersubjective milieu. Although there is undoubtedly a cultural component in any act of attention to one's own or another's body, it would be too imprecise to label any such act as an example of a somatic mode of attention. If you cut your finger while slicing bread, you'll attend to your finger in a way that is more or less culturally determined (Is it spiritually dangerous? Is it embarrassing? Must I see a doctor?) When you notice someone who weighs 275 pounds, your reaction is also culturally determined (that person looks fat, attractive, strong, ugly, friendly, nurturing). To define somatic modes of attention in such broad terms would probably only save to organize a variety of existing literatures into an over-broad category. I suspect, for example, that we could identify such loosely defined somatic modes of attention associated with a wide variety of cultural practices and phenomena. Marcel Mauss (1950) pointed out that there is what we are calling a somatic mode of attention associated with the acquisition of any technique of the body, but that this mode of attention recedes into the horizon once the technique is mastered. The imaginal rehearsal of bodily movements by athletes is a highly elaborated somatic mode of attention, as is the heightened sensitivity to muscle tone and the appetite for motion associated with health-consciousness and habitual exercise. The sense of somatic contingency and transcendence associated with meditation and mystic states would also be within our purview. There are certainly somatic modes of attention to basic bodily processes, such as pregnancy and menopause, in different cultures. On the pathological side the hyper-vigilance associated with hypochondria and somatization disorder, and the various degrees of vanity or tolerance for self-mortification associated with anorexia and bulimia, could be said to define particular somatic modes of attention.

It is evident that some of these examples suggest more or less spontaneous cultural elaboration, whereas others suggest modes that are consciously cultivated (compare Shapiro 1985). Some emphasize attending to the body and some with the body; some emphasize attending to one's own body, some attending to others' bodies, and some to others' attention to our bodies. My point is that the ways we attend to and with our bodies, and even the possibility of attending, are neither arbitrary nor biologically determined, but are culturally constituted. Leenhardt's (1979) classic study of the Canaques of New Caledonia described not only a way of conceptualizing the body radically distinct from our own, but the exclusion of the body per se as an object of consciousness until the people were introduced by missionaries to the objectified body of Christian culture. This suggests that neither attending to nor attending with the body can be taken for granted, but must be formulated as culturally constituted somatic modes of attention. I elucidate this construct with examples from the ethnographic record in the following discussion.

## Somatic Attention and Revelatory Phenomena

The somatic mode of attention I will delineate in this section is that of healers who learn about the problems and emotional states of their clients through bodily experiences thought to parallel those of the afflicted. I describe the phenomenon for both predominantly Anglo-American, middle-class Catholic Charismatic healers and for Puerto Rican spiritist mediums.

A variety of somatic experiences is cultivated in Charismatic ritual healing practice, but I shall focus on two types of experience reported by healers during their interaction with supplicants. One is called "anointing," the second, "word of knowledge." Although the physical act of anointing part of the body, typically the forehead or hands, with holy oil is a common form of blessing among Charismatics engaged in healing practice, a different use of the term is of interest in the present context. A healer who reports an "anointing" by God refers to a somatic experience that is taken to indicate either the general activation of divine power, or the specific healing of an individual. A conventional anthropology of ritual healing would say simply that the healer goes into trance, assuming trance to be a unitary variable or a kind of black box factored into the ritual equation, and perhaps assuming that somatic manifestations are epiphenomena of trance. The analysis would go no further than informants' reports that these epiphenomena "function" as confirmations of divine power and healing. Within the paradigm of embodiment, in contrast, we are interested in a phenomenology that will lead to conclusions both about the cultural patterning of bodily experience, and also about the intersubjective constitution of meaning through that experience.

The anointing is described by some healers as a general feeling of heaviness, or as a feeling of lightness almost to the point of levitation. The healer may experience tingling, heat, or an outflow of "power" similar to an electrical currant, often in the hands, but at times in other parts of the body. The hands of some healers visibly tremble, and I have felt this vibration as a healer laid a hand on my shoulder. Among healers themselves, however, the "authenticity" of this visible vibration as a manifestation of divine power is sometimes questioned, in the sense that the anointing may be feigned or sensationalized. In a large group healing service, when the healer moves from individual to individual, laying hands on each, the strength of the anointing may vary with each supplicant. One healer described an emotional complement of the anointing as a feeling of empathy, sympathy, and compassion. If this feeling were absent as he came to a particular person in line for his prayer, he might pass over that person, assuming that God did not plan to heal her at that moment.

The second Catholic Charismatic phenomenon in this somatic mode of attention is the "word of knowledge." It is understood as a "spiritual gift" from God by means of which healers come to know facts about supplicants through direct inspiration, without being told by the afflicted person or anyone else. The word of knowledge is sometimes experienced as an indeterminate "sense" that something is the case, but very often occurs in specific sensory modalities. The healer may see an afflicted body part in the "mind's eye" or hear the name of a body part or disease with "the heart." One healer distinguished clearly that when the problem is internal, she typically "sees" the organ, or cancer, appearing as a black mass, but when the problem is external, she typically "hears" the word naming the illness or the body part, such as arms and legs. Another healer reported that a snapping in his ear means someone in the assembly is undergoing an ear healing, and that intense pain in his heart means a heart healing. Still another reported heat in her elbow on one occasion, interpreting this as a sign of healing of an injury or arthritis. Some healers report being able to detect headache or backache in a supplicant through the experience of similar pain during the healing process.

Queasiness or confused agitation may indicate the activity of evil spirits, and an unexpected sneeze or a yawn may indicate that a spirit is passing out of the supplicant through the healer. One healer commonly reported an experience of "pain backup" from persons filled with resentment or previously engaged in occult activities. The pain would enter her arm as she laid hands on the person. It would be necessary to remove her arm and "shake out" the pain, while the supplicant would feel nothing. With one hand on the supplicant's chest and the other on his or her back, she claims to feel what's going on inside the person. For example, she can tell if the person is in bondage to Satan, and she gets an unspecified sensation as the person is set free. The odor of burning sulphur or of something rotting also indicates the presence of evil spirits, while the aroma of flowers indicates the presence of God or the Virgin Mary.

The most comprehensive phenomenological report was given by a healer who distinguished three components of word of knowledge. First was the sense of certainty that what he would say was actually happening. Second was a series of words that would come to him in abbreviated sequence, such as "heart ... of a lady N years old ... seated in the last pew ... ." He would call these words out to the assembly, much as one would read from a teleprompter, except that he heard rather than read them. Finally, at the same time he would feel a finger pressing softly on the part of his body corresponding to the afflicted part of the person being healed.

I will now turn to what I take to be essentially the same somatic mode of attention in a different healing tradition, Puerto Rican *espiritismo* (Harwood 1977). Two main cultural differences distinguish somatic attention in *espiritismo* and Charismatic healing. First, whereas for Catholic Charismatic anointings are direct experiences of divine power and words of knowledge are divinely empowered direct experiences of the supplicant's distress, for *espiritistas,* the corresponding experiences are the work of spirits that enter or possess the healer. These are either good guiding spirits, called "guias," or bad, distress-causing spirits, called "causas." The spirits dominate the healing process in that they are essential not only to diagnosis but also to treatment; and hence, the somatic experiences attended to are even more prominent than among Catholic Charismatic. Specific spirits may have distinct and recognizable voices, odor, or impact on the healer's body. However, the spirits themselves are more often seen and heard among spiritists than among Charismatic, and spiritist healers can distinguish between good *guias* and bad *causas.*

The second important cultural difference is with respect to conceptions of the body that go well beyond ritual healing. The ability to see spirits from in back of the eyes (*ojo oculto*) may be associated with the interpersonal salience of the eyes and the glance also found in the evil eye (*ojo malo*). The experience of a spirit entering through the stomach may be associated with the cultural emphasis on that organ not only as a seat of emotion, but also as an expressive organ with its own mouth (*boca del estomago*). The experience of spirits *asfluidos* coursing through the body may be associated with a humoral conception of how the body works. Although I would not rule out any of these experiences for Anglo-American Charismatics, it is doubtful that they would be cultivated within their somatic mode of attention.

Despite these differences, the experiences reported by the two types of healer are notably similar, although *espiritista* categories describing these experiences are even more explicit in distinguishing sensory modalities than the Charismatic anointings and words of knowledge. Based on writings of, and discussions with, leading researchers on *espiritismo* (Koss, Harwood, and Garrison), the phenomena appear to fall into four categories: seeing the spirits (*videncias*), hearing the spirits speak (*audiciones*), sensing immediately what is on the client's mind (*inspiraciones*), and feeling the pain and distress caused in the client by spirits (*plasmaciones*).

Most of the differences lie in visual experiences, since Charismatics typically see situations or images of problems, rather than problems objectified
as spirits. Perhaps most similar are the proprioceptive experiences, or *plasmaciones.* Koss (n.d.) cites use of the verb *plasmar* to refer to mediums' mold-
ing or forming clients' pain or emotional distress within their own bodies.
Harwood (personal communication) adds that *plasmaciones* are transmitted
though the medium of *plasma,* which in spiritist doctrine is a spiritual substance linking persons to spirits and to one another.

According to Harwood, the *plasmaciones* experienced by healers might
include pain, tingling, vibration, or a feeling of oration if possessed by a *guia*
spirit. Although Garrison (personal communication) does not recognize the
term *plasmaciones,* she acknowledges *sensations* that might include headache,
stomachache, or tension picked up from the client. Koss (n.d., 1992) pres-
ents the most elaborate inventory, including feeling of electrical charge,
accelerated heart rate, pain and other symptoms felt at the corresponding
body site, cool air blowing across the skin starting from the head, tingling,
energy entering the stomach and leaving the head or moving like a snake in
the body, *fluidos* like sexual energy, buzzing sounds, body lightness, rapid
thinking, feelings of contentment and relaxation in the presence of a good
spirit, feelings of nervousness, fatigue, or fear in the presence of a bad spirit.
Again, the principal differences appear to be associated with the role of spirits and with particular auditory, olfactory, or proprioceptive experiences
associated with particular *guias.* The elaboration of interaction with negative
spirits augments the *espiritista* repertoire of negative experiences and com-
pulsions to speak or hear involuntarily. Among Catholic Charismatic, evil
spirits are often ritually "bound" to prevent their manifestation in the form
of shrieking, writhing, vomiting, or challenging the proceedings. The acquiescence of spirits to this practice of binding is doubtless due in part to a class
habitus (Bourdieu 1977) that encourages behavioral moderation among
middle-class Charismatic. Protestant Pentecostals, typically of more working-
class provenance, tend to require some somatic manifestation as a sign of a
demon's departure from its host. In addition, evil spirits in the Charismatic
system are manifest only in the afflicted, not through the healer.

### Related Phenomena in Nonreligious Healing

The somatic mode of attention in both *espiritista* and Catholic Charismatic
systems is indigenously articulated in terms of religious revelation. I will
now briefly examine related phenomena in two healing systems that lack
such overtly religious character. Daniel (1984) describes the diagnostic taking of pulses by practitioners of Siddha medicine in South Asia as a three-
stage process that culminates with physicians making their own pulse
"confluent and concordant" with that of their patients. This final stage bears
the name *cama nilai,* the state of equipoise. Only after experiencing the
shared pulsations of *cama nilai* does the Siddha physician truly know the
patient's humoral disorder. In this instance, divinely inspired spontaneity is

replaced by cultivated diagnostic skill, but the somatic mode of attention remains characterized by its reference to another person's suffering.

Daniel's interpretation of Siddha pulse diagnosis also raises a method-ological issue, and requires us to return for a moment to the domain of semiotic analysis. Adopting the categories of Peirceian semiotics, Daniel describes the initial relation between the physician's passive fingertips and the patient's pulse as indexical—in their contact, they index each other as normal or abnormal. Also, the abnormal pulse of the patient indexes humoral imbalance, whereas the normal pulse of the physician indexes healthy humoral balance. As the physician's own pulse emerges and becomes confluent with that of the patient, the "indexical distance" between the signs decreases, until the relationship between the two pulses is transformed into an iconic one, and the two signs become one. According to Daniel, "At this moment of perfect iconicity, the physician may be said to have experienced in some sense the suffering as well as the humoral imbalance of the patient" (1984:120).

The semiotic analysis is of value in allowing Daniel to compare Siddha and similar traditional healing systems with Western biomedicine in terms of the relative power of indexicality or iconicity institutionalized within them (confer Kirmayer 1992 and Ots 1991). From the perspective of embodiment, however, the notion of indexical distance is too abstract, and the semiotic analysis allows only the conclusion that suffering is shared "in some sense." Daniel is forced into a neologism to express his understanding that, insofar as the process of taking the pulse neutralizes the divide between patient and physician, objectivity is replaced by "consubjectivity." The problematic of embodiment would pick up precisely at this point, with a phenomenological description of "consubjectivity" as characteristic of a particular somatic mode of attention.

A final example of this somatic mode of attention comes from contem-porary psychotherapy. Typically reported clinical experiences include a stir-ring in the penis in the male therapist's encounter with a "hysterical female," or a propensity to yawn when faced with an obsessive patient. Such phenomena occur spontaneously in psychotherapy, as in the religious set-tings described above, but the mode of attention to them is not consistently elaborated as indicative of something important about the patient or the condition being treated. Only certain schools, such as experiential, trans-personal, and analytical psychology, appear sympathetic to more explicit recognition of these phenomena. Samuels, for example, gives several exam-ples of countertransference as a "physical, actual, material, sensual expression in the analyst of something in the patient's psyche" (1985:52). He includes *bodily* and *behavioral responses,* such as wearing the same clothes as the patient, walking into a lamp-post, sensation in the solar plexus, pain in a particular part of the body; *affective responses,* such as anger, impatience, powerfulness, powerlessness; and *fantasy responses,* such as sudden delusional thoughts, mental imagery, or sensory distortions. Most important, he argues that such experiences are communication from patients, and against traditional theories of countertransference that impugn them as pathological reactions of the therapist.

This new example raises another methodological issue, that of the subject–object relationship as it pertains to the interpretive frameworks we bring to the objects of our analyses. Here I am not referring to our "objective" analysis of subjective phenomena, such as somatic modes of attention, but to the way our own interpretive subjectivity constitutes or objectifies the phenomena of interest. For the present discussion, work on countertransference from analytical psychology may appear to offer a valid interpretive framework. How can this be, however, when analytical psychology is itself the source of precisely the kind of data we wish to analyze under the heading of somatic mode of attention? Are we to place words of knowledge, *plasmaciones, cama nilai,* and embodied countertransference on an equal footing as phenomena to be interpreted, or can we justify using the last of these as a framework for interpreting the former three?

The nature of this problem is illustrated by the following vignette from my fieldwork. The setting was a Catholic Charismatic healing session conducted by a healer who was also a trained psychotherapist, and who made particular use of "bodywork" techniques. In this session, she asked the client, a 37-year-old man, to perform the postures of a technique known as "grounding," and to report what he felt in his body. In the context of ongoing therapeutic attention to the theme of overdiscipline and excessive need for control, it was not surprising that he observed that his fists were clenched and his knees locked. However, at the mentions of locked knees, my own crossed leg jumped as if it had been tapped by a doctor's hammer in a test of reflexes.

Insofar as my own somatic mode of attention was circumscribed by the motives of ethnography, I did not hesitate to use my own experience as an occasion for data collection. I later asked the healer how she would account for my knee jerk, and if it were possible for a nonbeliever to experience the divinely inspired word of knowledge. She responded that the experience could not be definitively interpreted, but that it could be one of three things: A somatic response caused by God, a consequence of my sharing some of the same personality issues as the client, or a natural result of deep attachment to another's experience. This "native exegesis" subsumes notions of divine agency, countertransference, and a psychosomatic understanding of empathy. In its postmodern juxtaposition of interpretive possibilities, it poses a challenge of reflexivity for the participant observer, and in so doing, it argues that the domain of interpretive possibilities is continuous between those of observer and those of observed.

It may be argued that, although a category such as countertransference may not be more correct, it may be more valuable for a comparative analysis of such phenomena, and that comparison itself is the source of validity. Nevertheless, this example reminds us that objective analytic categories become objective through a reflective movement within the process of analysis. I would argue that it is the perspective of embodiment itself that facilitates this insight. If the same insight can also be arrived at through other approaches, I would at least argue that embodiment offers a way to understand it in more depth. In any event, it is necessary to elaborate the finding that the attempt to define a somatic mode of attention decenters analysis such that no category is privileged, and all categories are in flux between subjectivity and objectivity.

## The Flux of Analytical Categories

All the examples we have called upon to illustrate the notion of somatic modes of attention are drawn from the domain of healing. If such modes of attention are general phenomena of human consciousness, we would expect that they can be identified in other domains as well. For example, Becker (1994, 1995) has observed that in Fijian culture the body is not a function of the individual "self" as in Euro-America, but of the community. An ongoing surveillance, monitoring, and commentary on body shape includes the changes that begin when a woman becomes pregnant. Fijians regard it as essential that a woman make her pregnancy known publicly, lest the power of its secrecy result in boats capsizing, contamination of food, and the spoiling of group endeavors. Unrevealed pregnancies can be manifest in the bodily experiences of others: Illness or weight loss caused by food cooked by the pregnant woman; loss of hair caused by cutting it; a lactating mother's milk drying up because of a glance. This phenomenon was fully cultivated as a somatic mode of attention by one woman who experienced an itch in her breast whenever a member of her family became pregnant. Such evidence typically led the head of the household to summon the family's young women and urge one of them to reveal her pregnancy before something untoward occurred.

An approach to cultural phenomena through embodiment should also make possible the reinterpretation of data already analyzed from other standpoints. We should then not only be able to discover undocumented somatic modes of attention as in the Fijian case, but also be able to recognize them right under our ethnographic noses in well-documented situations. I submit (based on observations made while my wife and I were expecting the birth of our twins) that such a reinterpretation of *couvade* is in order. The core of the phenomenon is that an expectant father experiences bodily sensations attuned to those of his pregnant mate. *Couvade* has been understood in one of two ways in the literature. On the one hand, it is thought of as a rather odd custom in which the man "simulates" or "imitates" labor (Broude 1988; Dawson 1929; Munroe et al. 1973). On the other, it is regarded as a medical phenomenon, or "syndrome" (Enoch and Trethowan 1991; Klein 1991; Schodt 1989). Thus, *couvade* is either exoticized as a primitive charade, or pathologized as a psychosomatic overidentification. Reconceived as a somatic mode of attention, it appears instead as a phenomenon of embodied intersubjectivity that is performatively elaborated in certain societies, while it is either neglected or feared as abnormal in others.

Pending additional empirical descriptions of somatic modes of attention, we can provisionally turn to the implications of the construct for a paradigm of embodiment. In outlining the phenomenology of somatic modes of attention in *espiritista* and Catholic Charismatic healing systems, I rigorously refrained from invoking any category other than "experience" and cast the description strictly in terms of sensory modalities. In the succeeding section, I showed that these modes of attention cannot be subsumed entirely under the category of religious experience, and that, in impinging on more conventional categories such as countertransference, they pose a challenge of reflexivity. The point I want to make now is about the poverty of our anthropological categories for going any further in understanding what it is to attend to one's body in a mode such as that described above. We operate with categories of cognition and affect, neither one of which alone can do justice to these phenomena, and between which there exists a nearly unbridgeable analytic gulf. The categories of trance and altered states of consciousness remain visual black boxes, and one colleague's suggestion of "proprioceptive delusion" is no help at all. To suggest that they are forms of "embodied knowledge" is provocative, but doesn't necessarily capture the intersubjective nature of the phenomena we have described. In his early programmatic work, Blacking referred to the existence of "shared somatic states" as the basis for a kind of "bodily empathy," but offered no specific examples of anything similar to what we have described above (1977:10).

I would like to go further here and briefly discuss these phenomena under four additional categories, if only to emphasize that we remain ill equipped to interpret them. These categories are intuition, imagination, perception, and sensation. I restrict the discussion in this section to the Charismatic and *espiritista* revelatory phenomena described above.

First, consider anointings, words of knowledge, *videncias,* and *plasmaciones* as kinds of intuition. The physician Rita Charon describes her practice of writing fiction to clarify her feelings when confused or distressed about a patient. She begins with known facts, tying together events, complaints, and actions of the patient, while making herself an actor in the story from the patient's point of view. She is "not surprised when details that I imagine about a patient turn out to be true. There is, after all, a deep spring of knowledge about our patients that is only slightly tapped in our conscious work" (Charon 1985:5). I think it is not difficult to conceive of intuition as embodied knowledge. Then why not conceive of revelatory phenomena as sensory intuition? Healers as well as physicians not only share with their patients a highly organized set of bodily dispositions summarized by Bourdieu (1977) under the term habitus, but also acquire a cumulative empirical knowledge of the range of human distress as they expand their experience.

Again, let us try to understand revelatory phenomena as forms of imagination. In current scholarship, imagination is discussed almost exclusively in terms of visual imagery, which is in turn readily thought of as "mental" imagery. So ingrained is the concept of mental imagery that the term physical imagery strikes one almost as an oxymoron. Yet if we allow the other sensory modalities equal analytic status with the visual, an expanded concept of sensory imagery would allow us to avoid the arbitrary dichotomy that tempts us to analyze Charismatic words of knowledge into distinct categories of mental images and physical sensations, and analytically to separate spiritist *videncias* from *plasmaciones*. We would then be taking a methodological step away from an empiricist conception of imagination as abstract representation to a phenomenological conception of imagination as a feature of the bodily synthesis, which Merleau-Ponty (1962) described as characteristic of a human consciousness that projects itself into a cultural world.

Once more, what if we take seriously the indigenous claim that these phenomena are forms of perception, if not of the divine, then of something else we can accept as concrete? This is a challenging proposition, and merits invoking Schwartz-Salant's (1987) attempt to integrate alchemical thinking into current psychotherapeutic theory. He suggests conceiving of an interactive field between two people that is "capable of manifesting energy with its own dynamics and phenomenology." This "in-between" field is palpable only on certain levels of perception in which the imagination itself can "become an organ that perceives unconscious processes" (1987:139). Samuels (1985), whose work has been discussed above, offers a related formulation, which, like that of Schwartz-Salant, is derived from analytical psychology. He elaborates Henry Corbin's concept of the *mundus imaginalis,* or imaginal world, as a distinct order of reality that exists both between two persons in therapeutic analysis, and between sense impressions and cognition or spirituality. Although the conception of imagination as a sense organ has its attraction, it creates methodological problems common to any model that tries to define "levels" of perception or consciousness. In addition, it does not address the problem that we have no independent way of "perceiving" unconscious processes so as to verify what is being perceived in revelatory phenomena.

Sensation is yet another category under which we might choose to subsume these phenomena. Sensation is inherently empiricist, however, and forces a conception of cultural meaning as referential meaning imposed on a sensory substrate. The relevant questions become whether the heat experienced by the healer is really the same as we feel when we blush, whether the tingling is really the same as the tingling of anticipation we feel in other highly meaningful situations, whether the "pain backup" in the healer's arm as she lays her hands on a person's shoulder is really the same feeling we have when our arm "falls asleep" after remaining too long in an uncomfortable position. All of these would be interesting determinations, but would not suit the aims of a cultural phenomenology. By reducing meaning to sensation or biological function, this approach requires a reconstitution of meaning that bypasses the bodily synthesis of sensory experience and the cultural synthesis of sacred experience.

The indeterminacy in our analytic categories is revealed when we encounter phenomena as essentially ambiguous as somatic modes of attention. This indeterminacy, it turns out, is an essential element of our existence.

Merleau-Ponty objected to conceiving perception as an intellectual act of grasping external stimuli produced by pregiven objects. Instead, he argued that the perceptual synthesis of the object is accomplished by the subject, which is the body as a fold of perception and practice (1964b:15–16). In effect, Merleau-Ponty's existential analysis collapses the subject-object duality in order to pose more precisely the question of how attention and other reflective processes of the intellect constitute cultural objects.

In taking up this enterprise, we find that the ambiguity between subject and object extends to our distinctions between mind and body, and between self and other. With regard to the first of these distinctions, if we begin with the lived world of perceptual phenomena, our bodies are not objects to us. Quite the contrary, they are an integral part of the perceiving subject. On the level of perception it is not legitimate to distinguish mind and body, since the body is itself the "general power of inhabiting all the environments which the world contains" (Merleau-Ponty 1962:311). Beginning from perceptual reality, however, it then becomes relevant to ask how our bodies may *become* objectified through processes of reflection. Likewise, in the lived world, we do not perceive others as objects. Another person is perceived as another "myself," tearing itself away from being simply a phenomenon in my perceptual field, appropriating my phenomena and conferring on them the dimension of intersubjective being, and so offering "the task of a true communication" (Merleau-Ponty 1964b:18). As is true of the body, other persons can become objects for us only secondarily, as the result of reflection.

It is in this embodied reality that we have had to begin the analysis of word of knowledge, *plasmacione, cama nilai,* and embodied countertransference. Originating in primordial experience characterized by the absence of duality between mind and body, self and other, the phenomena are objectified in reflective practice through a particular somatic mode of attention. Far from providing a causal account of these phenomena, our analysis has shown the difficulty of even finding adequate descriptive categories. What is revealed by a return to the phenomena—and the consequent necessity to collapse dualities of mind and body, self and other—is instead a fundamental principle of indeterminacy that poses a profound methodological challenge to the scientific ideal. The "turning toward" that constitutes the object of attention cannot be *determinate* in terms of either subject or object, but only *real* in terms of intersubjectivity.

## What's the Use of Indeterminacy?

Ironically, the approach through embodiment that has allowed us to elaborate somatic modes of attention as a construct with some demonstrable empirical value has also disclosed the rather slippery notion of the essential indeterminacy of existence. This is doubtless related to the discovery of existential and methodological indeterminacy in recent ethnographic writing (confer Favret-Saada 1980; Jackson 1989; Pandolfi 1991; Stoller 1989).

Inevitably, perhaps, when we try to give theoretical formulation to this indeterminacy, we easily slip back into the language of either textuality or embodiment, representation or being-in-the-world. In the present context, I can only point to this problem by briefly summarizing the principle of indeterminacy as formulated by Merleau-Ponty for perception, and by Bourdieu for practice. We thus return to the notion of indeterminacy, not to make it determinate as a concept that can be *applied* in our analyses, but to give some theoretical grounds for accepting it as an inevitable background *condition* of our analyses.

Merleau-Ponty, having demonstrated that all human functions (for example, sexuality, motility, intelligence) are unified in a single bodily synthesis, argues that existence is indeterminate:

> in so far as it is the very process by which the hitherto meaningless takes on meaning, whereby what had merely a [for example] sexual significance assumes a more general one, chance is transformed into reason; in so far as it is the act of taking up a de facto situation. We shall give the name "transcendence" to this act in which existence takes up, to its own account, and transforms such a situation. Precisely because it is transcendence, existence never utterly outruns anything, for in that case the tension which is essential to it [between objective world and existential meaning] would disappear. It never abandons itself. What it is never remains external and accidental to it, since this is always taken up and integrated into it. (1962:169)

The transcendence described by Merleau-Ponty is thus not mystical, but is grounded in the world, such that existential indeterminacy becomes the basis for an inalienable human freedom.

For Bourdieu, the synthesis of practical domains in a unitary habitus is likewise based on indeterminacy, but this variant of indeterminacy does not lead to transcendence. Instead of an existential indeterminacy, Bourdieu's is a logical indeterminacy, which "never explicitly or systematically limits itself to any one aspect of the terms it links, but takes each one, each time, as a whole, exploiting to the full the fact that two 'data' are never entirely alike in all respects but are always alike in some respect.... Ritual practice works by bringing the same symbol into different relations through different aspects or bringing different aspects of the same referent into the same relation of opposition" (Bourdieu 1977:111–112). Logical indeterminacy is the basis for transposition of different schemes into different practical domains, exemplified in his ethnography by the Kabyle application of the male-female opposition to outside-inside the house and, again, to different areas *within* the house. It is also the basis for the polysemy and ambiguity epitomized by the Kabyle cooking ladle that is sometimes male, sometimes female.

In sum, Merleau-Ponty sees in the indeterminacy of perception a transcendence that does not outrun its embodied situation, but that always "asserts more things than it grasps: when I say that I see the ash-tray over there, I suppose as completed an unfolding of experience which could go on ad infinitum, and I commit a whole perceptual future" (1962:361). Bourdieu sees in the indeterminacy of practice that, since no person has conscious mastery of the modus operandi that integrates symbolic schemes and practices, the unfolding of his works and actions "always outruns his conscious intentions" (1977:79). It would be convenient if we could pose these views of indeterminacy as perfectly complementary. Thus, we could say that human action is transcendent in taking up situations and endowing them with meaning that is open-ended and inexhaustible without ever outrunning those situations; and situations cannot be outrun because they are structured according to an enduring system of dispositions that regulates practices by adjusting them to other procedures, thereby creating the condition of possibility for the openendedness of action. However, there are serious conceptual differences between the two theorists that put this interpretation in doubt.

On Bourdieu's side, the locus of these differences is his rejection of the concepts of lived experience, intentionality, and the distinction between consciousness in itself and for itself. This rejection requires Bourdieu to ground the conditions for intelligibility in social life entirely on *homogenization* of the habitus within groups or classes (1977:80), and to explain individual variation in terms of *homology* among individuals, such that individuals' systems of dispositions are structural variants of the group habitus, or deviations in relation to a style (1977:86). Merleau-Ponty, on the other hand, insists on the *a priori* necessity of *intersubjectivity,* pointing out that any actor's adoption of a position presupposes his or her being situated in an intersubjective world, and that science itself is upheld by this basic *doxa.* This intersubjectivity is not an interpenetration of intentionalities, but an interweaving of familiar patterns of behavior: "I perceive the other as a piece of behavior, for example, I perceive the grief or the anger of the other in his conduct, in his face or his hands, without recourse to any 'inner' experience of suffering or anger, and because grief and anger are variations of belonging to the world, undivided between the body and consciousness, and equally applicable to the other's conduct, visible in his phenomenal body, as in my own conduct as it is presented to me" (Merleau-Ponty 1962:356). This analysis is echoed by Jackson: "To recognize the embodiedness of our being-in-the-world is to discover a common ground where self and other are one, for by using one's body in the same way as others in the same environment one finds oneself informed by an understanding which may then be interpreted according to one's own custom or bent, yet which remains grounded in a field of practical activity and thereby remains consonant with the experience of those among whom one has lived" (1989:135). Because body and consciousness are one, intersubjectivity is also a co-presence; another's emotion is immediate because it is grasped preobjectively, and is familiar insofar as we share the same habitus.

In the end, Bourdieu's principle of logical indeterminacy becomes the condition for regulated improvisation, whereas Merleau-Ponty's principle of existential indeterminacy becomes the condition for transcendence in social life. Each principle has a weakness, based on the implicit favoring of textuality or embodiment, representation or being-in-the-world. We will leave our discussion with a summary of these issues.

To Merleau-Ponty, authentic acts of expression "for themselves" constitute a world and are transcendent, but once a linguistic and cultural world is already constituted, reiteration of those acts is no longer transcendent, no longer projects itself into the world, and partakes more of being "in itself." For Merleau-Ponty this problem subsists primarily in the domain of speech, where the *speaking word* becomes sedimented as the *spoken word*. Here, Bourdieu's analysis of universes of practice subsisting alongside universes of discourse provides a corrective, forcing us to generalize this sedimentation from language to the rest of the habitus, and to acknowledge Merleau-Ponty's problem as endemic to his conception of existence. The problem, required by the (uncollapsed or uncollapsible) duality of the "in itself" (being) and "for itself" (existence), is having to distinguish genuine, transcendent expression from reiteration. This leads directly to the dilemma of having to specify conditions under which persons can become objects to others and to themselves, and under which socioeconomic classes can become objects to other classes and to themselves, as opposed to being subjects of their own action. *While existence is not text, it is essentially textualizable.*

Bourdieu, in rejecting the distinction between "in itself" and "for itself," can avoid this problem by conceptualizing the result of indeterminacy as regulated improvisation, open-ended yet circumscribed by the dispositions of the habitus. In this he is faced with a different problem, however: accounting for change, creativity, innovation, transgression, and violation. He claims that, "as an acquired system of generative schemes objectively adjusted to the particular conditions in which it is constituted, the habitus engenders all the thoughts, all the perceptions, and all the actions consistent with those conditions, and no others" (1977:95). This is difficult to conceive, he claims, if one remains locked in the dilemmas of determinism and freedom, conditioning and creativity. These are perhaps dualities that he is too quick to collapse, however, unless the "conditioned and conditional freedom" of the habitus's "endless capacity to engender products" includes the capacity for its own transformation (1977:95). Otherwise, the principle of indeterminacy becomes a disguise for lack of analytic specificity, and habitus loses its value as an analytic construct. *Although the habitus bears some of the schematism of a fixed text, it can be transcended in embodied existence.*

## Conclusion

Approaching cultural phenomena from the standpoint of embodiment has allowed us to define a construct of somatic modes of attention, which has in turn led us to a principle of indeterminacy that undermines dualities between subject and object, mind and body, self and other. In our concluding comparison of Merleau-Ponty and Bourdieu, we have seen that the relations between embodiment itself and textuality, and between representation and being-in-the-world, are indeterminate as well. These indeterminate relations constitute the shifting existential ground on which contemporary ethnography suggests we must increasingly situate cultural phenomena. Our attempts to objectify in analysis are analogous to the definitive gesture of the Senoufo diviner in striking his thigh (Zempleni 1988) to confirm his pronouncement. The act is not so much an invocation of the sacred as it is an embodied statement, in defiance of the wisdom that one never steps into the same river twice, that one has snatched a definitive outcome from the indeterminate flux of life, and that, once and for all, "This is the way it is."

It is this same principle of indeterminacy, inherent in social life, that has come to the fore in the conscious movement of postmodernism in art and the unconscious dissociation of signs and referents, symbols and domains, in contemporary culture. It is the fundamental indeterminacy of existence that is sensed as missing by those anthropologists attracted to the postmodernist methodological shift from pattern to pastiche, from key symbols to blurred genres. Their project has been begun in the semiotic paradigm of textuality, but a substantial contribution can also be made through elaboration of a phenomenological paradigm of embodiment. Yet, if indeterminacy is fundamental to existence, only careful elaboration of its defining features, such as Merleau-Ponty's transcendence and Bourdieu's improvisation, will allow it to become an awareness of our existential condition without becoming an excuse for analytical imprecision.

# CHAPTER TEN

# Shades of Representation and Being in Virtual Reality

One way to address the question of what it means to be human is to begin with the observation that we have a world and inhabit a world. The inquiry seems to unfold under its own weight from this point, with the next set of questions necessarily having to do with how worlds (for they are always multiple) are constituted, what it means to have them, and precisely how we inhabit them. In contemporary society biotechnology is increasingly implicated in transforming the very bodily conditions for having and inhabiting any world. This is doubly the case when biotechnology includes sophisticated computer applications, since computers and computer networks are recognized as having enormous transformative potential. Indeed, Sherry Turkle (1995) has suggested important modulations of the self are in the making, and the philosopher Michael Heim (1993) has suggested that the computer is leading to a major ontological shift—a modulation in the structure of human reality itself.

If this is the case, then what is at issue is the nature of our experience as social actors in a cultural world. What I mean by experience, and what is at issue in biotechnology applications of the computer, is the interplay between cultural representations in the form of speech, text, images, symbols, myths, dreams, and cultural modes of being-in-the-world, by which I mean the mode in which we are present in the world right now, at this moment in this situation, with the qualification that this moment is thoroughly immersed in temporality. I would prefer for the moment not to say more by way of abstract definition of representation and being-in-the-world. The theoretical challenge is not to abandon being-in-the world as inaccessible and confine ourselves to studying representations, nor to claim that being can be studied independently of representation, but to insist on the necessary intertwining of the two.

Elaborating the cultural consequences of biotechnology applications of the computer with respect to the having and inhabiting of worlds requires a cultural phenomenology that highlights this intertwining. Much recent cultural analysis privileges the pole of representation, with culture understood as constituted by symbols, signs, and images. From this standpoint, textuality is the most prominent metaphor guiding the interpretation of culture, and the world is not so much inhabited as represented in a way that can be read. While interpretively powerful, however, the notion of textuality is less apt for specifying cultural modes of being-in-the-world—that is, the kinds of engagement and participation of humans in our worlds—than is the complementary notion of embodiment. This notion places us at once at the most general and limiting condition of our existence. Our bodily existence, or embodiment, is from this standpoint understood to have a range of potential experiential modalities in relation to features of cultural and historical context.

The interplay between representation and being-in-the-world, and the complementarity between textuality and embodiment, is precisely at issue in biotechnology applications of the computer. First, the human body is the objective target of technology. By being taken up into the technological environment it is represented and, I would suggest, has its being-in-the-world altered. Second, the computer user is the embodied subjective manipulator of the technology. In this capacity a person encounters representations of the body and again, I would suggest along the lines of Turkle and Heim cited above, has its being-in-the-world altered. The example I offer in this paper involves the use of computer-generated virtual reality to create so-called virtual cadavers that are used for purposes such as the teaching of human anatomy and in computer-assisted surgery. The following discussion takes up these issues as they are being played out in the "Visible Human Project," and concludes with a reflection on their consequences for embodiment with respect to representation and being-in-the-world.

## The Creation of Computerized Cadavers

The creation of the "Visible Human" cadavers was an extraordinary technological feat, achieved with funding from the federal government. National Library of Medicine (NLM) Director Donald Lindberg notes that the Visible Human Project originated with his observation in 1987 that "the medical school community needed a better way to teach anatomy." In 1991 the NLM awarded a contract for development of the proposed data set to the University of Colorado Center for Human Simulation, with a subcontract for creation of three-dimensional volumetric visualizations of the computerized cadavers to the Visualization Group of the Scientific Computing Division of the National Center for Atmospheric Research. The first step was to find a suitable cadaver, beginning with a male. In 1993, after two and a half years of searching for a fresh cadaver that was " 'normal' and within guidelines of size and age," a qualified 39-year-old Texas death row inmate named Joseph P. Jernigan agreed to donate his body to science in exchange for being allowed to die by lethal injection rather than electrocution. This suited the researchers' purposes well, since electrocution alters the tissues in ways that would defeat the purpose of having as lifelike a body as possible.

The second step was creation of three matching sets of images composed of transverse sections or "slices" of Jernigan's body. The first two sets were obtained by magnetic resonance imaging (MRI) and computed tomography (CT). Next the body was encased in gelatin, frozen to minus 160 degrees Fahrenheit, and cut into four sections. It was then sectioned into 1,878 transverse slices each 1mm thick, corresponding exactly to the MRI and CT images. High-resolution color digital photographs were taken of the block after each slice. By the time a female cadaver—a 59-year-old woman who died of heart disease, and whose family insisted she remain anonymous—was subjected to the same procedure a year later, the researchers decided they could achieve greater detail and higher resolution by making the slices one-third the thickness. Consequently she was sectioned into 5,189 transverse slices. For both, the MRI and CT images were created from the whole body prior to freezing. The photographic images were of the face of each section not as it was planed away by a custom-designed laser-guided cryogenic macrotome. The physical remains thus became a collection of frozen shavings, which were then cremated, such that their digital remains now have, in some yet-to-be-understood sense that we will explore shortly, a more concrete existence.

Each of the three types of transverse images was digitally captured at 2,048 by 2,048 by 42 bits, and aligned with its companion images and with images of adjacent slices. The combined datasets are astronomically large: The male takes up a total of 15 gigabytes and the female occupies 39 gigabytes. These data are stored on an FTP site, and with a free license can be downloaded directly from the Internet.[1] Donald Lindberg states that, "With the Visible Human Project, we are returning to the idea that a library holds the knowledge of a profession—not just reprints, journals, and books. The advent of technology gives us the opportunity to store knowledge electronically and distribute it, virtually instantaneously, throughout the world." A NLM project report from 1996 notes that already the Visible Human data "are being applied to a wide range of educational, diagnostic, treatment planning, virtual reality, artistic, mathematical, and industrial uses," and by the end of the year over 700 licenses had been issued to users in 27 countries.

Let us explore the capabilities of the two virtual cadavers for producing computer images of the human body. The basic form is the transverse section (each in its own computer file), but because the images are precisely aligned, it is possible as well to produce vertical and horizontal sections through virtually any plane. More sophisticated programs are able to produce three-dimensional representations by stacking slices and isolating sites corresponding to particular internal or external anatomical structures. For example, the accompanying male and female heads, generated by William Lorenson of the General Electric Corporation, are not photographs but reconstructions of surface features from the slices (215 physical slices for the male, 209 CT slices for the female). Manipulating and combining these images using state-of-the-art visualization programs makes it possible to penetrate the body—giving the sense of walking or flying through (as in walking through walls, or superhero "x-ray vision"). Different levels of depth or systems (skin, muscles, skeleton) can be superimposed to be viewed simultaneously, and discrete anatomical structures can be isolated. Further, these images can be rotated to be viewed from different perspectives, not only in successive still images but in computer animations. Currently these have advanced to the point of allowing surgical simulations—to be discussed in somewhat more detail below—similar to flight simulations used in training pilots. In one easily available demonstration from the Center for Human Simulation, one can watch a computerized scalpel make an incision in a thigh sliced off from the body above and below the knee, the incision gradually opening to reveal muscle and fat, and the section then rotating in mid-screen and moving to a close-up to show several views of the incision. Developers of these methods promise that their animations—one is tempted to say reanimations—will eventually include blood coursing through veins and around organs, and breath pulsing through lungs. Beyond that is the prospect of transformations that will simulate the processes of aging (as well

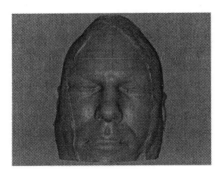

**Figure 10.1**    Visible Man, Frontal View

*Source*: Reproduced by permission of General Electric.

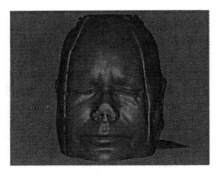

**Figure 10.2**    Visible Woman, Frontal View

*Source*: Reproduced by permission of General Electric.

as, of course, growing young again) and pathology (as well as its reversal). All this in the context of the technical aspiration to render future candidates into ever thinner slices to enter into ever larger databases. One can imagine a race of reanimated virtual cadavers faithful to their human form to the cellular level, infinitely mutable, able to be subject to many simultaneous surgical procedures and healing processes, capable of reanimation and regeneration in whole or in any part.

### Simuloids, Avatars, and Shades

Virtual cadavers are phenomena in and of cyberspace and virtual reality. Let me clarify my understanding of how these terms are related. A person who sends E-mail or peruses sites on the World Wide Web is tapping into cyberspace, but not in a strong sense entering a virtual reality. Here the computer is a communicative tool, a cybernetic enhancement of mail and media. A person outfitted with data glove and data goggles involved in an advanced computerized simulation is entering a virtual reality, but is not in cyberspace. Here the computer is an aesthetic tool, a cybernetic enhancement of dramatic and performative techniques by which we create imaginative terrains. To put it more formulaically, cyberspace is an intersubjective medium constituted socially (constituted by interaction among participants in the communicative medium), while virtual reality is the subjective sensory presence in that medium (constituted by interaction between individual user and technology). The intersection of the two is full sensory and bodily engagement in a virtual reality that is also fully networked and plugged into cyberspace—in other words, the full blown, Gibsonesque, science fiction version of virtual reality within cyberspace. This intersection is important to identify because, I would argue, it is precisely the cultural locus of the virtual cadavers.

What do we mean by "cultural locus" in this instance? Let us accept the metaphor of cyberspace with sufficient literalness to conceive it not as a "cultural domain," but as a distinct "ethnographic terrain," so that it is legitimate to talk about an ethnography of cyberspace. Indeed, there has very recently been a fluorescence of work along these lines in anthropology, and the number of sessions devoted to related topics at the last several meetings of the American Anthropological Association indicate a groundswell of interest. The most prominent concept in this area is that of "virtual communities," or networks of social interaction with fluid boundaries and varying degrees of apparent permanence, composed of actors whose agency is expanded to the point of controlling and manipulating their own identities as pure forms of representation, and who interact with ambiguous others whose identities are never certain. While there is thus a clearly articulated concern with the kind of relationships that exist in cyberspace, taking the metaphor of cyberspace literally as an ethnographic terrain allows us to formulate a parallel concern with the kind of beings that inhabit cyberspace. We can then propose a preliminary inventory of such beings, with the caveat that the "cyborg" is not among them, for a cyborg is by definition a creature of the technological interface. We are cyborgs whenever our bodily capacities are technologically altered or enhanced, including when we have our noses pressed up against the window of cyberspace by virtue of booting up. The goal here is an inventory of beings that exist wholly on the hither side of the interface, in the ethnographic terrain marked by what we just now identified as the intersection of cyberspace and virtual reality.

Accordingly, my preliminary inventory consists of three types of beings: simuloids, avatars, and shades. Simuloids are software-generated entities that have no sentient counterpart in actuality. In the language of the industry, these are referred to variously as humanoid technologies, virtual humans, human-modeling systems, computer-generated humanoids, or autonomous creatures. Simuloids are described as "autonomous" not in the sense that they have agency in their own right but in the sense that they are independent of human agency: They are software-controlled rather than human-controlled. They are also autonomous from any necessity to conform to concrete actuality, and thus their features may transcend the human—they can as easily be animal, machine, or monster. However, a great deal of attention is currently being devoted to the development of virtual humans per se, defined as "computer-generated people that live, work and play in virtual worlds, standing in for real individuals or carrying out jobs that real people cannot do." The news service story that carries this definition also quotes Sandra Kay Helsel, editor of *VR News,* as pronouncing that "Virtual Humans will be the growth industry of the 1990s!" Characters in computer games are simuloids, as are the characters Max Headroom, the computer known as "HAL" from the film *2001, A Space Odyssey,* and the villainous cyberman in the movie *Virtuosity.* The most advanced simuloids include the virtual humans developed by Nadia Thalmann with her "Marilyn" program that can simulate Marilyn Monroe and Humphrey Bogart as well, and includes characteristics such as emotional expression, speech, clothes, hair, and the ability to respond to computer users. Norman Badler's "Jack" system of human modeling is based on a figure designed to the specifications of the average American male that is capable of articulated motion including balance modification and collision avoidance, gesture and facial expressions, natural language processing, and transformation in size and color.

The term *avatar* is already in popular usage, and is sometimes applied to what I have called *simuloids,* but I want to restrict its meaning to virtual incarnations of human actors on the hither side of the interface that are directly controlled by those actors. It is worth playing out some of the cultural connotations of the notion of avatar because of the implications regarding self, agency, and being that can be read into it. The primary meaning of the term, of course is the incarnation of a Hindu deity in actual human form. By extension, on the one hand, the human computer operator is analogous to a kind of deity that manipulates the computerized avatar in virtual human form. On the other hand, the extension inverts the meaning of the Sanskrit term in a subtle way: Whereas the Hindu avatar is an incarnation of a deity into an earthly form, the computer avatar is a virtual apotheosis of an earthly being into an imaginative realm of fantastic powers and shape-shifting. A more secularized use of the term offered by Webster's dictionary suggest that an avatar can be an embodiment, as of a concept or philosophy, usually in human form, a definition which prompts the observation that what is being embodied by the computer avatar is the human form itself.

Another, more thoroughly abstracted definition from Webster's has the avatar as "a variant phase or version of a continuing basic entity." This one prompts the question of whether the avatar is best considered a representation of a person, a cybernetic extension of the person, a projection of a person into cyberspace, or indeed a "variant phase or version" of a person, for certainly the avatar is much more than a computerized double or simulation programmed to act like a person. Indeed, in February 1996 a virtual wedding took place in Los Angeles in which vows were exchanged on-screen via avatars while the participants remained in separate geographical locations in actuality. What varies from a technical standpoint is the degree of sensory engagement of operator in avatar. In practice, avatars can be imagined forms described to other users by typing in text, as in the interactive computer sex networks where people become animals or creatures endowed with the most amazing and creative types of sexual organs and multiple genders. Avatars can also be visual "body icons" that can be manipulated and come into the virtual presence of the body icons of other users, but are now often little more than "grim-looking peg-doll shapes" lacking faces or feet. The most advanced include multisensory feedback in which it is experientially the case that the avatars are not so much representations of the user as projections of the user into virtual space. Such projections are themselves customized computer-generated forms, but may in principle also be computer-animated video images as is pioneered in the "synthespian" technology developed by Jeff Kleiser and Diana Waczak.

To summarize the distinction between types of beings that I've drawn here, "simuloids" are computer-generated stand-ins for people with no connection to any actual person, while "avatars" are projections of living people as digitized persons. To draw on vivid popular-culture examples, the computer-generated villain of the movie *Virtuosity* is a simuloid that crosses the interface from virtuality into actuality and becomes embodied; the character Jobe in the movie *Lawnmower Man* is a human who crosses from actuality into virtuality to literally become his avatar and abandon his body. The purpose of this contrast is to introduce what is quite a distinct third type of being indigenous to cyberspace. The computerized virtual cadavers produced by the Visible Human Project of the National Library of Medicine are what I'll call "shades," derived from the use of that word to refer to a spirit in the netherworld. There are only two such beings at present, a male and a female, but their existence has profound consequences that we are just now barely beginning to work out. These shades are "in" cyberspace and virtual reality in a sense distinct from either simuloids or avatars. Like a simuloid, the shade can operate as a stand-in for a person, but unlike a simuloid it is a distillation of an actual person that can be digitally superimposed on another actual person. Like an avatar, the shade is a projection of a real person into cyberspace, but unlike an avatar that person is not only dead but has been dissolved as a physical being. It thus exists solely on the hither side of the interface, where it is not an animation but a reanimation—a new kind of being entirely. Let us elaborate this analysis by briefly examining the biotechnological and symbolic structure of shades.

### Adam and Eve in the Virtual World of the Dead

When the first images of the Visible Human male were presented, the regents of the NLM broke out in spontaneous applause, prompting a comment by project director Ackerman quoted by the *Denver Post:* "It was sort of like applauding at the end of a movement in a concert. It was inappropriate, and the decorum is not that way. But it was a clear indication of how excited people are" (Schrader 1994). Given the excitement generated by this biotechnological juggernaut, let us pose the following question: What is the relation between Joseph Jernigan, the housewife from Maryland, and the datasets they have become? In other words, what constitutes their cultural status (being) as shades? For a preliminary answer I take as data representations of the event in newsletters of the National Library of Medicine and approximately 55 articles from the popular media.

One way of thinking about the change in cultural status is most evident with Jernigan, whose fate was the more public. Here there is a sense of "through the looking glass" with respect to representation of his personhood. The first mention of Jernigan is in retrospect rather eerily incognizant of his subsequent posthumous celebrity. It is a typical article furnished by Reuters in the *New York Times,* August 6, 1993, reporting that he was executed by lethal injection after having admitted to killing a homeowner who surprised him during a burglary. Such deaths by execution are routinely reported in the news media, and the article's significance does not go beyond the observation that because the reinstitutionalization of capital punishment in the United States remains at least a back-burner issue of public discussion, executions remain newsworthy. By April of 1994 his previous existence was a distorted footnote. A bylined article on the forthcoming release of the Visible Human Male data set picked up by several Knight-Ridder papers described the cadaver as a "drug overdose victim" (this incredibly bad article also referred to the film *Fantastic Voyage* as *Incredible Journey*), as did articles in 1994 and 1995 in the *Denver Post.* The *Denver Post* in 1994 mentions the project in an article on "cryonics," the practice of freezing dead diseased individuals in hopes of bringing them back to life when medical technology has advanced sufficiently to cure them, with the concluding comment that "the project doesn't aim to revive the subjects." Nevertheless one is left with the titillating image of a person, whose personal identity is rather beside the point, who is in some sense capable of being reanimated.

The sense of "through the looking glass" that defines the cultural status of shades is most evident in a series of metaphors invoked to describe them.

**Figure 10.3**   Visible Man Torso Minus Skin

*Source*: Reproduced by permission of Karl Heinz Höhne at the Institute of Mathematics and Computer Science in Medicine, University of Hamburg, Germany.

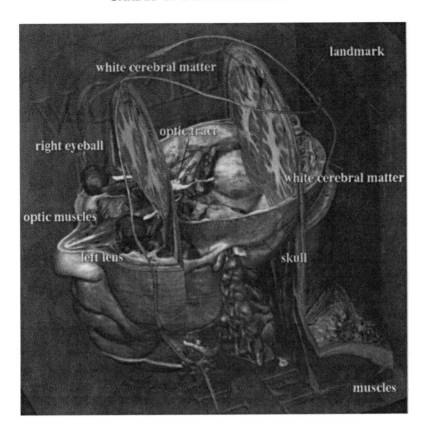

**Figure 10.4**   Visible Man Head Showing Internal Organs

*Source:* Reproduced by permission of Karl Heinz Höhne at the Institute of Mathematics and Computer Science in Medicine, University of Hamburg, Germany.

There are four sets of metaphors, one describing the shades as Adam and Eve, one in images of birth or immortality, another as if they were a virtual terrain to be mapped, and one that invokes Leonardo da Vinci. The Adam and Eve metaphor is doubtless the most symbolically charged. This is evident in that it was the original intent of the project directors to use these as the formal titles for their shades instead of the markedly more awkward "Visible Human Male" and "Visible Human Female." The plan had to be abandoned in the face of threatened legal conflict with a company that had already named itself and its interactive anatomy software program "A.D.A.M.," an acronym for Animated Dissection of Anatomy for Medicine. The competition—never mind that the A.D.A.M. company is now itself licensed to use the Visible Human shades in its product development—indicates that something rather existential might indeed be at stake in the advent of shades.

A sampler of these metaphors will give a flavor of what I mean. The *Philadelphia Inquirer* (April 14, 1994) refers to the shades as a " 'perfect couple'—an Adam and Eve for computer immortality." The *Denver Post* (June 6, 1994) calls them the "first couple." The *Baltimore Sun* (November 29, 1995)

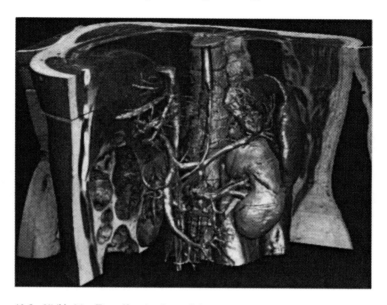

**Figure 10.5**   Visible Man Torso Showing Internal Organs
*Source:* Reproduced by permission of Karl Heinz Höhne at the Institute of Mathematics and Computer Science in
Medicine, University of Hamburg, Germany.

announces that the female shade is a "Partner for 'Visible Man,'" and the
*New York Times* (November 29, 1994) reports that NLM was "on the look-
out for a female donor to share his life on the wire." *The Economist* (March
5, 1994) goes yet further in its reference to "A new family—Adam, Eve, and
their embryonic offspring," pairing the Visible Humans with the Visible
Embryo Project at the Armed Forces Institute of Pathology. Elsewhere in the
piece they are referred to as "Adam, Eve, and little Cain," and the National
Library of Medicine as their "electronic Garden of Eden." Referring to plans
to increase the inventory of shades, the *San Diego Union-Tribune* (March 15,
1995) notes that "The future might also bring Visible progeny, Grandpa and
Grandma, or a younger pre-menopausal woman."

The Adam and Eve metaphor cannot be written off as either tritely cute
or opportunistically commercial—the metaphor is simply too good a fit with
contemporary gender symbolism (see also Treichler 1998). Although there
was no more of a relation between Jernigan and the Maryland housewife than
between any two cadavers donated to medical research, there is evident appeal
in transforming them into a mythical first couple in a new virtual world, digitally reanimated and capable of necrosexual procreation of a family or species
of shades. Conveniently, this time around Adam and Eve are both white.
Naturally the male was created first, and the female second, though this time
not from Adam's rib. As is the case generally in contemporary society, the male
is guaranteed an identity (though fortuitously because his cause of death was
public, court-ordered), while the female remains anonymous (though on the
hither side of the looking-glass that anonymity was understood as a right).
The image of Adam rendered in 1 mm slices is rougher, while Eve is more

refined or, in a pun spun by one journalist, she "looks sharper" than her male counterpart. Stated otherwise, in the Foucauldian idiom of bodily surveillance, if the male can be scrutinized, the female can be scrutinized more thoroughly. The male was an evil victimizer killed by lethal injection, the woman was an innocent victim who died of a heart (as in "bless her heart") attack. Adam is sufficient in his own right, indeed a magnificent specimen who pumped iron, while word is that Eve needs to be supplemented by a premenopausal counterpart—we can expect polygyny in cyberspace.

The second set of metaphors has to do with birth and immortality. On the one hand, a 1996 World Wide Web self-description of the University of Colorado Health Sciences Center (UCHSC) Center for Human Simulation by its staff observed that their anatomical imaging laboratory has "given birth to the Visible Human—Male and Female," while the *New York Times* (November 29, 1994) reported "Executed killer reborn as 'visible man' in Internet." On the other hand, the National Library of Medicine newsletter (1995:50:6) reported that the visible humans are "immortalized on the Internet." *The Independent* (November 29, 1994) announced that "A killer was yesterday let loose on the Internet computer network," and *NetGuide* magazine (April 1, 1995) that "A killer ... has been immortalized ... ." *Life* magazine (February 1997) refers to "electronic afterlife," the *Baltimore Sun* (November 29, 1995) stated that the shade "... has won a measure of computerized immortality," and the *Denver Post* (June 6, 1994) that the project "promises eternal life for the participants." These metaphors are neither idle nor contradictory, but reflect views from different sides of the looking glass. The humans are in a sense immortal in their new form, their shades are in a sense born again beings of a new space.

Two rather less developed metaphors that yet indicate something of the cultural status of shades can be identified. One is the elevation of the shades in a celebration of the human form that places them alongside the renderings of Leonardo. A group at University Hospital in Hamburg, Germany, that has developed an impressive 3-D Atlas called VoxelMan draws a direct historical line from Leonardo to the development of x-rays, on to the invention of CT and MRI technology, and thence to the Visible Humans. A group of artists in Japan is explicitly juxtaposing Leonardo's representations with representations of the Visible Human shades. Implicit is the ideal of approaching reality via virtuality, a connotation that is also evident in the slightly jarring phrase of the *Baltimore Sun* (November 29, 1995) referring to "real bodies on a computer." The final image, which I found only once in a quote from one of the project coordinators, is telling in its appeal to the inanimate. This was in a reference by project director Ackerman to the need to label each site and segment of the Visible Humans, since "Right now, looking at them is like looking at a road map with no street names." Here the shades are understood as a terrain to be traversed—not an unknown virgin territory, but a map not yet useful because not yet labeled.

Taken together, I would suggest that these sets of metaphors disclose the workings of a deep essentialism that constitutes the cultural status of shades.

The Adam and Eve metaphors point to gender essences defined by the heterosexual reproductive couple—the two shades could have been defined as siblings, or even, given their age difference, as mother and son. The immortality metaphors define a moral essence, whether conceived as an untainted being prior to the Fall or to a redeemed being in the guise of a born-again convict. The Leonardo metaphor outlines an aesthetic essence of the apotheosized ideal man—notably without female counterpart. The map metaphor implies a cosmic essence by assimilating the body to a terrain, but in particular one that still needs to be charted. The positing of essences in these popular metaphors is implicitly a strategy of identity, made all the more compelling by two features. First, positing the essence of the other (the not-me, the shade) is a double-edged act of self-definition either by denial (me as the opposite of not-me) or by desire (the wished-for me). Second, the force of this double-edgedness is enhanced by the paradoxical condition of the shades as virtuality in actuality, their apparent existence as "real bodies on the Internet." But how can all this be so?

## The Meaning of Metaphor: Virtual or Actual?

Two methodological points can help assess the consequences of the foregoing discussion of popular metaphors. First, the discussion assumes that such metaphors offer an interpretive opening to cultural meaning. The analysis can only be valid, for example, if the metaphoric description of the first shades as "Adam and Eve" is accepted as nontrivial, more than an eye-catching journalistic ploy. It was certainly common enough—only the sober British *Daily Telegraph* consistently reported on the project without recourse to metaphor. Once accepted, this assumption allows the implications of the metaphors to be spun out and regarded as data about the "cultural imaginary," which is further assumed to be as consequential as what we could call the "culturally literal," that is, the language of technology and its application. In this context, the cultural imaginary is the realm of possibility, desire, and fear in which we participate passively insofar as it lurks anxiously beneath conscious awareness[2] and in which we participate actively by the exercise of imagination, the "capacity to articulate what used to be separate ... which allows one either to make a new move or change the rules of the game" (Lyotard 1984:52). Looking at a text from the standpoint of the culturally literal, one could argue that a metaphor is only a colorful analogy to help clarify an objective relation or function—the metaphor is discursively subordinate to the function. Regarding metaphor as an opening into the cultural imaginary grants it a far more important role, one in which the cultural imaginary has an equivalent status with—indeed is in a dialectical relation with—the culturally literal.[3] From the glass half-full perspective this dialectic is one in which they mutually constitute one another. From the glass half-empty perspective it is one in which they mutually destabilize one another. At the very least, one could say that the cultural imaginary provides the context by means of which the existential implications of the technical innovation can be examined, while the culturally literal leaves discussion at the level of policy implications.

The second point is the importance of identifying the social origins and destinations of the cultural meanings that are spun into the gossamer fabric of a cultural imaginary. In the case of computerized cadavers, the sense-making process that accompanies the technological development originates principally from the Visible Human Project coordinators, from groups developing project applications, and from public media that disseminate information about the project. Still another source is the metadiscourse of cultural analysts—we must reflexively include, for example, the metaphorical offering of "shades" to define the computerized cadavers. Members of each group participate both passively and actively in the cultural imaginary, but each has a socially positioned stake in the relation between the cultural imaginary and cultural literality. The appeal to the literal through the use of technical and policy language is more or less pronounced depending on whether it is being made by project coordinators whose audience is government funding sources, potential database users, and the media; by developers of applications whose audience is a potential market for those applications; by the media whose audience is the general public; and by cultural analysts whose audience is academia. Each is thoroughly ensconced in the dialectic between imaginary and literal, and that dialectic is constituted by the sum of their social positionalities.

This understanding of the relation between the cultural imaginary and the culturally literal remains somewhat strained unless it takes into account what I regard as an orthogonal distinction between representation and being-in-the-world. Cultural analysis will always be subject to suspicions of whether it is dealing with reality if it is cast entirely in terms of representation, or analysis of representation. The popular metaphors surrounding the Visible Human shades, and the images themselves, are necessarily of limited consequence if they are analyzed on the representational level alone. The metaphors are indeed frivolous unless they are interpreted as hinting at or disclosing a subtle shift in our mode of being-in-the-world, and the remarkable images with all their combinatorial possibilities bear no more intimate connection to the original people-cum-cadavers than a photograph torn up and taped together again unless they do the same. My intent in introducing the notion of a "shade" is to push us into thinking beyond representation and toward being.

Yet when dealing with such a general notion as being, it is important to think globally in order to avoid essentializing a cultural particularity. Thus the disposition of the actual bodies might not appear much more radical to a North American than that of a medical cadaver or a cremated dead person. Consider the integrity of being required in a Buddhist society like Japan, where one cannot take one's place among the ancestors or expect a higher reincarnation if one goes to the grave lacking a part of one's body. From that standpoint, would one react with complaisance regarding the moral import of creating shades? Final cremation aside, what might be the cosmological status of a human who has been ground into dust? On the other side of the existential looking-glass, once transformed into a shade, what might be the status of one who can be divided into chunks repeatedly? On a more mundanely North American ethical level, what about anonymity and privacy? A conventional cadaver used in anatomy training is both anonymous and lacks identity. A shade is not anonymous, for even the woman from Maryland might be recognized by an acquaintance who saw her reconstituted visage. Project director Ackerman has said "We're hopeful that if she is recognizable, that people will respect her anonymity. There is nothing we can do" that wouldn't compromise the data (*Baltimore Sun,* November 29, 1995). Moreover, a shade retains identity, for even Jernigan at least in some sense remains who he was whether or not he comes to be called "Adam" instead of Joseph.

All things considered, the argument about the being of shades would remain rather disingenuous if we were not clear that the primary concern was our own being in relation to them, or in other words how the technological innovation induces a subtle modulation in our own embodiment and hence in our own culturally situated being-in-the-world. Subjectivity and intersubjectivity are bodily phenomena, and thus the question becomes the potential transformation of subjectivity on the part of those who use the technology, and especially with regard to physicians, in the intersubjective relation formed with those bodily beings who are their patients. In this light, let us briefly consider the two most immediate sites of impact, namely anatomy training for medical students and computer-aided surgery.

### "It Empowers Us"

At the fourth annual conference on "Medicine and Virtual Reality" in January of 1996, Michael Ackerman and Victor Spitzer, director of the project at NLM and director of the contracting group at UCHSC, received the Satava Award, named for Colonel Richard Satava, a pioneer in VR telesurgery. Helene Hoffman, herself director of the anatomy curriculum using Visible Human data at UCSD medical school, observed that "This data set has become the new standard for human physiology education. For example, 30 to 40 percent of the papers presented at this year's conference alone relied on this data set." A variety of medical schools are actively developing anatomy curricula based on Visible Human data.[4] The primary debate is over whether these methods are to be used to complement or to replace conventional dissection in anatomy education. Traditionalists are resistant to the idea that medical students would not have the hands-on experience of work with real bodies in what is implicitly sacrosanct as a rite of passage in medical training. Innovators point out that actual cadavers are increasingly short in supply in comparison to the infinitely reusable shades, and that in any case, most physicians other than surgeons will never have occasion to work on the insides of their patients.

The potential consequences with respect to embodiment for both medical students and their future patients must be understood with respect to the already profound phenomenological transformation wrought in conventional training. In his ethnographic study of medical students, the anthropologist Byron Good has observed that the dimensions of the world of experience built up in their training were "more profoundly different from my everyday world than nearly any of those I have experienced in other field research" (1994:71). Reminiscent of the map without street signs metaphor I mentioned above, students learning anatomy are "as geographers moving from gross topography to the detail of microecology" (1994:72). Good repeatedly refers to the intimacy with which medical students come to know the body, and describes the anatomy lab as a kind of ritual space in which the reconstitution of experience takes place. The accustomed body surfaces that define personhood are drawn back, revealing an "interior" that consists not of a person's inner emotional life but a complex three-dimensional space with planes of tissue that are separated to distinguish the boundaries of gross forms and fine structures. As one student said, "Emotionally a leg has such a different meaning after you get the skin off" (1994:72). The new way of seeing beneath the surface that is central to the medical gaze can usually be turned on and off, but also bleeds over into the student's everyday perception of other persons, as they are constituted and reconstituted—translated and retranslated—between the perceptual languages of medicine and everyday life. Good observes that this training is profoundly visual, and the profundity of phenomenological transformation can only be enhanced by the new anatomy curricula based on virtual reality. The penetration to increasingly minute levels of biological hierarchy (epidemiological—clinical—histological—cellular—molecular/genetic) will be complemented by a penetration based on transparency, the sense of "x-ray vision."

A preliminary glimpse of the potential change comes from a reflection by a medical student who attended the introduction of the Visible Human Female in 1995. Referring to the Visible human as a prime example of high performance computing as applied to biomedical science, he writes "It empowers us. We students know that a world of information is out there at the touch of our fingertips" (Roberts 1996). He was impressed by the fact that brain sections would not fall apart as they sometimes do in dissection of a real cadaver, that one can isolate sections of the body rather than "deal with the whole daunting thing," that the circulatory system would appear like a real, three-dimensional loop rather than flat as in a textbook, that the database could be reformatted to change body characteristics, that the images could be rotated, dissected, and resected, and that some day he would be able to call up these images in his offices to help educate patients about illnesses and procedures.

Several questions arise concerning the ultimate experiential consequences of applying this technology. What will be the consequences of isolating body parts for detailed, intensive work? Will it enhance the sense of intimacy noted by Good or will it initiate a more fragmented, objectified sense of the body? What will be the consequences of digital dissection that is both exceedingly neat and comfortably reversible in comparison to actual dissection, in much the way word processing allows easy deletion and substitution in comparison to writing or typing? Will it introduce a sense of arbitrariness of biological process, or enhance the understanding of meticulous detail? Finally, what will be the consequence of empowerment as it is alluded to by the medical student rapporteur. Will it be the power of humanizing intimacy and compassion, or that of apotheosizing omnipotence and objectification of one's fellow beings? Will it refine the sensibilities of physicians as a flower blossom unfolding to reveal its intimate recesses rather than having to be sliced open or peeled apart petal by petal?

## "From Blood and Guts to Bits and Bytes"

Beyond the training of medical students, shades will increasingly play a role in the development of surgical training and what is called "telepresence surgery." The title of this section is a favorite phrase of Colonel Richard Satava, M.D., one of the leading figures in this area, in referring to a major paradigm shift in which the blood and guts of conventional surgery is replaced by the bits and bytes that will facilitate the work of a new generation of "digital physicians" and "Nintendo surgeons." Virtual reality surgical simulations are already available for prostrate, eye, leg, and cholycystectomy procedures. Telepresence surgery allows the physician to project himself or herself to a remote location via video and audio monitors, with computer-controlled instruments controlled from the remote site with dummy handles able to provide "force feedback" that gives the surgeon—or surgeons collaborating by network from different geographical locations—a sense of tactile immediacy.

Satava distinguishes between artificial and natural virtual reality, the former completely synthetic and imaginary as in the simulation of being inside a molecule, the latter a situation that could physically exist as in surgery on a realistic recreation of a human body (1992:360–361). Both surgical simulation and telepresence surgery are forms of natural virtual reality, though obviously only the latter is performed on actual patients. Yet Satava says that "the day may come when it would not be possible to determine if an operation were being performed on a real or computer generated patient ... the threshold has been crossed; and a new world is forming, half real and half virtual" (1992:363). He and his colleagues are working on just such a system, in which the operator can fly around the organs and travel through the digestive system (Key Words: 935), and use of shade data is allowing them to move from a cartoon-like visual display to an increasingly life-like one. Likewise, the overlay and enhancement of live CT/MRI data with shade data promises to augment the vividness of telepresence surgery, as the immediacy of the electronic image and remote manipulation come to "dissolve time and space" (Key Words: 939).

The development of shade-enhanced telepresence surgery has consequences for embodiment with respect to the skills it requires of the surgeon—as what Marcel Mauss (1950) called a "technique of the body"—and with respect to its applications on the bodies of patients. With respect to the former, the emerging field of "Human Interface Technology" dictates that a system have sensory intuitiveness—that it "should feel and be used as naturally as possible." As Satava observes, telepresence surgery has the same eye-hand axis as open surgery insofar as the surgeon looks down at a monitor, thus preserving the correspondence of visual with proprioceptive and kinesthetic senses. Contemporary laparoscopy requires visually looking up at a video monitor, while surgical simulation requires wearing a virtual-reality helmet such that the surgeon must learn the tool rather than the tool accommodating the surgeon (1994:819–820). With respect to patient care, the new technology will allow comparing normal and abnormal organs by substituting images, simulating the biomechanics of muscles and joints to make more effective replacement joints, demonstrating projected treatment courses for patients. Military applications—one of Colonel Satava's ultimate interests—of shade-enhanced simulation would include plotting the path of a bullet before treating a bullet wound, and applications of telepresence surgery would include "to metaphorically project a surgeon into every foxhole" (n.d.:12).

At least two questions are posed by these developments. The first comes from considering that both surgical simulation and telepresence surgery pose a paradox of simultaneously increased remoteness and enhanced intimacy. Simulation is remote from living persons and telesurgery is geographically remote; both partake of the intimacy afforded by the technologically enhanced medical gaze. What will be the consequences of this paradox, and what the limits of access to the inner recesses of biological process? The second question arises in considering Drew Leder's analysis of the typical *disappearance* of the body from awareness in everyday life as it "not only projects outward in experience but falls back into unexperiencable depths" (1990:53). Leder argues that it is the body's own structure that leads to its self-concealment and to a notion of the immateriality of mind and thought that is reified as mind–body dualism. Could it be on the culture-technological horizon that shade-enhanced virtual reality will make the intimate core of bodily processes accessible in a new way, offering the possibility of transcending this Cartesianism of the natural attitude?

## Frozen Representation and Virtual Being-in-the-World

I want to return to the broader question of the cultural significance of shades not in terms of the relation between the cultural imaginary and cultural practice, but in terms of the relation between representation and being-in-the-world. The notion of representation holds a virtual hegemony over contemporary cultural analysis, hand in hand with the associated methodological metaphor of textuality. This extends to cultural analysis of the body, so that scholarly works are filled with phrases like the body as text, writing on the body, bodies of writing, the inscription of meaning on/in the body, representations of the body, reading the body. A less prominently articulated tradition understands culture from the standpoint of embodiment as our fundamental and culturally conditional mode of being-in-the-world. As bodily beings we inhabit the world in terms of the space and extension of our bodies, we engage in movement and experience resistance to that movement we incorporate and explore the world via our senses, we interact with others or find ourselves in solitude. The modes of representation and being-in-the-world are intimately intertwined in practice, for example in the way their relation can be superimposed on the relation between subject and object: If the body is conceived as an object, representations of the body are the site of subjectivity; if the body is conceived as subject, representations are objectifications of the body.

I would argue that understanding the interaction between the body as representation and the body as being-in-the-world is critical to cultural analysis in general (see Csordas 1994b), and furthermore that this interaction defines the cultural process that is critically at stake in the existential analysis of the shades created by the Visible Human Project. From the standpoint of Jernigan and the Maryland housewife, are their shades no more than hypertext versions of a photographic representation, no more connected to their particular essences than a snapshot that could be torn to bits then reassembled with tape and glue? Or is there something of the transformation of quantity into quality in the degree of specificity with which their physical beings have been digitalized, some way in which they have gone "through the looking-glass?"

It is possible to indulge a debate about whether even a simple photograph captures something essential about a person (and anthropologists know that in some societies this is thought quite literally to be so), or is better understood as an arbitrary and momentary simulation that can be repeated without limit to the ultimate degradation of meaning, similar to what might happen to the meaning of the word "egg" if it is repeated a hundred times. However, the question of the shades' being-in-the-world in itself is academic insofar as, the tool of science fiction placed aside, there is no question of personal subjectivity for them. What is all the more at issue is the subjectivity of the rest of us—specialist medical students and surgeons to be sure, but also the coming generation. Indeed, UCHSC's Spitzer has said "I think in the future, kids will grow up with him" (certainly an improvement on Barney). More importantly, while by the same token there is no question of defining intersubjectivity between shades and users, there is all the more a question of how, given the premise that intersubjectivity is also grounded in our bodily being, it may become transformed, enhanced, or distorted by the existence and application of shades. What will interpersonal relations be like when I can casually visualize your skeleton as we converse, and you can feel your way around inside my brain?

Finally, if the biotechnological innovations in virtual reality, of which shades are only one example, are indeed pointing toward a modulation of embodiment, it may be so only because of the historical condition in which culture now exists. Daniel Boorstin wrote in 1961 that the contemporary world is already one "where fantasy is more real than reality, where the image has more dignity than its original. We hardly dare face our bewilderment, because the solace of belief in contrived reality is so thoroughly real" (quoted in Kearney 1988:252). This is to say that what we are describing is not a technological determinism of embodiment, but a highly specific way of incorporating a technological development into the postmodern condition of culture. Understanding this process will require a cultural phenomenology that can capture the essence of the particular in an embodiment constituted in the existential space between virtual and actual, between the cultural imaginary and culturally literal, between remoteness and intimacy, and between representation and being-in-the-world.

## Epilogue 2001

Since the first version of this chapter was prepared in 1996/1997, a variety of scholarly treatments of the Visible Human project have appeared (Cartwright 1998, n.d.; Curtis 1999; Kember 1998, Van Dijk n.d.; Waldby 1997a, 1997b, 1999). Each of these makes an important contribution using the project as symptomatic of issues in the cultural studies of science and feminist theory. At this juncture, rather than summarizing or analyzing these works, I want to update the preceding discussion by presenting some additional material that highlights the biopolitical—and religious—implications of the Visible Human as it has found a home in cyberspace and in technoscientific research. Notably, from the end of 1996 to the end of 2000 the number of licenses to access the Visible Human data stored on a government FTP approximately doubled, from about 700 to 1,419. My discussion singles out one project making use of these data, the implications of which are truly noteworthy. It has to do with biopolitics both in the sense of the institutional competition for resources for biotechnological research, and in Foucault's sense of the creation of knowledge that institutes control over individual bodies and populations.

In January of 2000 I attended the annual Medicine Meets Virtual Reality conference, in order to get a sense of the profile of the Visible Human work among people in this field, and of what other kinds of potentially self-transformative technologies were in the works. The group I decided to follow has as its goal the development of what they call the Virtual Human Project, intended to be a quantum leap beyond the Visible Human with respect to the capacity for simulation. The group is led by several people from Oak Ridge National Laboratories, and prominently includes the leader of the group at the University of Colorado who actually converted the Visible Humans from real cadavers into datasets, and telesurgery pioneer Richard Satava, formerly of the U.S. Army and currently of Yale Medical School. This is where the institutional form of biopolitics is in evidence in the form of interagency competition and cooperation within the world of high-stakes government research that draws in both the university and the private sector. Notably absent from the conference session was

Michael Ackerman, NLM's project director for the Visible Human. In my interview with him, Ackerman said that in his view, the Oak Ridge researchers were being opportunistic and trying to ensure their positions in the face of an unsettled situation in which a changing of the guard in government contractors was underway at the national labs. He was also emphatic that in the long run a project of such scope should be carried out under the aegis of the National Institutes of Health. There is a story in this that needs to be pursued in future work on this initiative.

The goal of the Virtual Human initiative is indeed massive, and I quote from project material:

> The Virtual Human will integrate data, biophysical (and other) models, and advanced computational algorithms into a research environment used to investigate human responses to stimuli. This effort will go far beyond the visualization of anatomy produced by the Visible Human Project to incorporate physics (such as mechanical and electrical properties of tissue) and biology (from physiological to biochemical information) into a platform so that responses to varied stimuli (biological, chemical, physical, and—it is hoped—psychological) can be predicted and results viewed [in three dimensions].

This work is expected to require simultaneous analysis of hundreds to thousands of measured variables and multidisciplinary collaboration of thousands of researchers. The key word for my purposes in this excerpt is "responses," which occurs twice. The prototypes include simulations that can, for example, predict the results of blunt trauma—these images, frozen here, are animated in the computer displays. Applications? Well, the Marine Corps wants to simulate the response of the body to rubber bullets and other kinetic weapons used to control unruly civilians; the Army wants to be able to predict whether a soldier's lung has collapsed as the result of a puncture wound. Implications? I think we can conceive the very real possibility of a simulation ultimately so sophisticated, including the psychological dimension, that it responds as a form of life—and I use the word "as" intentionally to exploit its ambiguity, since "as" can be taken both literally and in the sense of "as if" or "like" a form of life. To be specific, this development portends the creation of an entirely new figure of knowledge in the life sciences: At present biological research can be conducted either in vivo or in vitro—from this point it will also be able to be conducted *in virtuo*.

There is more, however. I quote again from the Oak Ridge material:

> The beauty of a computer model of a human is that it can be customized for a specific person at any time .... By using equations and changing some parameters, we can make the heart smaller or larger. We can make a human model or phantom grow with age. Use of a customized model—a computerized clone of you that includes your genetic makeup—will make it possible to predict how you might

respond to different doses of radiation, chemicals, and drugs, or what damage you might suffer if you were in an automobile accident or airplane crash .... Looking ahead five to ten years from now, [we see] a "human model on a chip" ... [after all,] The more you know about your body ... the more you can take charge of your well-being.

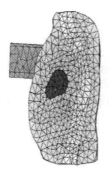
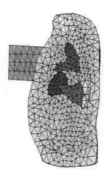

**Figure 10.6**    Finite-element model of a wood block moving at constant velocity of 1 m/s impacting a finite-element model of the human torso and deforming the torso. The impact lasts for 0.0001 second. This is a qualitative simulation to test the finite-element models and the simulation software. At this stage of research, the tissue elastic constants used were only very approximately related to real tissue elasticity.

*Source*: Image and caption reproduced by permission from research sponsored by the Laboratory Directed Research and Development Program of Oak Ridge National Laboratory (ORNL), managed by UT-Battelle, LLC for the U.S. Department of Energy under Contract No. DE-AC05-00OR22725.

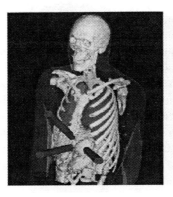

**Figure 10.7**    The beams are qualitative representations of radiation therapy treatment beams coming from different directions and focusing on a spot in the human torso that represents a tumor. The image demonstrates the capability to display the human body in three dimensions with the radiation therapy beams. The software allows the body to be rotated in three dimensions for easy evaluation of the treatment scenario.

*Source*: Image and caption reproduced by permission from research sponsored by the Laboratory Directed Research and Development Program of Oak Ridge National Laboratory (ORNL), managed by UT-Battelle, LLC for the U.S. Department of Energy under Contract No. DE-AC05-00OR22725.

There is a question here of who in fact will "take charge" of this knowledge, and who will have the capacity to take charge—the individual patient, patient advocacy organizations, physicians, insurance companies, the government. But let me approach this question from within the argument I've started to develop in this chapter.

The terms I introduced for types of beings in cyberspace might be useful in understanding the transformation that is occurring between the Visible Human and the Virtual Human. The former I called a "shade" in the sense that it is the reconstituted digital remain of a specific person. The virtual phantoms being prototyped now are what I called simuloids, in the sense that they are generic representations built on the anatomical platform of the Visible Human, but constructed in a way that is a highly diffuse and distributed composite of databases and computational resources. The projected customized computer chips will be avatars in the sense of being stand-ins for real, live people. A clue to the cultural and ontological ambiguity of these entities can be taken from the appearance in tandem of the terms "model" and "clone." A model is a kind of representation, bearing an abstract and neutral relationship to that of which it is a model. A clone implicates a certain kind of being-in-the-world, one that is not abstract but concretely parallel to the being from which it has been cloned.

Before concluding with my reflection on representation and being-in-the-world, let me extend the comparison between the Visible Human and the Virtual Human in Foucault's terms. The Visible Human raises issues of what Foucault would call anatomo-politics of the body, but in the quite specific sense of the deployment of anatomy as a discourse about life. The conventionalized bio-ethical problems being discussed at this level are those of the rights of those who become virtual cadavers to anonymity as medical subjects, and the pedagogical ones are about the value of simulations in anatomy education and surgical training. The Virtual Human crosses the threshold into another concern of Foucault's, that of the biopolitics of populations. The philosopher Giorgio Agamben has posed the following question about a certain ambiguity in Foucault: "... where, in the body of power, is the zone of indistinction (or at least, the point of intersection) at which techniques of individualization [characteristic of anatomo-politics] and totalizing procedures [characteristic of biopolitics] converge?" (1998:6). If this can be understood as an ethnographic question, then one possible answer is "right here." The customized human model or clone on a chip is an end point of biotechnical individualization, and the fact that such a chip could be generated for everyone is an end point of totalization through a governmental/private/academic initiative, the power of which is diffuse in that like any big science it is highly distributive and collaborative. The question opened for reflection here is whether this moment of intersection points to the possibility for a new form of human existence, or at least a new form of understanding human life, or whether it is explicitly one of the potential for a surveillance that penetrates to the very sinews of the self. In the terms I introduced above, the answer may have to do with whether our participation in the cultural imaginary is in the active or the passive mode.

These considerations lead directly to the ethnography of imagination, and back to religion. My thinking about representation and being-in-the-world has been developed in my ongoing ethnographic studies of the religious imagination. To me, the prototypic ethnography of imagination was the one carried out by the English poet and engraver William Blake, as he traveled imaginatively among the fires of hell, inventing the ethnography of speaking as he "... collected some of their Proverbs; thinking that as the sayings used in a nation mark its character, so the Proverbs of Hell shew the nature of Infernal wisdom better than any description of buildings or garments" (Erdman 1988:35). I've tried to follow Blake's lead in my work with Catholic Charismatics on religious healing, religious language in performance, and religion in the context of globalization. In becoming interested in the question of technology, I have also become aware that I am once again, or still, very emphatically in the zone of imagination. Stated another way, if on the one hand I have encountered, as I expected, a field of study with its own imperatives, agendas, and problematic, on the other hand, I find that the issues are not so far removed from a concern with religion and the religious imagination as I had once expected.

Blake, in addition to grounding his understanding of imagination in bodily experience, recognized that imagination and religion are one, and that when they are separated, imagination is reduced to mere fantasy, and religion becomes a claim to absolute truth. Both thereby become denials of human life and energy, relegating their creative power to the realm of mystery. But if we are to understand this, our work needs to be broader than an invocation of the liminal, and more specific than a definition of the imaginary as an analytic space or zone of cultural analysis, a psychological process, or a modality of being-in-the-world. We need instead to identify imaginaries in the plural as ethnographically specifiable zones of human creativity, possibility, fantasy, desire, horror, alterity, holiness—but also planning, research and development, product design. An understanding of the imaginary must be pluralizing in the sense of identifying cultural portals to imagination. For example, each of the constituencies of the Visible Human Project I have discussed in this chapter has its own subject position within, or its own socially circumscribed mode of access to, the cultural imaginary. The general public exposed to media images inhabits an imaginary of spectacle in relation to the computerized cadavers, reminiscent of the body's display in public execution and dismemberment, or in the freakishness of the carnival side show. Medical students and surgeons training for telesurgery inhabit an imaginary of initiation, in which they are inculcated with dispositions toward anatomy and incorporate novel techniques of the body. The originators of the Virtual Human inhabit an imaginary of creation, aspiring to the production of new forms of knowledge about life and response.

Insofar as a turn to the question of technology is thus not at all a turn away from the study of religion, it is also to be taken in the sense of

Martin Heidegger's remark in his essay on technology that we typically assume that we have technology "spiritually in hand." Heidegger wants us to be aware that the kind of revealing of reality and being that we achieve through technology is not the kind that is based on a poesis that brings forth something, but is based on an enframing that, in his somewhat odd phrase (at least in translation), challenges forth. This enframing transforms things and objects into a kind of standing-reserve for technical application. Heidegger (1977b) says that "The coming to presence of technology threatens revealing with the possibility that everything will present itself only in the unconcealedness of standing-reserve." He would likely be concerned that the Human Genome Project portends that the phrase "gene pool" is at risk of becoming transformed into something analogous to "motor pool"—not a standing-reserve of cars and trucks, but of DNA. He would likely be concerned that what the Virtual Human Project offers spiritually is to create a standing-reserve not of DNA but of human responsiveness. What we can ask is whether Heidegger's distinction between poesis and enframing condemns our engagement with technology to the passive mode of the imaginary—we can ask whether, to the extent that technology reveals not only bare life but also raw existence, it can be the subject of poesis and the active imaginary. And we must ask ourselves for a critical vigilance with respect to the consequences.

In another sense, the work of Agamben reminds us that the original Visible Human male, Richard Jernigan, meets the criterion of the original meaning of *homo sacer*—that is, "sacred man" in archaic Roman law. This is the man condemned for a crime, who may be killed with impunity by anyone and yet may not be sacrificed. Resolving the paradox in this statement is a central aim of Agamben's work, but it also speaks to something particularly uncanny about an origin myth for a race of shades whose progenitor is this kind of sacred man. And if we were to pursue this line of thinking, we would have to ask if indeed what has happened is that he was killed (not through impunity, but through the sovereign power of the state of Texas) and not sacrificed. But is the immolation of a body willed to science by slicing it into sections a sacrifice? Is the act of sacrifice one that is carried out by a technoscientific priesthood, or one that was achieved in the condemned man offering up of himself? Was the condemned man's intent one of sacrifice or of self-enshrinement? At this point my remarks end at the tangled religious roots of our civilization, roots constituted in part by questions about the nature of life and the problem of death. This is what we see in the images of Adam and Eve in the cultural imaginary framed by the popular press, the transformed consciousness of the body and perception of life and death in the transformation of consciousness undergone by medical students and practitioners of telesurgery, the creation of life *in virtuo* in the biopolitical simulation of response by the Virtual Human advocates, and the implicit question of the sanctity and inviolability of life in all of them.

# Notes

## Chapter One

1. A few studies have been done on the compatibility of religious and scientific belief systems in the treatment of emotional distress (Cox 1973), and on the effect of religious belief on the work of practicing psychotherapists (Gaines 1982). Unlike Gaines's informants, who were psychiatrists introducing religious ideas into their therapy, the informant in this example is a religious healer introducing her knowledge of psychotherapy into her ritual work. On these issues, see also Csordas (1990), Schumaker (1992), and Koenig (1998).
2. A notable exception to this tendency is the discussion in Good et al. (1982) of transference and countertransference in a collaboration between traditional healers and mental health professionals.
3. Deliverance is regarded as appropriate only in cases of demonic oppression; cases of full-scale possession are to be referred to a priest for treatment with the official Church rite of Exorcism, and possibly, in addition, to a trained psychotherapist.
4. Catholic Pentecostals also make use of the Sacrament of the sick (formerly Extreme Unction) in conjunction with healing, and encourage participants to have recourse to the Eucharist as a source of spiritual well-being.
5. Bourguignon (1976), for purposes of a broad review, chooses not to distinguish between religious healing movements and religious healing in other contexts. Yet, in practice, the fact that participation in a healing movement is the product of secondary socialization may represent a significant difference from situations in which a particular form of religious healing is a taken-for-granted option in a society's hierarchy of therapeutic resort. In addition, the fact that (in Kroeber's terms) religious movements constitute "part-societies," with techniques of healing relevant only to movement participants, may represent a significant difference from situations in which religious healing is available to whole societies.
6. In this context the term "to walk" is a shortened form of "to walk with the Lord," a phrase that, in the Charismatic vernacular, is synonymous with "to lead the Christian life."
7. The reader familiar with psychotherapeutic techniques will recognize the clinically trained informant's reference to both behavior modification and, in the notion of statements she makes to herself, to elements of Albert Ellis's rational-emotive therapy. An intriguing glimpse at the interface of the sacred and secular is offered in her explanation of why, at the same time, she dislikes gestalt techniques: "I question it [gestalt] because of some of the fruits that come from it. Much of the fruit that comes from it turns out to either border on sin or be sin; [that is,] in terms of how a lot of gestalt therapists get into a lot of sensitivity stuff where there's a lot of freedom, which really borders on encounter groups, which can even border on TM. A lot of gestalt therapists have gotten into that. They get into a lot of stuff that is transcendental in nature that doesn't 'test the spirit' as to whether that's really the spirit of Jesus or another spirit."
8. The fact that this supplicant was himself a healer may explain his willingness to attempt the self-healing, but does not diminish the significance of its solitary nature.
9. *Discourse* can be defined simply as linguistic performance in contrast with competence, or *parole* in contrast with *langue,* though it does include nonlinguistic semiotic modes and forms of communication (Ricoeur 1979). In its strongest formulation, the structure of discourse is the locus of the very conditions of possibility of knowledge (Foucault 1970, 1972).

The notions of discourse elaborated by Ricoeur and Foucault are derived from textual analysis rather than from analysis of oral tradition or cultural practice; hence the autonomy of discourse is based on the observation of the independence of text from author subsequent to composition. Anthropologists who distrust analogies from textual analysis should bear in mind that, in all but the most rigidly mentalistic formulations, culture is conceived to have systemic properties that make it semiautonomous of individual culture-bearers. In this light the contribution of discourse to the study of culture is twofold: (1) While maintaining an emphasis on culture as a semiautonomous system, discourse is by definition a type of process, thus overcoming the tendency of culture to be conceived as a static system and the consequent conceptual difficulties in explaining cultural change, and allowing an unstrained analysis of culture in action; (2) since discourse is by definition a social (communicative) product that in turn influences social action in a dialectical fashion, the semiautonomy of discourse can be postulated without an appeal to a "superorganic" ontological level à la Kroeber. The concept of culture associated with the notion of discourse is neither superorganic nor mentalistic, but closer to that of Sapir, for whom the locus of culture was in the interaction *between* individuals.

10. As an aspect of discourse, rhetoric can be conceived as its "cutting edge"—the means by which participants in discourse are convinced of its validity and relevance. A concern with rhetoric has been introduced into the human sciences in part through the work of Kenneth Burke (1970a, 1970b), and is evident in the "performance-centered" approach in folklore (Abrahams 1968) and sociolinguistics (Hymes 1975), and the "cultural performance" approach in cultural anthropology (Fernandez 1972, 1974; Geertz 1973; Singer 1958; Turner 1974; confer. Csordas 1980a).

11. It is best to summarize Schutz's approach to meaning in his own terms:

It is misleading to say that experiences have meaning. Meaning does not lie *in* the experience. Rather, those experiences are meaningful which are grasped reflectively. The meaning is the way in which the ego regards its experience .... (1967:69)

[A] meaning is not really attached to an action. If we say it is, we should understand that statement as a metaphorical way of saying that we direct our attention upon our experiences in such a way as to constitute out of them a unified action. (1967:63)

The meaning is merely the special way in which the subject attends to his lived experience; it is this which elevates the experience into an action. It is incorrect, then, to regard meaning as if it were some kind of predicate which could be "attached" to an action. (1967:215)

12. Ness makes the important observation that this form of interaction may in the long run have greater therapeutic effect than the healing rituals themselves (1980:178).

13. Mills defines *motives* as follows: "Motives are words. Generally, to what do they refer? They do not denote elements 'in' individuals. They stand for anticipated situational consequences of conduct. Intention or purpose (as stated as a 'program') is awareness of anticipated consequences; motives are names for consequential situations, and surrogates for actions leading to them ..." (1940:905).

14. This is not to argue that the magical element is totally lacking (confer McGuire 1982), but that the gesture is much more complex than a magical transfer. Even when a Charismatic lays hands on the hood of his car when it won't start in the morning, the "magical" gesture is accompanied by the "religious" act of prayer. It might seem that the person in this situation lacks technical knowledge about his car in a manner similar to the "magical" thinkers of traditional society described by Malinowski. Yet the intent of the prayer is as likely to be that the day's activities not be upset by having to summon a mechanic as that a miraculous process be worked in the mysterious engine. In addition, performing a gesture such as this has the meaning of an exhibition of faith to the point of being overtly "foolish for the Lord."

15. I thank Jay Geller for this observation about the hands as a shield. Geller also suggests that some of the rhetorical force of the laying on of hands may derive from an inversion whereby gesture, which is normally marginal and ancillary to discourse narrowly defined as the spoken word, becomes central within the ritual setting. Indeed, this may contribute to the sense of mystical power conveyed by the gesture. It is not necessarily the case that language loses its central place, however; and there is simultaneously a rhetorically powerful inversion within the linguistic component itself, namely, the inversion of intelligibility achieved in glossolalia.

16. An interesting historical aside is the strong belief of one of the early eighteenth-century originators of literal immersion (hydrotherapy) as a medical treatment in its simultaneous spiritual efficacy: "I have made the Immersion almost an Universal Remedy for our infirm Bodies, as well as a

miraculous Purifier and Cleanser of the Soul by its Supernatural virtue .... [A]ll Divine Institutions have such large and diffusive Virtue, as to remedy the Disorders both of Body and Mind" (Floyer, quoted in Gabbay 1982:38).

17. Not directly related to the Healing of Memories or Deliverance, but an important example of the link between experience of the holy and that of physical/emotional well being, is the *technique du corps* known as "Slaying in the Spirit" or "resting in the Spirit." A person slain in the Spirit is in a state of motor dissociation, defined by loss of control of voluntary muscle activity for a period of from ten minutes to half an hour, followed by a feeling of relaxation and rejuvenation. In the release of the Holy Spirit's power triggered by the laying on of hands the oft-cited mechanism of suggestion (Calestro 1972) is undoubtedly called into play to activate the endogenous process of dissociation. From informants' descriptions of the blissful rapture of this state, it appears that its physiological correlate might possibly be a release of endorphins in the central nervous system (confer Prince [1982] for analyses of the role of endorphins in ecstatic religion). Yet neither the psychological mechanism of suggestion nor the physiological mechanism of endorphins is sufficient to an interpretation of how Slaying in the Spirit contributes to healing—that is, why people are susceptible to this experience and what benefit they derive from it. In fact, the experience is understood as a gratuitous infusion of divine power for the purpose of enhancing general well-being. The extent of rhetorical control over the endogenous physiological process is evident in supplicants' claim that, in spite of the dissociation, they never lose consciousness. The reasons, consistent with Catholic Pentecostal notions of personhood and spirituality, are that the experience is meant to be appreciated and enjoyed, and that God would never deprive those He wants as Servants of their faculties. A Slaying in the Spirit in which a person lost consciousness would be interpreted as demonically inspired. The total somatic experience is a physiological symbolization of mystical union, in which the supplicant is literally overwhelmed with divine power.

18. An easily observed, but less easily explained feature of rhetoric is its remarkable immunity to contradiction. In the present case Spontaneity remains a potent motive in spite of the facts that prayer meetings have become subject to conformity with increasingly authoritative norms, and the "holy hug" has become a ritualized greeting.

19. Kapferer (1979c) constructs an equally convincing account of transformation in healing, starting from, as it were, the opposite direction from that taken in this chapter. Whereas our focus is on the rhetorical *means* of transformation, and analysis shows how attention is redirected and a new world and selfhood are created, Kapferer's focus is on the *subject* of transformation, and his Meadian analysis shows how the self in illness and healing enters into a variety of relationships with the other and generalized other. The primary differences are his greater emphasis on interaction vis-à-vis our greater emphasis on endogenous processes and the re-creation of the phenomenological world (predisposition and empowerment) as well as the self (transformation). The latter difference may be due less to methodology than to research setting: transformative social movement for us, everyday social practice for Kapferer.

20. The importance of presence and present in Catholic Pentecostal practice was originally pointed out in a dissertation by Mawn (1975), and in a rather different context from that in which they are presented in this paper.

21. Each patient recruited agreed to allow me to be present during the session; given the intimacy and painfulness of many issues dealt with in their healing session, this was a privilege for which I remain deeply grateful. For each participant, I recorded up to five healing sessions on audiocassette tape. During a subsequent interview, I asked each participant to identify the most important or meaningful event within the session. I played back the recording of each event and elicited commentaries from each person, using an adapted form of the Interpersonal Process Recall (IPR) method developed by psychotherapy process researchers (Elliott 1984, 1986). An additional background interview covered basic life history and medical/psychiatric history, nature and level of involvement in the Charismatic Renewal, and attitudes and expectations of religious healing. In order to confirm presence or absence of psychiatric disorder, this interview included an adapted and shortened form of the Schedule of Affective Disorders and Schizophrenia (SADS). I contacted participants prior to the first session observed for purposes of explaining the study, obtaining informed consent, and initiating rapport, so that my presence would not be perceived as disruptive. I presented healing ministers with the option of recording sessions without the researcher's presence, but all invariably declined. Although several participants were somewhat nervous in early sessions, none chose to terminate involvement with the project. Except on rare occasions, one of which is reported here, I was not drawn into the proceedings in any way. The importance of this method is

that it allows observation of nonverbal behavior in the healing sessions and adds to informant-researcher rapport. The researcher with access to the intimacy of the healing session is, however, no mere fly on the wall. Indeed, inevitably some topics discussed in research interviews are recycled back into the healing process. The act of listening to the tapes of the sessions raises thoughts and emotions that may otherwise remain unexamined. These issues of reflexivity are too large for adequate discussion here. This research was supported by NIMH grant 1 RO1 MH 40473-03.

22. Her medications at the time of the study, a tranquilizer (Clonipan) and an antidepressant (Nardil), appear to confirm this diagnosis.

23. Father Felix never explored the possible experiential connection between the frightening presence and the earlier death of Margo's brother, since the occurrence of that event never emerged in the sessions.

24. According to Ralph and his parents, various physicians disagreed over whether he was in fact diagnosable as paranoid schizophrenic and whether he in fact had a brain lesion, but his primary medications included an antipsychotic (Mellaril), an anticonvulsant (Tegretol), and an antidepressant (Elavil).

25. It is more than likely that this delusion of reference with fear of homosexuality is the basis for the diagnosis he reports for himself, but it is also the case that he exhibits no other typical symptoms of schizophrenia. He mentions the term "grandiosity" with reference to his self-conception as a potentially great poet but admits that another poet has literally complimented him at that level. In any case, adolescent grandiosity about the stature of one's poetry cannot be considered pathological; chances are that his use of the term "grandiosity" is based on an interpretation by a mental health professional.

26. Ralph brought up the fact that other doctors had warned him against hypnosis for fear he might develop the idea that the doctor was controlling his mind, but he denied having any such delusions.

27. However, it is uncertain to what extent Margo shared this attitude with her mother. Ironically, as a psychologist Father Felix was opposed to the use of ETC.

28. Personal communication from movement leaders indicates that since its inception in 1967, 10 million North American Catholics have been exposed to the movement, along with 20 million Catholics in other parts of the world. Active participation at the time of this writing, however, is estimated at 162,500. This remains a substantial number, especially when placed alongside the larger number of Protestant Charismatics and Pentecostals.

29. For this reason, despite the preliminary nature of their analyses, the works of Richard Noll (1983), Larry Peters (1981), and Larry Peters and Douglass Price-Williams (1980) on shamanism are important steps toward understanding altered states as loci for personal transformation, as social rites of passage, and as the simultaneous activation of physiological and symbolic processes.

30. One can go beyond the relativistic starting point only in the context of a broad and thoroughgoing cultural critique; in the present connection, see for example LaBarre's (1972) analysis of the impact of St. Paul's sexual neuroses on Christian culture.

31. The problem of guilt looms large for conventional as well as for religious psychotherapy in the Judaeo-Christian sphere of influence. As Cox notes, guilt may be interpreted quite differently in the two settings (1973:9). Precise comparison of how these interpretations differ should be the focus of further study, for they raise the fundamental question of whether sacred and secular therapies are in fact alternate ways of dealing with the same problems, or whether they address quite different ranges of cognition and affect.

32. McGuire's (1982) recent book on Catholic Pentecostalism includes a general discussion of healing that focuses primarily on physical illness. Her views will be reviewed in detail elsewhere.

33. This clarification of motives seems important in light of Ohnuko-Tierney's rather tight-lipped response to Comaroff's (1983) review of her book. Whereas the reviewer attempts to demonstrate the relevance of a wider range of issues to medical phenomena, one gets the impression that Ohnuko-Tierney regards her as an interloper raising questions of peripheral relevance. The lessons to be learned from medical systems are too important to allow such an attitude to become entrenched; certainly they are too important to allow the field to degenerate into another battleground between symbolists and cognitivists.

34. Good and Good's (1981) method really involves a double hermeneutic: One is carried out by the student interpreting the medical systems, and the other by the therapist as part of the treatment. The clinical training described involves raising the therapist's awareness of his own inevitable hermeneutic from that of taken-for-granted "native exegesis" to that of systematic interpretive method.

# Chapter Two

*Acknowledgments:* The research reported in this paper was supported by NIMH grant 1 RO1 MH 40473-03. I am grateful to participants in the weekly seminar in Clinically Relevant Medical Anthropology at Harvard University directed by Arthur Kleinman and Byron Good, for creating an environment that has stimulated the development of this work at various stages. Comments on an earlier version of the paper were offered by Pierre Maranda and Byron Good during a symposium organized by Gilles Bibeau and Ellen Corin at ICAES XII in Zagreb, Yugoslavia. Gananath Obeyesekere, Robert LeVine, and Nancy Scheper-Hughes proffered the greatest encouragement by selecting the paper as winner of the Stirling Award. The argument was further refined in response to a helpful critique by Richard Shweder. Finally, I am grateful to Janis Jenkins, whose theoretical dialogue and editorial pen contributed much to what, by any standards, is an experimental argument.

1. In addition to works cited in the text, several major theorists have developed perspectives on the body (Douglas 1973: Foucault 1973, 1977; Ong 1967; Straus 1963). Anthropologists have periodically examined the social and symbolic significance of the body and the senses (for example, Hertz 1960; Leach 1958; Benthall and Polhemus 1975; Blacking 1977; Obeyesekere 1981; Howes 1987; Hanna 1988; Tyler 1988). Particular fields that have made recent contributions include medical and psychiatric anthropology (Devisch 1983; Scheper-Hughes and Lock 1987; Frank 1986; Good 1988; Martin 1987; Kleinman 1980, 1986; Kirmayer 1984; Favazza 1987), social anthropology (Jackson 1981), sociology (Armstrong 1983; Turner 1984), philosophy (Johnson 1987; Levin 1985; Tymieniecka 1988), history (Bell 1985; Bynum 1987; Feher 1989), and literary criticism (Scarry 1985; Berger 1987, Suleiman 1986). This is naturally only a sampling of relevant works, and the list continues to expand. [For additional bibliography since the original version of this chapter was published in 1990, see Lock (1993) and Csordas (1999a, 1999b).]

2. The argument I am developing about the body as existential ground of culture is to be distinguished from that of Johnson (1987), who analyzes the body as cognitive ground of culture.

3. These distinctions roughly presage the empirical delineation of a continuum of person-concepts between egocentric and sociocentric by Shweder and Bourne (1982).

4. Whereas empiricism erroneously posits a world of impressions and stimuli in itself, the antithetical error of intellectualism posits a universe of determining, constituting thought. Intellectualism (epitomized by Descartes) confuses perceptual consciousness with the exact forms of scientific consciousness. Both positions start with the objective world rather than sticking closely to perception, and neither can express, as Merleau-Ponty wrote, the "peculiar way in which perceptual consciousness constitutes its object." Intellectualism is weakened, he says, by its lack of "contingency in the occasions of thought," and its requirement of an abstract capacity of judgment that transforms sensation into perception (Merleau-Ponty 1962:26–51).

5. Merleau-Ponty's reference to the unequal lines of an optical illusion is to the well-known Muller-Lyer diagram. Cross-cultural studies suggest that both shaping of geometric perception within the behavioral environment (the carpentered-world hypothesis) and psychophysiological factors (variations in retinal pigmentation) may play a role in whether the diagram is perceived as illusory (Cole and Scribner 1974). It is these very differences that make it important to begin with the perceiving subject rather than the analytically constituted object in the study of perception as a psychocultural process, especially when we move from visual perception to self-perception.

6. Hallowell (1955) makes a similar point that environmental resources are not objectified as "resources" until they are recognized as such by a people and until there is a technology developed to exploit them.

7. The first umpire declares, "I calls 'em as they are." The second replies, "I calls 'em as I sees 'em." The third announces, "They ain't nothing till I calls 'em."

8. Bourdieu rejects phenomenology in the guise of Schutz and the ethnomethodologists on the one hand, and Sartre on the other, while including favorable citations of Merleau-Ponty's (1942) early work on behavior.

9. The distinction between existence and being is essential to the thought of Merleau-Ponty and, in general, to phenomenology and existential psychology. In anthropological terms it can be roughly translated as the distinction between intentional action and constituted culture.

10. I do not believe that Bourdieu's reference to a generative principle implies a search for a "deep grammar of practices" reminiscent of Chomsky's linguistics. Insofar as Bourdieu's generative and unifying principle is the socially informed body, it must be considered as *given* in an existential sense

rather than as *innate* in the sense of cognitive hard-wiring. Bourdieu explicitly includes Chomsky in his critique of the objectivist conception of rule in social and linguistic theory (1977:10–30) The critical distinction is that the habitus and its constituent dispositions are *nonrepresentational, as* opposed to the objectivist model and its constituent rules. In accounting for practices governed by rules unknown to agents and thus outside their experience, it thus avoids the "fallacy of the rule which implicitly places in the consciousness of the individual agents a knowledge built up against that experience" (1977:29).

11. On the relation between Merleau-Ponty and structuralism proper, see Edie (1971). Boon offers a brief but insightful analysis of parallelism between the mutual attempts by Levi Strauss and Merleau-Ponty to overcome the subject-object duality promulgated by Sartre: "For Levi-Strauss totemisms institutionalize reciprocal object-object relations from the viewpoint of the totalizing classification system (langue). For Merleau-Ponty pronouns, art, and so forth institutionalize reciprocal subject-subject relations (artists and pronouns 'view' objects as subjects) from the viewpoint of intersubjectivity" (Boon 1982:281).

12. I avoid the term "mental imagery" because it begs the question of our problematic distinctions between body and mind, because it tends to imply a focus on visual imagery rather than the integration of the senses in imagery processes (confer Ong 1967 on the "sensorium"), and because it belies the need to examine the relation of image and emotion.

13. This is not the place to discuss cultural concepts of power, but it can be said that the concept invoked here has much more in common with ethnologically familiar notions of spiritual power such as *mana, orenda,* or *manitou,* than with current North American ethnopsychological notions of "personal empowerment."

14. Field's account can be compared with the outlawing of drums among African slaves in the antebellum United States. Here was a situation where the threat was not explicitly linguistic, but was semantically a more complete form of embodied communication insofar as actual messages can be sent via "talking drums." From the slave owners' perspective the drumming was both unintelligible and a concrete threat to social order.

15. The cultural language of self-objectification is here preferable to the psychoanalytic language of "regression in service of the ego" (Kris 1952), because the latter is less attuned to what kind of ego—in this case, one constituted in religious terms—is in question.

16. Because the ritual systems of different branches of Charismatic Christianity vary somewhat, for the sake of consistency the discussion in this section is restricted to the Roman Catholic Charismatic Renewal.

17. This level of indeterminacy made glossolalia a key symbol in the postmodernist fiction of Pynchon, who not only constantly invokes Pentecost and speaking in tongues, but impregnates his pages with a multitude of languages and pseudo-languages. For Pynchon, "Pentecost is a version of the state of entropy which takes what is, and celebrates it. Pentecost is entropy with value added—the value of communication" (Lhamon 1976:70). 1 have not used Pentecost as an image of an entropic postmodernist world in which everything refers to everything else, but would argue that the principle of indeterminacy essential to embodiment makes such a world possible.

18. See note 4 on Merleau-Ponty's parallel critique of empiricism and intellectualism. For a contemporary critique of empiricist language in medical science see Good and Good (1981).

19. An additional example is provided by Fernandez (1990), who points out that the drug-induced bodily experience of Fang participants in the Bwiti religion is misrecognized precisely as its opposite, a state of disembodiment, and thematized as an approximation of the serene, purified disembodiment of the ancestors.

20. Bourdieu is perhaps less successful in going beyond dialectic to the collapse of dualities, remaining bound to apparently oxymoronic articulations about spontaneous dispositions, regulated improvisation, or intentionless invention. Accordingly, discussion in this section leans more heavily on the work of Merleau-Ponty.

21. I have offered Bourdieu's concept of the habitus to forestall the lapse of phenomenology into the microanalysis of individual subjectivity, and to emphasize the social and cultural background that Merleau-Ponty requires but does not sufficiently elaborate. I have confronted Bourdieu's antiphenomenological bias with preobjective intentionality and the transcendent constitution of cultural objects, in order to compensate for his inadequate provision for self-motivated change within the habitus.

22. The ramifications are too great to broach here. Consider only the reliance of cognitive developmental theory, which owes much to Piaget, on the objective notion of *representation* intervening

between stimulus and response (Kohlberg 1969). A phenomenology of the body does not posit this kind of object and concentrates not on intervening reference and representation, but on immediate relation and rapport of the body with the world (Hottois 1988).

23. The very distinction between hard and soft is imbued with machismo, for there is no doubt about its cultural connotation that hard data are more tough-minded and hence better. To the extent that our attitudes are shaped by conventional metaphors, and as someone who has worked in both modes, I would propose that we experiment with replacing "soft and hard" data by "flexible and brittle" data.

24. The most vivid example of the constitution of the real as an indefinite series of perspectival views is Merleau-Ponty's (1964a) essay on "Cezanne's Doubt," which he begins with the observation that the painter required 100 working sessions for a still life, and 150 sittings for a portrait.

25. Undoubtedly the most fruitful attempt to date to deal with indeterminacy is Fernandez's elaboration of the notion of the inchoate as "the underlying (psychophysiological) and overlying (sociocultural) sense of entity (entirety of being or wholeness) which we reach for to express (by predication) and act out (by performance) but can never grasp" (1982:39). For Fernandez, the inchoate is the ground of emotional meaning, moral imagination, identity, and self-objectification. That the principle of indeterminacy elaborated in the paradigm of embodiment may contribute to understanding the inchoate is suggested by Fernandez's (1990) attempt, in dialogue with Werbner, to rethink earlier analyses of religious experience from the perspective of bodily experience.

It may also be this principle of indeterminacy, inherent in social life, that has come to the fore in postmodernist anthropology's shift from pattern to pastiche, from key symbols to blurred genres. Anthropologists such as Tyler (1988) have launched a critique of empiricist theories of the senses and called for an approach to language as incarnate, but the postmodernist critique remains committed to the idiom of semiotics and textuality. The perspective of embodiment may provide psychological anthropology with its own analytic purchase on postmodern processes of culture and self.

## Chapter Three

1. For comprehensive treatments of Catholic Charismatic healing, see Csordas (1994a) and McGuire (1982, 1983).

2. For a cultural analysis of the loss of wanted pregnancy that includes religious and symbolic responses see Layne (1992).

3. I am grateful to Susan Sered for drawing my attention to the Japanese case.

4. My discussion of *mizuko kuyo* and abortion in Japan relies heavily on the excellent account provided by LaFleur (1992).

5. Necessity is sometimes conceived under the metaphor of "culling of seedlings," is performed in order to enhance the viability of those that survive (LaFleur 1992:99) The notion of *tatari*, that spirits of those who die untimely, unnatural, or unjust deaths may seek revenge on the living, is an old one in Japan, and is currently rather controversial with respect to the practice of *mizuko kuyo* (LaFleur 1992:55, 163–172).

## Chapter Four

*Acknowledgments:* Earlier versions of this chapter were presented to the Seminar in Clinically Relevant Medical Anthropology at Harvard University, where valuable comments were offered especially by Arthur Kleinman, Byron Good, and Janis Jenkins. A version was presented to the invited symposium on The Dialectic of Medical and Sacred Realities at the 1986 annual meetings of the American Anthropological Association, where valuable comments were added by Jean Comaroff, Stanley Tambiah, and Atwood Gaines. The paper was completed under support from NIMH grant 2ROI-MH40473-04.

1. In all, I conducted three interviews with Peggy during the spring and summer of 1986. These were followed up with periodic phone conversations that continued for two and a half years. Although it was clear that I offered no therapeutic help or direct contribution to the religious resolution of the problem, Peggy remained open to my questions on the grounds that, at the least, the account of Martin's trials could help other similarly afflicted people in the future. On these grounds she also (unsuccessfully) encouraged Martin to complete a standard psychiatric symptom checklist (SCL-90).

Although she understood that it was designed to assess symptoms of psychopathology, she was firmly convinced that his was a religious, instead of a psychiatric, problem.

2. In the Charismatic religious tradition demons typically have names drawn from the cultural repertoire of negative affects, personality traits, and behaviors (see Chapter One).

3. Peggy rejected the notion that the spirit could be negotiated with or "converted" on the grounds that it is one of Satan's minions and as such is irredeemably diabolical. She also rejected the notion that the spirit was Martin's deceased father, although such an identification would be acceptable to some Catholic Charismatic healers who practice "generational" or "ancestral" healing (see Chapter One).

4. Nevertheless, in these curses it never directly utters the name "Jesus." This is a notable element of cultural and religious shaping or modeling of the spontaneous audition and can be interpreted according to the belief that Jesus is so powerful that a demon fears to use his name even in a curse.

5. Compare the discussion of superimposition of images in hallucination by Merleau-Ponty (1962:334–345).

6. Compare the theory of emotions proposed by William James (1967) in which emotional experience is a response to prior biological and physical changes.

7. The formal Church rite of Exorcism differs from Deliverance prayer in two important respects. First, it implies a full-blown possession in which the demon is understood to be inside a person and in control of all that person's faculties. In a situation requiring Deliverance, the demon is typically outside the person, "harassing" and "oppressing" rather than possessing him. This distinction in degree of severity is crucial to the claim of legitimacy for Deliverance prayer, which is borrowed from the protestant Pentecostal tradition, within the Catholic setting. It thus bears directly on the second difference between the two ritual forms. That is, exorcism must be performed by a priest with the formal consent of the local bishop, and requires an eligibility procedure in which all other causes, including psychopathology, must be systematically excluded before causality is attributed to an evil spirit. Deliverance prayer in theory, because it is assumed to be less serious, is often performed by laypeople. The format of the prayer is much more flexible than that of exorcism, and the presence of an evil spirit is established not through a formal procedure but through the discernment of the healer or healers. (Deliverance prayer is often performed in teams of several healers whose "spiritual gifts" are complementary.)

8. There was no noticeable manifestation of the evil spirit during that portion of the baptismal rite that includes a prayer for exorcism; neither was there any evidence that the spirit's hold was weakened by the rite. Although I questioned Peggy specifically about this, she appeared not to have regarded it as a significant moment. With regard to overt behavioral "manifestations," however, it is noteworthy that the evil spirit made its presence felt in any situation that might have resulted in public embarrassment for those involved.

9. The situation is more complex than is evident from the cases that typically achieve notoriety in which medical treatment is objected to on religious grounds by the parents of a minor afflicted with life-threatening illness (see, for example, Redliner and Scott 1979). Within any health-care system, the relation between any two forms of healing may be characterized as compatible alternatives, conflicting or contradictory alternatives, complementary forms addressing different aspects of a problem, or coexistent and non-interacting forms.

10. This discussion could be fruitfully compared to the analysis of cultural, narrative, and experiential issues relevant to the understanding of chronic pain in Good, Brodwin, Good, and Kleinman (1992).

11. The distinction between "I have" and "I am" in an illness has recently been discussed specifically with respect to schizophrenia by Estroff (1989).

# Chapter Five

*Acknowledgments:* The section subtitled "Reflections on a Mystery Illness" is based on a paper presented in 1993 to the symposium on "Symbol and Performance in Healing: The Contributions of indigenous Medical Thought," a preconference of the XIII International Congress of Anthropological and Ethnological Sciences in San Cristabal de las Casas, Chiapas, Mexico. I am grateful to the colleagues who offered comments and questions during presentation of earlier versions of this article to the Scholars Seminar at the Russell Sage Foundation, to the Seminar in Clinically Relevant Medical Anthropology at Harvard University, and to the 1997 meetings of the International Society for the Sociology of Religion, particularly John Logan, Michael Hout, and Arthur Kleinman. Special thanks go to Janis Jenkins for her insightful critique of my argument. The article was completed during my tenure as a Visiting Scholar at the Russell Sage Foundation in 1997.

I am also grateful to the project staff and all those associated with the Navajo Healing Project: Mitzie Begay, Beulah Allen, Mick Storck, Don and Steph Lewis-Kraitsik, John Garrity, Thomas Walker, Jr., Elizabeth Lewton, Victoria Bydone, Nancy Maryboy, David Begay, Derek Milne, Wilson Howard, Nancy Lawrence, Deborah Diswood, Mary Diswood, Helen Curley, Ray Begaye, Alyse Neundorf, Theresa Cahn-Tober, Elizabeth Ihler, Chris Dole, Meredith Holmes, Cindy Retzer, Matt Strickland, and Heather Rushcamp.

1. This chapter is based on data from a five-year project on ritual healing in contemporary Navajo society funded by National Institute of Mental Health grant MH50394-05. The Navajo Healing Project was carried out under Navajo Nation Cultural Resources Investigation Permit C9708-E and with the endorsement of five Community Health Advisory Boards in regions of the Navajo reservation in which the project was conducted. This chapter was reviewed and approved by the Navajo Nation Health Research Board on October 12, 1999. Research was conducted by four teams, each consisting of an ethnographer and an interpreter, and by a psychiatrist with substantial clinical experience with Navajo patients. As principal investigator, I supervised all research conducted. The initial phase of the project consisted of ethnographic interviews with 95 healers distributed across Traditional, Native American Church, and Christian forms of healing. Working with a smaller selected group of healers, we followed a total of 84 patients for a minimum of four to six months through ethnographic and clinical interviews, as well as observation of healing ceremonies and domestic environments. See also Csordas (2000). Interviews with the three patients discussed in this chapter were conducted by the team of Elizabeth Lewton and Victoria Bydone.
2. Anglo is the generic English term used by Navajos for Euro-Americans; the corresponding Navajo term is Bilagaana. African Americans and Mexicans are recognized as distinct groups.
3. As Keith Basso (1979) has documented among the neighboring Apache, such stories constitute a major genre of contemporary Navajo expressive culture.
4. Such rare public hierophanies may be particularly associated with moments of collective stress such as the 1996 drought. Clyde Kluckhohn (1942:59–60) reports two apparitions of Holy People in 1936, during the traumatic period of U.S. government - forced livestock reduction on the reservation. In these instances, the divine message also included instructions that ceremonial activity be carried out.
5. There is evidence, however, that some Navajo ceremonials have originated in dreams or visions (Haile 1940; Kluckhohn 1942).
6. This request for silence brings the Navajo politics of identity face to face with the politics of ethnographic representation in that, although at the time individual Navajos were willing to discuss the apparition with me, it was unclear in what way I could respectfully write about the incident in an ethnographic article. In the present discussion, I take my lead from the tribal newspaper article cited in the text, maintaining cautious respect for the sacred by not publishing the names of the particular Holy People who appeared, or the details of their mode of appearance.
7. A comprehensive account of the Navajo health care system would have four components, including biomedical care practiced in facilities of the Indian Health Service and private or publish hospitals both on and off the reservation (Csordas and Garrity 1992). Interaction of the spiritual traditions with biomedical care is beyond the scope of the present argument.
8. Some Navajos are critical of the term "Evilway" as the English rendering of the *Hochxoo'ji* ceremony. In their opinion, the adverse effects of exposure to spirits of the deceased are not well described by a word that in English connotes profound malevolence and even demonic influence.

## Chapter Six

*Acknowledgments:* This chapter is newly prepared for this volume.

1. These patients were participants in the Navajo Healing Project. See Chapter Five and Csordas (2000).

## Chapter Seven

1. The research reported on in this paper was supported by grants from the National Center for American Indian and Native Alaskan Mental Health Research, the Milton Fund of Harvard Medical

School and the Arnold Center for Pain Research and Treatment. Thanks are due to Margaret Jose, Michael Storck, Beulah Allen, Martha Austin, Roseann Willink, Babette Daniels, Arthur Kleinman, Spero Manson, Jerrold Levy, Louise Lamphere, Oswald Werner, Mary Jo Good, Stuart Lind, David Begay, Mike Mitchell, Andy Natonabah, Wilson Arnoleth, Frank Isaac, and the staffs of the Indian Health Service hospitals at Fort Defiance and Tuba City. I am especially grateful to the patients on the Navajo Reservation and in Boston who spoke with me about their illness and regret the passing of those who have not survived. This chapter is dedicated to the memory of Gregor Allen.

2. This position grows out of Kleinman's (1980) formulation of the "explanatory model" of an illness episode, which included not only causal attributions per se, but the afflicted person's understanding of the pathophysiology and course of the illness as well. These aspects have been downplayed by researchers professing to take up Kleinman's model.

3. The table shows neither prevalence nor incidence rates, but the actual prevalence of cases at a particular point in time when the present research was carried out. Care should thus be taken, for example, in interpreting the relatively inflated number of cases of cervical cancer. Differences in the prevalence of different types of cancer may in large part be attributable to different durations (survival times) and cure rates, the comprehensiveness of screening programs, and/or the percentage of false positives in diagnostic tests.

4. For purposes of this research, I developed an open-ended instrument elaborating on the explanatory model interview for specific illness episodes developed by Kleinman (1980). The interview covered a variety of aspects of illness experience and patient–doctor communication and provided data on causality for both groups. The original English version was translated into Navajo and subsequently revised in light of the adaptations in concept and phrasing that became necessary in preparing the Navajo version. Navajos who chose to be interviewed in English received the same version of the interview as the members of the Boston group did.

5. The Boston study, using parallel interview protocols, was carried out by a team headed by Mary-Jo Delvecchio-Good and including the present author. The results of this work have been reported in Good et al. (1990), Good et al. (1993), and Good et al. (1994).

6. Although it has been shown that elderly Navajos use Shootingways significantly more frequently than the control group (Levy, personal communication), this factor cannot account for the results among cancer patients, since the performance of Shootingways was distributed across the age range of our sample.

7. Indeed, one could hypothesize that a Navajo understanding of the effects of ozone depletion would be cast in these terms. This is certainly the case with respect to the connection between uranium and cancer, as is evident in the following statement by one of our bicultural medicine man informants: "People use to say, 'That mountain right there is harmful, don't bother it.' They'll say, 'What's harmful about it?' 'Nothing harmful unless you fall off of it.' But sure enough, you dig in there, there is uranium, so powerful so dangerous you don't mess with it. That's what they [traditional Navajos] were talking about." Also relevant is the notion of electromagnetic pollution current in Anglo-American popular culture, particularly with respect to possible negative health consequences for people living in proximity to high-tension electrical transmission lines.

8. The means of 2.7 and 2.1 were calculated by dividing the total number of responses (68 for Navajos, 85 for Anglos, combining the Related and Caused categories) by the total number of respondents (25 for Navajos, excluding 3 who offered no response, and 40 for Anglos, excluding 10 who offered no response). The percentage of those who offered no causal construal is then 3 of 28 (11 percent) for Navajos and 10 of 50 (20 percent) for Anglos (compare Table 7.5).

9. Lamphere's discussion of natural phenomena contrasts what she takes to be a Navajo emphasis on features of the external environment, such as animals and meteorological phenomena, with Victor Turner's (1966) description of Ndembu emphasis on bodily phenomena. In particular, she argues that Navajo color symbolism is associated with such external natural phenomena rather than with body substances. However, data from the bicultural medicine men interviewed in the present study indicate that while colors come from the sun, the rainbow colors are the same as the colors of sandstone *and* of the body. The following scheme of seven colors and seven body organs (not body emissions as with the Ndembu) was offered: white—bone, yellow—marrow, glitter—fat, brown—skin, gray—internal organs, red—blood, and black—hair. These seven internal organs correspond with seven external body parts (the order of relation is uncertain), namely foot, leg, waist, trunk, arm, head, and nose, and with a series of seven herbs used to cure ailments of the corresponding organs and members.

# Chapter Eight

*Acknowledgments:* The research reported on in this chapter was supported by grants from the National Center for American Indian and Native Alaskan Mental Health Research, the Milton Fund of Harvard Medical School, and the Arnold Center for Pain Research and Treatment of the New England Deaconess Hospital.

1. This raises the issue of cultural assumptions about rehabilitation and recovery, a topic as yet very inadequately examined. It is possible that the Anglo-American assumption that some form of formal rehabilitation should begin as early as possible is here contradicted by a Navajo assumption that one should wait until one is rehabilitated, or until one's capacities "come back" before entering any formal retraining, even if that retraining is itself oriented toward not over-exerting the patient.
2. The transcriptions of Dan's words give evidence of continued illness-related linguistic disability, and for that reason they have not been edited to increase their fluency.
3. Most authors report a latency of several years between onset of epilepsy and emergence of the behavioral syndrome; Dan's case study was completed between 12 and 30 months after his surgery.
4. My use of the term *schema* is loosely related to the notion of image schema put forward by Mark Johnson (1987). In this context, I use the term out of convenience rather than conviction, for Johnson's approach is essentially cognitive rather than phenomenological. However, not wishing to open a large area of theoretical debate, I would suggest that, appropriately elaborated as part of a cultural phenomenology, the notion of *habit* would perform an equivalent theoretical role, and for our purposes be more apt than that of *schema*.
5. On the concept of indeterminacy in embodiment theory, see Chapter Nine.
6. This is evidently different from the traditional Navajo interpretation of such trembling as a sign that the person is a candidate for initiation as a diagnostician or "handtrembler."
7. As Ortner (1974) has argued with respect to the universality of male dominance and the symbolic assimilation of males to culture and females to nature, the existence of similar strategies across cultures and religions does not place such strategies outside culture. Even less does it place them outside "being-in-the-world."

# Chapter Nine

*Acknowledgments:* Portions of this article were presented in 1988 to the symposium "Beyond Semantics and Rationality," organized by Gilles Bibeau and Ellen Corin at the Twelfth International Congress of Anthropological and Ethnological Sciences in Zagreb. A version was presented to the session "Embodied Knowledge," organized by Deborah Gordon and Jean Lave at the 1988 Annual Meeting of the American Anthropological Association in Phoenix. Since that time, I am grateful to Janis Jenkins for ongoing scholarly discussion that challenged me to refine my argument. Thanks as well to the two anonymous reviewers for *Cultural Anthropology*. Fieldwork among Catholic Charismatics was supported by NIMH grant R01 MH40473.

# Chapter Ten

*Acknowledgments:* This chapter was written while I was a Visiting Scholar at the Russell Sage Foundation. I received valuable comments from Stefania Pandolfo and Lawrence Cohen when I presented this material at the University of California, Berkeley.

1. Visible human images presented here are available on the Internet at the following addresses: General Electric (http://www.crd.ge.com/esl/cgsp/projects), UC San Diego (http://cybermed.ucsd.edu), and University of Hamburg (http://www.uke.uni-hamburg.de). Each of these can be accessed via the National Library of Medicine: (http://www.nlm.nih.gov). An impressive array of images based on the Visible Human can be found in Tsiaras (1997).
2. Compare Ragland-Sullivan (1986:138–162) on Lacan's notion of the imaginary.
3. A similar point can be made with respect to the addition of vivid color to the anatomical images generated from the Visible Human data. On the level of literality, those who produce the images say that they are using color for purposes of contrast, so that different organs and tissues can be seen better. But it is also the case that on the imaginary level for the viewer of the images this coloration

produces a kind of alterity, making the image seem real and other-worldly at the same time, with a kind of dramatization in a comic-book color way, and perhaps adding a sense of the heroic to the spectacle of it all.

4. Among these are the University of California at San Diego Medical School, Loyola University Stritch School of Medicine, Johns Hopkins in collaboration with the National University of Singapore, SUNY Stony Brook, University of Pennsylvania Medical Center, Washington University Medical School, the University of Chicago in collaboration with Argonne National Laboratory, and Columbia University Medical School in collaboration with the Stephens Institute of Technology. Outside the United States, projects are underway at the University of Hamburg School of Medicine, the Keio University School of Medicine in Tokyo, Australian National University, and the Queensland University of Technology.

# Bibliography

Aberle, David. 1982. *The Peyote Religion Among the Navajo.* University of California Press, Chicago.

Abrahams, Roger. 1968. "Introductory Remarks to a Rhetorical Theory of Folklore." *Journal of American Folklore* 81:143–148.

Ackerman, Susan E. 1981. "The Language of Religious Innovation: Spirit Possession and Exorcism in a Malaysian Catholic Pentecostal Movement." *Journal of Anthropological Research* 37:90–100.

Adair, John, Kurt Deuschle, and Walsh McDermott. 1957. "Patterns of Health and Disease Among the Navajos." In *American Indians and American Life,* pp. 80–94. American Academy of Political and Social Science, Philadelphia.

Agamben, Giorgio. 1998. *Homo Sacer: Sovereign Power and Bare Life.* Stanford University Press, Stanford.

Alchon, S. A. 1991. *Native Society and Disease in Colonial Ecuador.* Cambridge University Press, Cambridge.

American Psychiatric Association. 1987. *Diagnostic and Statistical Manual (DSM-III).* American Psychiatric Association, Washington, D.C.

———. 1994. *Diagnostic and Statistical Manual of Mental Disorders (DSM-IV).* American Psychiatric Association, Washington, D.C.

Anderson, R. 1991. "The Efficacy of Ethnomedicine: Research Methods in Trouble." *Medical Anthropology* 13:1–18.

Appadurai, Arjun. 1986. "Theory in Anthropology: Center and Periphery." *Comparative Studies in Society and History* 28:356–361.

———. 1996. *Modernity at Large: Cultural Dimensions of Globalization.* University of Minnesota Press, Minneapolis.

Arieti, Silvano. 1974. *Interpretation of Schizophrenia.* Basic Books, New York.

Armstrong, David. 1983. *Political Anatomy of the Body: Medical Knowledge in Britain in the Twentieth Century.* Cambridge University Press, Cambridge.

Assagioli, Roberto. 1965. *Psychosynthesis: A Manual of Principles and Techniques.* Hobbs Dorman, New York.

Atwood, Margaret. 1985. *The Handmaid's Tale.* McClelland and Stewart, Toronto.

Austin, Martha. n.d. "Cancer Patient and its Terminology." Unpublished manuscript in author's possession.

Bales, Fred. 1994. "Hantavirus and the Media: Double Jeopardy for Native Americans." *American Indian Culture and Research Journal* 18:251–263.

Barthes, Roland. 1986. *The Rustle of Language.* Hill and Wang, New York.

Basso, Keith H. 1979. *Portraits of "The Whiteman": Linguistic Play and Cultural Symbols among the Western Apache.* Cambridge University Press, New York.

Becker, Anne E. 1994. "Nurturing and Negligence: Working on Others' Bodies in Fiji." In *Embodiment and Experience,* edited by Thomas J. Csordas. Cambridge University Press, Cambridge.

———. 1995. *Body, Self, and Society: The View from Fiji.* University of Pennsylvania Press, Philadelphia.

Behar, Ruth, and Deborah A. Gordon (editors). 1995. *Women Writing Culture.* University of California Press, Berkeley.

Bell, Rudolph. 1986. *Holy Anorexia.* University of Chicago Press, Chicago.

Benedict, Ruth. 1934. *Patterns of Culture*. Houghton Mifflin, New York.

Benedikt, Michael (editor). 1994. *Cyberspace: First Steps*. MIT Press, Cambridge.

Benthall, Jonathan, and Ted Polhemus (editors). 1975. *The Body as a Medium of Expression*. E. P. Dutton, New York.

Berger, Harry, Jr. 1987. "Bodies and Texts." *Representations* 17:144–166.

Berman, Morris. 1990. *Coming to Our Senses: Body and Spirit in the Hidden History of the West*. Simon and Schuster, New York.

Birdwhistell, Ray L. 1970. *Kinesics and Context: Essays on Body Motion Communication*. University of Pennsylvania Press, Philadelphia.

Black, Richard. 1975. "The Chronic Pain Syndrome." *Surgical Clinics of North America* 55:999–1011.

Blacker, Carmen. 1989. "The Seer as Healer in Japan." In *The Seer in Celtic and Other Traditions*, edited by Hilda Ellis Davidson, pp. 116–123. John Donald, Edinburgh.

Blacking, John (editor). 1977. *The Anthropology of the Body*. Academic Press, London.

Bloch, Marc. 1973. *The Royal Touch: Sacred Monarchy and Scrofula in England and France*. Routledge and Kegan Paul, London.

Bock, Phillip. 1988. *Rethinking Psychological Anthropology*. Freeman, San Francisco.

Boon, James. 1982. *Other Tribes, Other Scribes: Symbolic Anthropology in the Comparative Study of Culture, Histories, Religions, and Texts*. Cambridge University Press, Cambridge.

Boorstin, Daniel J. 1961. *The Image: A Guide to Pseudo-events in America*. Harper & Row, New York.

Bordo, Susan. 1993. "Reading the Slender Body." In *Body/Politics: Women and the Discourses of Science*, edited by Mary Jacobus, Evelyn Fox Keller, and Sally Shuttlesworth, pp. 83–112. Routledge, New York.

Bouckoms, Anthony, R. E. Litman, and L. Baer. 1985. "Denial in the Depressive and Pain-Prone Disorders of Chronic Pain." *Advances in Pain Research and Therapy* 9:879–887.

Bourdieu, Pierre. 1977. *Outline of a Theory of Practice*. Cambridge University Press, Cambridge.

———. 1984. *Distinction*. Harvard University Press, Cambridge.

———. 1987. *Choses Dites*. Editions de Minuit, Paris.

———. 1990. *The Logic of Practice*. Stanford University Press, Stanford.

Bourguignon, Erika. 1964. "The Self, the Behavioral Environment, and the Theory of Spirit Possession." In *Context and Meaning in Cultural Anthropology*, edited by Melford Spiro. Free Press, New York.

———, ed. 1973. *Religion, Altered States of Consciousness, and Social Change*. Ohio State University Press, Columbus.

———. 1976. "The Effectiveness of Religious Healing Movements: A Review of the Literature." *Transcultural Psychiatric Research Review* 13:5–21.

———. 1979. *Psychological Anthropology*. Holt, Rinehart, and Winston, New York.

———. 1982–3. "Belief and Experience in Folk Religion: Why do Women Join Possession Trance Cults?" *Papers in Comparative Studies* 2:1–16.

———. 1983. "Multiple Personality, Possession Trance, and the Psychic Unity of Mankind." In *Die Wilde Seele/The Savage Soul*, edited by Hans Peter Duerr. Syndikat, Frankfurt.

Bowden, Henry Warner. 1981. *American Indians and Christian Missions: Studies in Cultural Conflict*. University of Chicago Press, Chicago.

Brandt, Jason, Larry Seidman, and Deborah Kohl. 1984. "Personality Characteristics of Epileptic Patients: Controlled Study of Generalized and Temporal Lobe Cases." *Journal of Clinical and Experimental Neuropsychology* 7.

Brena, Steven, and Stanley Chapman. 1985. "Acute V8. Chronic Pain States: The Learned Pain Syndrome." *Clinics in Anesthesiology* 3:41–55.

Brodwin, Paul. 1996. *Medicine and Morality in Haiti: The Contest for Healing Power*. Cambridge University Press, Cambridge.

Brody, Howard. 1980. *Placebos and the Philosophy of Medicine*. University of Chicago Press, Chicago.

Brook, James, and Iain Boal (editors). 1995. *Resisting the Virtual Life: The Culture and Politics of Information*. City Lights Books, San Francisco.

Broude, Gwen. 1988. "Rethinking the Couvade: Cross-Cultural Evidence." *American Anthropologist* 90:902–911.

Broudy, D. W., and P. A. May. 1983. "Demographic and Epidemiological Transition among the Navajo Indians." *Social Biology* 30:1–16.

Brown, George, and Tirril Harris. 1978. *The Social Origins of Depression: a Study of Psychiatric Disorder in Women.* Free Press, New York.

Browner, Carole, Bernard R. Ortiz de Montellano, and Arthur J. Rubel. 1988. "A Methodology for Cross-Cultural Ethnomedical Research." *Current Anthropology* 29:681–702.

Burke, Kenneth. 1970a. *The Rhetoric of Religion.* University of California Press, Berkeley.

————. 1970b [1966]. *Language as Symbolic Action.* University of California Press, Berkeley.

Butler, Judith. 1990. *Gender Trouble: Feminism and the Subversion of Identity.* Routledge, London.

————. 1993. *Bodies that Matter: On the Discursive Limits of "Sex."* Routledge, London.

Bynum, Carolyn Walker. 1987. *Holy Feast and Holy Fast: The Religious Significance of Food to Medieval Women.* University of California Press, Berkeley.

————. 1989. "The Female Body and Religious Practice in the Later Middle Ages." In *Fragments for a History of the Human Body, Part One,* edited by Michel Feher, pp. 160–219. Zone, New York.

Calestro, Kenneth. 1972. "Psychotherapy, Faith Healing, and Suggestion." *International Journal of Psychiatry* 10:83–113.

Calhoun, Craig (editor). 1994. *Social Theory and the Politics of Identity.* Blackwell, Cambridge.

Campbell, E. J. M. 1976. "Clinical Science." Paper presented at Symposium on Research and Medical Practice, London.

Cannon, W. B. 1942. "Voodoo Death." *American Anthropologist* 44:169–181.

Caputo, John D. 1991. "Incarnation and Essentialization." *Philosophy Today* 35:32–42.

Carneiro, Edison. 1940. "The Structure of African Cults in Bahia." *Journal of American Folklore* 53:271–278.

Cartwright, Lisa. 1998. "A Cultural Anatomy of the Visible Human Project." In *The Visible Woman: Imaging Technologies, Gender, and Science,* edited by Paula Treichler, Lisa Cartwright, and Constance Penley, pp. 21–43. New York University Press.

————. n.d. "The Real Life of Biomedical Body Images." Unpublished manuscript.

Cassel, Eric. 1992. "The Body of the Future." In *The Body in Medical Thought and Practice,* edited by Drew Leder, pp. 233–250. Kluwer Academic Publishers, Dordrecht.

Centers for Disease Control and Prevention. 1993. "Update: Hantavirus Disease—United States, 1993." *Morbidity and Mortality Weekly Report* 42:612–614.

Charon, Rita. 1985. Commencement Address.

Chesebro, James. 1982. "Illness as a Rhetorical Act: A Cross-Cultural Perspective." *Communication Quarterly* 30:321–331.

Classen, Constance. 1993. *Worlds of Sense: Exploring the Senses in History and Across Cultures.* Routledge, New York.

Clifford, James, and George E. Marcus (editors). 1986. *Writing Culture: The Poetics and Politics of Ethnography.* University of California Press, Berkeley.

Cole, Michael, and Sylvia Scribner. 1974. *Culture and Thought.* John Wiley, New York.

Comaroff, Jean. 1980. "Healing and the Cultural Order: the Case of the Barolong Boo Ratshidi." *American Ethnologist* 7:637–657.

————. 1982. "Medicine: Symbol and Ideology." In *The Problem of Medical Knowledge: Examining the Social Construction of Medicine,* edited by Peter Wright and Andrew Treacher, pp. 49–68. Edinburgh University Press, Edinburgh.

————. 1983. "The Defectiveness of Symbols or the Symbols of Defectiveness? On the Cultural Analysis of Medical Systems." *Culture, Medicine, and Psychiatry* 7:3–20.

————. 1985. *Body of Power, Spirit of Resistance: The Culture and History of a South African People.* University of Chicago Press, Chicago.

Connor, Linda. 1982. "The Unbounded Self: Balinese Therapy in Theory and Practice." In *Cultural Concepts of Mental Health and Therapy,* edited by Anthony Marsella and Geoffrey White. D. Reidel, Dordrecht.

Corbin, Alain. 1986. *The Foul and the Fragrant: Odor and the French Social Imagination.* Harvard University Press, Cambridge.

Corin, Ellen. 1990. "Facts and Meaning in Psychiatry: An Anthropological Approach to the Life-World of Schizophrenics." *Culture, Medicine, and Psychiatry* 14:153–188.

Cox, Richard (editor). 1973. *Religious Systems and Psychotherapy.* Charles C. Thomas, Springfield, IL.

Crandon, Libbet. 1989. "Changing Times and Changing Symptoms: The Effects of Modernization on Mestizo Medicine in Rural Bolivia." *Medical Anthropology* 10:255.

Crapanzano, Vincent. 1973. *The Hamadsha: A Study in Moroccan Ethnopsychiatry.* University of California Press, Berkeley.

Crapanzano, Vincent, and Vivian Garrison (editors). 1977. *Case Studies in Spirit Possession.* Wiley, New York.

Crawford, Robert. 1984. "A Cultural Account of 'Health': Control, Release, and the Social Body." In *Issues in the Political Economy of Health Care,* edited by John McKinlay. Tavistock Publications, New York.

Csordas [Chordas], Thomas. 1980. "Catholic Pentecostalism: A New Word in the New World." In *Perspective on Pentecostalism: Case Studies from the Caribbean and Latin America,* edited by Steven Glazier. University Press of America, Washington, D.C.

———. 1985. "Medical and Sacred Realities: Between Comparative Religion and Transcultural Psychiatry (Review)." *Culture, Medicine, and Psychiatry* 9:103–111.

———. 1987. "Health and the Holy in African and Afro-American Spirit Possession." *Social Science and Medicine* 24:1–11.

———. 1990. "The Psychotherapy Analogy and Charismatic Healing." *Psychotherapy* 27:79–80.

———. 1994a. *The Sacred Self: A Cultural Phenomenology of Charismatic Healing.* University of California Press, Berkeley.

———, ed. 1994b. *Embodiment and Experience: the Existential Ground of Culture and Self.* Cambridge University Press, Cambridge.

———. 1997. *Language, Charisma, and Creativity: the Ritual Life of a Religious Movement.* University of California Press, Berkeley.

———. 1999a. "Embodiment and Cultural Phenomenology." In *Perspectives on Embodiment,* edited by Gail Weiss and Honi Haber, pp. 143–162. Routledge, New York.

———. 1999b. "The Body's Career in Anthropology." In *Anthropological Theory Today,* edited by Henrietta Moore, pp. 172–205. Polity Press, Cambridge.

———. 2000. "The Navajo Healing Project." *Medical Anthropology Quarterly* 14:463–475.

———. 2001. "Notes for a Cybernetics of the Holy." In *Essays in Honor of Roy A. Rappaport,* edited by Michael Lambek and Ellen Messer, pp. 227–243. University of Michigan Press, Ann Arbor.

Csordas, Thomas, and John Garrity. 1992. "Co-Utilization of Biomedicine and Religious Healing: A Navajo Case Study." In *Year Book of Cross-Cultural Medicine and Psychotherapy, 1992,* edited by W. Andritzky, pp. 241–252. Verlag fur Wissenschaft und Bilung, Berlin.

Csordas, Thomas, and Arthur Kleinman. 1996. "The Therapeutic Process." In *Medical Anthropology: Contemporary Theory and Method,* edited by Carolyn F. Sargent and M. T. Johnson, pp. 3–20. Praeger, London.

Csordas, Thomas, and Elizabeth Lewton. 1998. "Practice, Performance and Experience in Ritual Healing." *Transcultural Psychiatry* 35:435–512.

Curtis, Neal. 1999. "The Body as Outlaw: Lyotard, Kafka, and the Visible Human Project." *Body and Society* 5:249–266.

D'Andrade, Roy, and Claudia Strauss (editors). 1992. *Human Motives and Cultural Models.* Cambridge University Press, Cambridge.

Daniel, E. Valentine. 1984. "The Pulse As an Icon in Siddha Medicine." *Asian Studies* 18:115–126.

Dawson, Warren R. 1929. *The Custom of Couvade.* Manchester University Press, Manchester.

DelVecchio Good, Mary Jo, Paul Brodwin, Byron Good, and Arthur Kleinman (editors). 1992. *Pain as Human Experience: An Anthropological Perspective.* University of California Press, Berkeley.

Denber, Herman. 1955. "Studies on Mescaline III. Action in Epileptics: Clinical Observations and Effects on Brain Wave Patterns." *Psychiatric Quarterly* 29:433–438.

Deren, Maya. 1953. *The Divine Horseman: The Living Gods of Haiti.* Thames and Hudson, New York.

Desjarlais, R. 1992. *The Aesthetics of Illness and Healing in the Nepal Himalayas.* University of Pennsylvania Press, Philadelphia.

Devisch, Rene. 1983. "Le Corps Sexue et Social ou les Modalites D'Echange Sensoriel chez les Yaka du Zaire." *Psychopathologie Africaine* 19:5–31.

———. 1993. *Weaving the Threads of Life: the Khita Gyn-Eco-Logical Healing Cult among the Yaka.* University of Chicago Press, Chicago.

Devisch, Rene, and Antoine Gailly (editors). 1985. *Symbol and Symptom in Bodily Space-Time.* Special issue of the *International Journal of Psychology* 20:389–663.

Dewhurst, Kenneth, and A. W. Beard. 1970. "Sudden Religious Conversions in Temporal Lobe Epilepsy." *British Journal of Psychiatry* 117:497–507.

Dieterlin, Germaine. 1971. "L'Image du Corps et les Compsantes de la Personne chez les Dogon." In *La Notion de Personne en Afrique Noir,* edited by Germaine Dieterlin, pp. 205–229. Editions du Centre National de la Recherche Scientifique, Paris.

Douglas, Mary. 1966. *Purity and Danger.* Routledge and Kegan Paul, London.

———. 1973. *Natural Symbols.* Pelican, New York.

Dow, James. 1986a. *The Shaman's Touch: Otomi Indian Symbolic Healing.* University of Utah Press, Salt Lake City.

———. 1986b. "Universal Aspects of Symbolic Healing: A Theoretical Synthesis." *American Anthropologist* 88:56–69.

Dubos, Rene. 1959. *Mirage of Health: Utopias, Progress, and Biological Change.* Harper, New York.

Durkheim, Emile. 1965 [1915]. *The Elementary Forms of the Religious Life.* Free Press, New York.

Edie, James M. 1971. "Was Merleau-Ponty a Structuralist?" *Semiotica* 4:297–323.

Eisenberg, Leon. 1977. "Disease and Illness." *Culture, Medicine and Psychiatry* 1:9–23.

Ekman, Paul. 1982. *Emotion in the Human Face.* Cambridge University Press, Cambridge.

Eliade, Mircea. 1958. *Patterns in Comparative Religion.* World Publishing, Cleveland.

Elliott, Robert. 1984. "A Discovery-Oriented Approach to Significant Events: Interpersonal Process Recall and Comprehensive Process Analysis." In *Patterns of Change: Intensive Analysis of Psychotherapy Process,* edited by Laura Rice and Leslie Greenberg, pp. 249–286. Guilford Press, New York.

———. 1986. "Interpersonal Process Recall (IPR) as a Process Research Method." In *The Psychotherapeutic Process,* edited by Laura Greenberg and William Pinsoff, pp. 180–211. Guilford Press, New York.

Enoch, M. David, and Sir William Trethowan. 1991. "The Couvade Syndrome." In *Uncommon Psychiatric Syndromes,* edited by M. David Enoch and Sir William Trethowan, pp. 92–111. Butterworth-Heinemann, Oxford.

Erdman, David V. 1988. *The Complete Poetry and Prose of William Blake.* Doubleday, New York.

Estroff, Sue E. 1989. "Self, Identity, and Subjective Experiences of Schizophrenia: In Search of the Subject." *Schizophrenia Bulletin* 15:186–196.

Evans-Pritchard, Edward E. 1937. *Witchcraft, Oracles, and Magic Among the Azande.* Clarendon Press, Oxford.

Farella, John R. 1984. *The Main Stalk: A Synthesis of Navajo Philosophy.* University of Arizona Press, Tucson.

Fausto-Sterling, Anne. 1992. *Myths of Gender: Biological Theories about Women and Men.* Basic Books, New York.

Favazza, Armando. 1982a. *Bodies Under Siege.* Johns Hopkins University Press, Baltimore.

———. 1982b. "Modern Christian Healing of Mental Illness." *American Journal of Psychiatry* 139:728–735.

Favret-Saada, Jeanne. 1980. *Deadly Words: Witchcraft in the Bocage.* Cambridge University Press, Cambridge.

Featherstone, Michael. 1991. "The Body in Consumer Culture." In *The Body: Social Process and Cultural Theory,* edited by Michael Featherstone, Michael Hepworth, and Bryan S. Turner. SAGE, London.

Featherstone, Michael, Michael Hepworth, and Bryan S. Turner (editors). 1991. *The Body: Social Process and Cultural Theory.* SAGE, London.

Feher, Michel (editor). 1989. *Fragments for a History of the Human Body.* Special issues of *Zone* 3, 4, and 5.

Feld, Steven. 1982. *Sound and Sentiment: Birds, Weeping, Poetics and Song in Kaluli Expression.* University of Pennsylvania Press, Philadelphia.

Fernandez, James. 1972. "Persuasions and Performances: Of the Beast in Every Body ... and the Metaphors of Every Man." *Daedalus* 101:39–60.

———. 1974. "The Mission of Metaphor in Expressive Culture." *Current Anthropology* 15:119–145.

———. 1982. *Bwiti: An Ethnography of the Religious Imagination in Africa.* Princeton University Press, Princeton.

———. 1990. "The Body in Bwiti: Variations on a Theme by Richard Werbner." Unpublished manuscript.

Field, Karen. 1982. "Charismatic Religion as Popular Protest." *Theory and Society* 11:305–320.

Field, Margaret. 1937. *Religion and Medicine of the Ga People.* Oxford University Press, London.

———. 1960. *Search for Security: An Ethnopsychiatric Study of Rural Ghana.* Northeastern University Press, Evanston, Ill.

Finkler, Kaja. 1980. "Non-Medical Treatments and Their Outcomes, Part 1." *Culture, Medicine, and Psychiatry* 4:271–310.

———. 1981. "Non-Medical Treatments and Their Outcomes, Part 2." *Culture, Medicine, and Psychiatry* 5:65–103.

———. 1985. *Spiritualist Healers in Mexico.* Bergin and Garvey, South Hadley, Mass.

Fogelson, Raymond, and Richard Adams (editors). 1977. *The Anthropology of Power: Ethnographic Studies from Asia, Oceania, and the New World.* Academic Press, New York.

Fogelson, Raymond D. 1982. "Person, Self and Identity: Some Anthropological Retrospects, Circumspects and Prospects." In *Psychosocial Theories of the Self,* edited by Benjamin Lee, pp. 67–109. Plenum Press, New York.

Fortes, Meyer. 1987. "The Concept of the Person." In *Religion, Morality, and the Person: Essays on Tallensi Religion,* edited by Meyer Fortes. Cambridge University Press, Cambridge.

Foucault, Michel. 1965. *Madness and Civilization: A History of Insanity in the Age of Reason.* Vintage, New York.

———. 1970. *The Order of Things: An Archaeology of the Human Sciences.* Vintage, New York.

———. 1972. *The Archaeology of Knowledge and the Discourse on Language.* Harper Colophon, New York.

———. 1973. *The Birth of the Clinic: An Archaeology of Medical Perception.* Vintage, New York.

———. 1977. *Discipline and Punishment: the Birth of the Prison.* Vintage, New York.

———. 1978. *The History of Sexuality: An Introduction.* Vintage, New York.

———. 1985. *The Use of Pleasure.* Vintage, New York.

———. 1986. *The Care of the Self.* Vintage, New York.

Frank, Gelya. 1986. "On Embodiment: A Case Study of Congenital Limb Deficiency in American Culture." *Culture, Medicine, and Psychiatry* 10:189–219.

Frank, Jerome. 1973 [1961]. *Persuasion and Healing.* Johns Hopkins University Press, Baltimore.

———. 1978. *Psychotherapy and the Human Predicament.* Schocken, New York.

Frank, Jerome D., and Julia B. Frank. 1991. *Persuasion and Healing.* Johns Hopkins University Press, Baltimore.

Frankenberg, Ronald. 1986. "Sickness as Cultural Performance: Drama, Trajectory, and Pilgrimage Root Metaphors and the Making Social of Disease." *International Journal of Health Services* 16:603–626.

Friedman, Jonathan. 1992. "The Past in the Future: History and the Politics of Identity." *American Anthropologist* 94:837–859.

Gabbay, John. 1982. "Asthma Attacked? Tactics for the Reconstruction of a Disease Concept." In *The Problem of Medical Knowledge,* edited by Peter Wright and Andrew Treacher, pp. 23–48. Edinburgh University Press, Edinburgh.

Gaines, Atwood. 1982a. "Cultural Definitions, Behavior, and the Person in American Psychiatry." In *Cultural Conceptions of Mental Health and Therapy,* edited by A. J. Marsella and G. M. White. D. Reidel, Dordrecht.

———. 1982b. "The Twice-Born: 'Christian Psychiatry' and Christian Psychiatrists." *Culture, Medicine, and Psychiatry* 6:305–324.

———. 1992. "Medical/Psychiatric Knowledge in France and the United States: Culture and Sickness in History and Biology." In *Ethnopsychiatry: The Cultural Construction of Professional and Folk Psychiatries,* edited by Atwood Gaines, pp. 171–201. SUNY Press, Albany, New York.

Garrison, V. 1977. "Doctor, Espiritista, or Psychiatrist?: Health-Seeking Behavior in a Puerto Rican Neighborhood of New York City." *Medical Anthropology* 2:65–91.

Garrity, John F. 1998. *The Ethos of Power: Navajo Religious Healing of Alcohol and Substance Abuse.* Ph.D. diss., Case Western Reserve University, Cleveland, Ohio.

Geertz, Clifford. 1973. *The Interpretation of Cultures.* Basic Books, New York.

———. 1984. " 'From the Native's Point of View': on the Nature of Anthropological Understanding." In *Culture Theory,* edited by Richard Shweder and Robert LeVine. Cambridge University Press, Cambridge.

Gellner, D. N. 1994. "Priests, Healers, Mediums, and Witches: The Context of Possession in the Kathmandu Valley, Nepal." *Man* 29:27–48.

Gerlach, Luther, and Virginia Hine. 1970. *People, Power and Change*. Bobbs-Merrill, Indianapolis.

Giddens, Anthony. 1990. *Modernity and Self-Identity*. Stanford University Press, Stanford.

Ginsberg, Faye. 1989. *Contested Lives: The Abortion Debate in an American Community*. University of California Press, Berkeley.

Good, Byron. 1977. "The Heart of What's the Matter: The Semantics of Illness in Iran." *Culture, Medicine, and Psychiatry* 1:25–58.

———. 1988. "A Body in Pain: The Making of a World of Chronic Pain." Paper presented at Wenner Gren Conference on Medical Anthropology, Lisbon, Portugal.

———. 1992. "A Body in Pain." In *Pain as Human Experience: An Anthropological Perspective*, edited by Mary-Jo DelVecchio Good, Paul Brodwin, Byron Good, and Arthur Kleinman, pp. 29–48. University of California Press, Berkeley.

———. 1994. *Medicine, Rationality, and Experience: an Anthropological Perspective*. Cambridge University Press, Cambridge.

Good, Byron, and Mary-Jo DelVecchio Good. 1980. "The Meaning of Symptoms: A Cultural Hermeneutical Model for Clinical Practice." In *The Relevance of Social Science for Medicine*, edited by Leon Eisenberg and Arthur Kleinman. D. Reidel, Dordrecht.

———. 1981. "The Semantics of Medical Discourse." In *Sciences and Cultures: Anthropological and Historical Studies of the Sciences*, edited by Everett Mendelsohn and Yehuda Elkana. D. Reidel, Dordrecht.

———. 1982. "Toward a Meaning-Centered Analysis of Popular Illness Categories: 'Fright-Illness' and 'Heart Distress' in Iran." In *Cultural Conceptions of Mental Health and Therapy*, edited by A. J. Marsella and G. M. White, pp. 141–166. D. Reidel Dordrecht.

Good, Byron, Henry Herrera, Mary-Jo DelVecchio Good, and James Cooper. 1982. "Reflexivity and Countertransference in a Psychiatric Cultural Consultation Clinic." *Culture, Medicine, and Psychiatry* 6:281–303.

Good, Mary-Jo DelVecchio. 1995. *American Medicine: the Quest for Competence*. University of California Press: Berkeley.

Good, Mary-Jo DelVecchio, Paul E. Brodwin, Byron J. Good, and Arthur Kleinman. *Pain as Human Experience: An Anthropological Perspective*. University of California Press: Berkeley.

Good, Mary-Jo DelVecchio, Byron J. Good, Cynthia Schaffer, and Stuart E. Lind. "American Oncology and the Discourse on Tape." *Culture, Medicine, and Psychiatry*. 14:59–79.

Good, Mary-Jo DelVecchio, Linda Hunt, Tseunetsugu Munakata, and Yasuki Kobayashi. "A Comparative Analysis of the Culture of Biomedicine: Disclosure and Consequences for Treatment in the Practice of Oncology." In *Sociological Perspectives in International Health*. Edited by Peter Conrad and Eugene Gallagher. Temple University Press. Philadelphia, Pa.

Good, Mary-Jo DelVecchio, Tseunetsugu Munakata, Yasuki Kobayashi, Cheryl Mattingly, and Byron J. Good. 1994. "Oncology and Narrative Time." *Social Science and Medicine*. 38(6):855–862.

Goodenough, Ward. 1957. "Cultural Anthropology and Linguistics." In *Report of the Seventh Annual Round Table Meeting in Linguistics and Language Study*, edited by P. Garvin, pp. 141–166. Georgetown University, Washington, D.C.

Goodman, Felicitas. 1972. *Speaking in Tongues*. University of Chicago Press, Chicago.

———. 1981. *The Exorcism of Anneliese Michel*. Doubleday, New York.

Gossen, Gary. 1976. "Language as Ritual Substance." In *Language in Religious Practice*, edited by William Samarin. Newbury House, Rowley.

Gottlieb, L., and L. Husen. 1982. "Lung Cancer Among Navajo Uranium Miners." *Chest* 81:449–452.

Grady, Denise. 1993. "Death at the Corners." *Discover* 14:82–90.

Gray, Chris Hables (editor). 1995. *The Cyborg Handbook*. Routledge, New York.

Grosz, Elizabeth. 1994. *Volatile Bodies: Toward a Corporeal Feminism*. Indiana University Press, Bloomington.

Guimares de Magalhaes, E. 1974. *Orixas de Bahia*. S. A. Artes Grafias, Salvador.

Guimera, Louis Mallart. 1978. "Witchcraft Illness in the Evuzok Nosological System." *Culture, Medicine, and Psychiatry* 2:373–396.

Gupta, Madhulika A. 1986. "Is Chronic Pain a Variant of Depressive Illness? A Critical Review." *Canadian Journal of Psychiatry* 31:241–248.

Hahn, Robert. 1984. "Rethinking 'Disease' and 'Illness.' " In *South Asian Systems of Healing,* edited by E. Valentine Daniels and Judy Pugh. E. J. Brill, Leiden.

Hahn, Robert, and Arthur Kleinman. 1983. "Belief as Pathogen, Belief as Medicine." *Medical Anthropology Quarterly* 14:3, 16–19.

Halifax, J., and H. Weidman. 1973. "Religion as a Mediating Institution in Acculturation: The Case of Santeria in Greater Miami." In *Religious Systems and Psychotherapy,* edited by R. Cox Thomas, Springfield.

Hall, Edward T. 1959. *The Silent Language.* Fawcett World Library, Greenwich, Conn.

———. 1966. *The Hidden Dimension.* Anchor Books, New York.

Hallowell, Irving A. 1955. "The Self in Its Behavioral Environment." In *Culture and Experience,* pp. 75–110. University of Pennsylvania Press, Philadelphia.

Hanks, William. 1989. "Text and Textuality." *Annual Review of Anthropology* 18:95–127.

———. 1990. *Referential Practice: Language and Lived Space Among the Maya.* University of Chicago Press, Chicago.

Hanna, Judith Lynne. 1988. *Dance, Sex, and Gender: Signs of Identity, Dominance, Defiance, and Desire.* University of Chicago Press, Chicago.

Haraway, Donna J. 1990. "Investment Strategies for the Evolving Portfolio of Primate Females." In *Body/Politics: Women and the Discourses of Science,* edited by Mary Jacobus, Evelyn Fox Keller, and Sally Shuttlesworth, pp. 139–162. Routledge, New York.

———. 1991. *Simians, Cyborgs, and Women: The Reinvention of Nature.* Routledge, New York.

———. 1997. *Modest-Witness@Second-Millennium.FemaleMan-Meets-OncoMouse: Feminism and Technoscience*/Routledge, New York.

Harwood, Alan. 1977. *Rx: Spiritist as Needed: A Study of a Puerto Rican Mental Health Resource.* John Wiley, New York.

———. 1987. *Rx: Spiritist as Needed: a Study of a Puerto Rican Community Mental Health Resource.* Cornell University Press, Ithaca.

Heidegger, Martin. 1977a. "Letter on Humanism." In *Basic Writings,* edited by David Farrell Krell, pp. 230. Harper and Row, New York.

———. 1977b. *The Question Concerning Technology and Other Essays.* Translated and introduced by William Lovitt. New York: Harper and Row.

Heim, Michael. 1993. *The Metaphysics of Virtual Reality.* Oxford University Press, New York.

Henderson, James. 1982. "Exorcism and Possession in Psychotherapy Practice." *Canadian Journal of Psychiatry* 27:129–134.

Henderson, W. 1982. *Healing and Affliction.* Beacon Press, Boston.

Hertz, Robert. 1960 [1909]. "The Preeminence of the Right Hand." In *Death and the Right Hand,* translated by Rodney Needham and Claudio Needham. Choehn & West, Aberdeen.

Hinde, R. A. (editor). 1972. *Non-Verbal Communication.* Cambridge University Press, Cambridge.

Hodge, W. 1969. "Navajo Pentecostalism." *Anthropological Quarterly* 37:73–93.

Hoeppner, Jo-Ann, David Garron, Robert Wilson, and Margaret Koch-Weiser. 1987. "Epilepsy and Verbosity." *Epilepsia* 28:35–40.

Holland, Dorothy, and Naomi Quinn (editors). 1987. *Cultural Models in Language and Thought.* Cambridge University Press, Cambridge.

Horm, John, A. Asire, J. Young, and E. Pollock (editors). 1984. *Cancer Incidence and Mortality in the United States, 1973–81.* National Cancer Institute, Bethesda, Md.

Hottois, Gilbert. 1988. "De L'Objet de la Phenomenologie ou la Phenomenologie Comme Style." In *Maurice Merleau-Ponty, la Psychique et le Corporel,* edited by A. T. Tymieniecka. Aubier, Paris.

Howes, David. 1987. "Olfaction and Transition: An Essay on the Ritual Uses of Smell." *Canadian Review of Sociology and Anthropology* 24:390–416.

Howes, David (editor). 1991. *The Varieties of Sensory Experience: A Sourcebook in the Anthropology of the Senses.* University of Toronto Press, Toronto.

Hymes, Dell. 1975. "Breakthrough into Performance." In *Folklore: Communication and Performance,* edited by Kenneth Goldstein and Dan Ben-Amos. Mouton, The Hague.

Ingleby, David. 1982. "The Social Construction of Mental Illness." In *The Problem of Medical Knowledge*, edited by Peter Wright and Andrew Treacher, pp. 123–143. Edinburgh University Press, Edinburgh.

Insel, Thomas, Theodore Zahn, and Dennis C. Murphy. 1985. "Obsessive-Compulsive Disorder: An Anxiety Disorder?" In *Anxiety and the Anxiety Disorders*, edited by A. H. Tuma and J. O Maser. Erlbaum Publishers, Hillsdale, N.J.

Irigaray, Luce. 1985a. *The Sex Which is Not One*. Cornell University Press, Ithaca, New York.

———. 1985b. *Speculum of the Other Woman*. Cornell University Press, Ithaca, New York.

———. 1993. *An Ethics of Sexual Difference*. Cornell University Press, Ithaca, New York.

Jackson, Michael. 1981. "Knowledge of the Body." *Man* 18:327–345.

———. 1989. *Paths Toward a Clearing: Radical Empiricism and Ethnographic Inquiry*. Indiana University Press, Bloomington.

Jacobus, Mary, Evelyn Fox Keller, and Sally Shuttlesworth (editors). 1990. *Body/Politics: Women and the Discourses of Science*. Routledge, New York.

James, William. 1961 [1902]. *The Varieties of Religious Experience*. Macmillan, New York.

———. 1967. "The Emotions." In *The Emotions*, edited by Carl Georg Lange and William James. Jafner, New York.

Janet, P. 1925. *Psychological Healing*. Macmillan, New York.

Janzen, J. M. 1992. *Ngoma: Discourses of Healing in Central and Southern Africa*. University of California Press, Berkeley.

Jenkins, Janis. 1988. "Ethnopsychiatric Interpretations of Schizophrenic Illness: The Problem of *Nervios* within Mexican-American Families." *Culture, Medicine, and Psychiatry* 12:301–330.

———. 1991. "Expressed Emotion and Schizophrenia." *Ethos* 19:387–431.

———. 1994. "Culture, Emotion, and Psychopathology." In *Emotion and Culture: Empirical Studies of Mutual Influence*, edited by Shinobu Kitayama and Hazel Markus, pp. 307–335. American Psychological Association, Washington, D.C.

Jenkins, Janis H., and Marvin Karno. 1992. "The Meaning of Expressed Emotion: Theoretical Issues Raised by Cross-Cultural Research." *American Journal of Psychiatry* 149(1):9–21.

Jenkins, Janis H., and Valiente Martha. 1994. "Bodily Transactions of the Passions: *El Calor* among Salvadoran Women Refugees." In *Embodiment and Experience*, edited by Thomas J. Csordas. Cambridge University Press, Cambridge.

Jilek, Wolfgang. 1974. *Salish Indian Mental Health and Cultural Change: Psychohygienic and Therapeutic Aspects of the Guardian Spirit Ceremonial*. Holt, Rinehart, and Winston of Canada, Montreal.

Johnson, Mark. 1987. *The Body in the Mind: The Bodily Basis of Meaning, Imagination, and Reason*. University of Chicago Press, Chicago.

Kane, Steven M. 1982. "Holiness Ritual Fire Handling: Ethnographic and Psychophysiological Considerations." *Ethos* 10:369–384.

Kapferer, Bruce. 1976. "Mind, Self, and Other in Demonic Illness: The Negation and Reconstruction of Self." American Ethnologist 6:1–23.

———. 1979a. "Emotion and Healing in Sinhalese Healing Rites." *Social Analysis* 1:153–176.

———. 1979b. "Introduction: Ritual Process and the Transformation of Context." *Social Analysis* 1:3–19.

———. 1979c. *Transaction and Meaning: Directions in the Anthropology of Exchange and Symbolic Behavior*. Institute for the Study of Human Issues, Philadelphia.

———. 1983. *A Celebration of Demons: Exorcism and the Aesthetics of Healing in Sri Lanka*. University of Indiana Press, Bloomington.

———. 1997. *The Feast of the Sorcerer: Practices of Consciousness and Power*. University of Chicago Press, Chicago.

Kaplan, Bert, and Dale Johnson. 1964. "The Social Meaning of Navajo Psychopathology and Psychotherapy." In *Magic, Faith, and Healing: Studies in Primitive Psychiatry Today*, edited by Ari Kiev, pp. 203–229. The Free Press, New York.

Kearney, Richard. 1988. *The Wake of Imagination*. University of Minnesota Press, Minneapolis.

Kehoe, Alice, and Dody Giletti. 1981. "Women's Preponderance in Possession Cults: The Calcium-Deficiency Hypothesis Extended." *American Anthropologist* 83:549–561.

Kember, Sarah. 1998. *Virtual Anxiety: Photography, New Technologies and Subjectivity*. Manchester: Manchester University Press.

Kennedy, John G. 1967. "Nubian Zar Ceremonies as Psychotherapy." *Human Organization* 26:185–194.

Kenny, Michael. 1986. *The Passion of Ansel Bourne: Multiple Personality in American Culture.* Smithsonian Institution, Washington, D.C.

Kiev, A. (editor). 1964. *Magic, Faith, and Healing.* Free Press, New York.

Kirmayer, Laurence. 1984. "Culture, Affect, and Somatization." *Transcultural Psychiatric Research Review* 21:160–187, 237–261.

———. 1989. "Mind and Body as Metaphors." In *Biomedicine Examined,* edited by Margaret Lock and Deborah Gordon, pp. 57–94. Kluwer Academic Publishers, Dordrecht.

———. 1992. "The Body's Insistence on Meaning: Metaphor as Presentation and Representation in Illness Experience." *Medical Anthropology Quarterly* 6:323–346.

———. 1993. "Healing and the Invention of Metaphor: the Effectiveness of Symbols Revisited." *Culture, Medicine, and Psychiatry* 17:161–195.

Klein, Hilary. 1991. "Couvade Syndrome: Male Counterpart to Pregnancy." *International Journal of Psychiatry in Medicine* 21:57–69.

Kleinman, Arthur. 1973. "Medicine's Symbolic Reality." *Inquiry* 16:206–213.

———. 1980. *Patients and Healers in the Context of Culture: An Exploration of the Borderland Between Anthropology, Medicine, and Psychiatry.* University of California Press, Berkeley.

———. 1982. "Neurasthenia and Depression." *Culture, Medicine, and Psychiatry* 7:97–99.

———. 1983. Editor's Note. *Culture, Medicine, and Psychiatry* 7:97–99.

———. 1986. *Social Origins of Distress and Disease.* Yale University Press, New Haven.

———. 1988. *The Illness Narratives: Suffering, Healing, and the Human Condition.* Basic Books, New York.

———. 1995. *Writing at the Margin: Discourse Between Anthropology and Medicine.* University of California Press, Berkeley.

———. 1997. "Depression, Somatization and the 'New Cross-Cultural Psychiatry.' " *Social Science and Medicine* 11:3–10.

Kleinman, Arthur, Veena Das, and Margaret Lock (editors). 1997. *Social Suffering.* University of California Press, Berkeley.

Kleinman, Arthur, and James Gale. 1982. "Patients Treated By Physicians and Folk Healers: A Comparative Outcome Study in Taiwan." *Culture, Medicine, and Psychiatry* 6:405–423.

Kleinman, Arthur, and Byron Good (editors). 1985. *Culture and Depression.* University of California Press, Berkeley.

Kleinman, Arthur, and Joan Kleinman. 1991. "Suffering and its Professional Transformation: Toward an Ethnography of Experience." *Culture, Medicine, and Psychiatry* 15:275–302.

Kleinman, Arthur, and Lilias Sung. 1979. "Why Do Indigenous Practitioners Successfully Heal?" *Social Science and Medicine* 13:7–26.

Kluckhohn, Clyde, and Dorothea Leighton. 1946. *The Navajo.* Doubleday, New York.

Kluver, Heinrich. 1966. *Mescal and Mechanisms of Hallucination.* University of Chicago Press, Chicago.

Koenig, Harold G. 1998. *Handbook of Religion and Mental Health.* Academic Press, San Diego.

Kohlberg, Lawrence. 1969. "Stage and Sequence: The Cognitive-Developmental Approach to Socialization." In *Handbook of Socialization Theory and Research,* edited by D. Goslin, pp. 347–480. Rand McNally, Chicago.

Koss, Joan. n.d. "The Experience of Spirits: Ritual Healing as Transactions of Emotion." *Unpublished manuscript.*

Koss-Chioino, Joan. 1992. *Women as Healers, Women as Patients: Mental Health Care and Traditional Healing in Puerto Rico.* Westview Press, Boulder, Colo.

Kris, Ernst. 1952. *Psychoanalytic Explorations in Art.* International Universities Press, New York.

Kroker, Arthur, and Marilouise Kroker. 1987. *Body Invaders: Panic Sex in America.* St. Martin's Press, New York.

Kunitz, Stephen J. 1983. *Disease Change, and the Role of Medicine: The Navajo Experience.* University of California Press, Berkeley.

Kunitz, Stephen, and J. E. Levy. 1981. "Navajos." In *Ethnicity and Medical Care,* edited by Alan Harwood, pp. 131–167. Harvard University Press, Cambridge.

LaBarre, Weston. 1969. *The Peyote Cult.* University of Oklahoma Press, Norman.

———. 1972. *The Ghost Dance: Origins of Religion.* Delta, New York.

———. 1975. *The Peyote Cult.* Schocken, New York.

Laderman, C. 1991. *Taming the Wind of Desire: Psychology, Medicine, and Aesthetics in Malay Shamanistic Performance.* University of California Press, Berkeley.

Laderman, Carol, and Marina Roseman (editors). 1996. *The Performance of Healing.* Routledge, New York.

LaFleur, William R. 1992. *Liquid Life: Abortion and Buddhism in Japan.* Princeton University Press, Princeton.

Lakoff, George. 1987. *Women, Fire, and Dangerous Things: What Categories Reveal about the Mind.* University of Chicago Press, Chicago.

Lakoff, George, and Mark Johnson. 1980. *Metaphors We Live By.* University of Chicago Press, Chicago.

Lamphere, Louise. 1969. "Symbolic Elements in Navajo Ritual." *Southwestern Journal of Anthropology* 25:279–305.

———. 1977. *To Run After Them: Cultural and Social Bases of Cooperation in a Navajo Community.* University of Arizona Press, Tucson.

Langer, Suzanne. 1957. *Philosophy in a New Key.* Harvard University Press, Cambridge.

Laplantine, Francois. 1987. *L'Anthropologie de la Maladie.* Payot, Paris.

Larson, David, and Susan S. Larson. 1994. *The Forgotten Factor in Physical and Mental Health: What Does the Research Show?* John Templeton Foundation, seminar text.

Lash, Scott, and Jonathan Friedman (editors). 1992. *Modernity and Identity.* Blackwell, Cambridge.

Layne, Linda. 1992. "Of Fetuses and Angels: Fragmentation and Integration in Narratives of Pregnancy Loss." *Knowledge and Society* 9:29–58.

Leach, Edmund. 1958. "Magical Hair." *Journal of the Royal Anthropological Institute* 88:147–164.

Leacock, Seth. 1972. *Spirits of the Deep.* Doubleday, New York.

———. 1978. "Review of Crapanzano and Garrison: Case Studies in Spirit Possession with reply by Crapanzano and Garrison." *Reviews in Anthropology* 5:399–409, 420–425.

Leder, Drew. 1990. *The Absent Body.* University of Chicago Press, Chicago.

Leenhardt, Maurice. 1979 [1947]. *Do Kamo: Person and Myth in a Melanesian World.* University of Chicago Press, Chicago.

Leighton, Alexander H., and Dorothea C. Leighton. 1941. "Elements of Psychotherapy in Navajo Religion." *Psychiatry* 4:515–524.

Leiris, Michel. 1958. *La Possession ses Aspect Theatraux chez les Ethiopiens de Gondar.* Plon, Paris.

Leslie, C. M. 1980. "Medical Pluralism." *Social Science and Medicine* 1.

Lessa, William, and Evon Z. Vogt (editors). 1979. *Reader in Comparative Religion: an Anthropological Approach.* Harper and Row, New York.

Leuba, James H. 1896. "A Study in the Psychology of Religious Phenomena." *American Journal of Psychology* 7:309–385.

Levin, David Michael. 1985. *The Body's Recollection of Being.* Routledge and Kegan Paul, London.

Levin, Jeffrey, and H. Vanderpool. 1987. "Is Frequent Religious Attendance Really Conducive to Better Health? Toward an Epidemiology of Religion." *Social Science and Medicine* 24:589–600.

Levine, Jeffrey, N. C. Gordon, and H. L. Fields. 1978. "The Mechanism of Placebo Analgesia." *Lancet* ii:656–657.

Levi-Strauss, Claude. 1964. *Le Cru et le Cuit.* Plon, Paris.

———. 1968. *L'Origine des Manieres de Table.* Plon, Paris.

Levy, Jerrold E. 1983. "Traditional Navajo Health Beliefs and Practices." In *Disease Change and the Role of Medicine: the Navajo Experience,* edited by S. J. Kunitz, pp. 118–145. University of California, Berkeley.

Levy, Jerrold, Raymond Neutra, and Dennis Parker. 1987. *Hand Trembling, Frenzy Witchcraft, and Moth Madness: a Study of Navajo Seizure Disorders.* University of Arizona Press, Tucson.

Levy, Robert. 1973. *Tahitians.* University of Chicago Press, Chicago.

Lewis, I. M. 1971a. *Ecstatic Religion: an Anthropological Study of Spirit Possession and Shamanism.* Penguin, Middlesex.

———. 1971b. "Spirit Possession in Northern Somaliland." In *Spirit Mediumship and Society in Africa,* edited by John Beattie and John Middleton. Africana, New York.

Lewis, I. M., A. Al-Safi, and S. Hurreiz (editors). 1991. *Women's Medicine: the Zar-Bori Cult in Africa and Beyond.* Edinburgh University Press, for the International African Institute, Edinburgh.

Lewton, Elizabeth. 1997. *Living Harmony: the Transformation of Self in Three Navajo Religious Healing Traditions.* Ph.D. dissertation, Case Western Reserve University, Cleveland, Ohio.

Lex, Barbara. 1974. "Voodoo Death: New Thoughts on an Old Explanation." *American Anthropologist* 76:818–823.

Lhamon, W. T. 1976. "Pentecost, Promiscuity, and Pynchon's *V*: from the Scaffold to the Impulsive." In *Mindful Pleasures: Essays on Thomas Pynchon,* edited by G. Levine and D. Leverenz. Little, Brown & Co., Boston.

Lindenbaum, Shirley. 1979. *Kuru Sorcery: Disease and Danger in the New Guinea Highlands.* Mayfield, Mountain Yew, Calif.

Linn, M., B. Linn, and S. Stein. 1982. "Beliefs about Causes of Cancer in Cancer Patients." *Social Science and Medicine* 16:835–839.

Linn, Matthew, Dennis Linn, and Sheila Fabricant. 1985. *Healing the Greatest Hurt.* Paulist Press, New York.

Lock, Margaret. 1993. "Cultivating the Body: Anthropology and Epistemologies of Bodily Practice and Knowledge." *Annual Review of Anthropology* 22:133–155.

Low, Setha. 1994. "Embodied Metaphors: Nerves as Lived Experience." In *Embodiment and Experience,* edited by Thomas J. Csordas. Cambridge University Press, Cambridge.

Lowie, Robert H. 1924. *Primitive Religion.* Grosset and Dunlap, New York.

Luckert, Karl. 1975. *The Navajo Hunter Tradition.* University of Arizona Press, Tucson.

Lutz, Catherine. 1982. "The Domain of Emotion Words on Ifaluk." *American Ethnologist* 9:113–128.

Lyon, Margot L. 1990. "Order and Healing: The Concept of Order and Its Importance in the Conceptualization of Healing." *Medical Anthropology* 12:249–268.

————. 1993. "Psychoneuroimmunology: the Problem of the Situatedness of Illness and the Conceptualization of Healing." *Culture, Medicine, and Psychiatry* 17:77–97.

Lyotard, Jean-Francois. 1984. *The Postmodern Condition: A Report on Knowledge.* University of Minnesota Press, Minneapolis.

MacNutt, Francis. 1974. *Healing.* Ave Maria Press, Notre Dame.

Marcus, George, and Michael M. J. Fischer. 1986. *Anthropology as Cultural Critique: An Experimental Moment in the Human Sciences.* University of Chicago Press, Chicago.

Martin, Emily. 1987. *The Woman in the Body: a Cultural Analysis of Reproduction.* Beacon Press, Boston.

————. 1992. "The End of the Body?" *American Ethnologist* 19:121–140.

————. 1994. *Flexible Bodies: The Role of Immunity in American Culture from the Days of Polio to the Age of AIDS.* Beacon Press, Boston.

Mathews, Dale, David Larson, and Constance Barry. 1993. *The Faith Factor: An Annotated Bibliography of Clinical Research on Spiritual Subjects.* National Institute for Healthcare Research, Bethesda, Md.

Mauss, Marcel. 1950 [1934]. "Les Techniques du Corps." In *Sociologie et Anthropologie.* Presses Universitaires de France, Paris.

————. 1950 [1938]. "Une Categorie de L'Esprit Humain: la Notion du Personne, Celle du 'Moi.' " In *Sociologie et Anthropologie.* Presses Universitaires de France, Paris.

Mawn, Benedict. 1975. *Testing the Spirits.* Ph.D. diss., Boston University, Boston.

May, L. Carlyle. 1956. "A Survey of Glossolalia and Related Phenomena in Non-Christian Religions." *American Anthropologist* 58:75–96.

Mayeux, Richard, Jason Brandt, Jeff Rosen, and Frank Benson. 1980. "Interictal Memory and Language Impairment in Temporal Lobe Epilepsy." *Neurology* 30:120–125.

McCall, Kenneth. 1982. *Healing the Family Tree.* Sheldon Press, London.

McDonnell, Kilian. 1976. *Charismatic Renewal and the Churches.* Seabury Press, New York.

McGuire, Meredith. 1982. *Pentecostal Catholics: Power, Charisma, and Order in a Religious Movement.* Temple University Press, Philadelphia.

————. 1983. "Words of Power: Personal Empowerment and Healing." *Culture, Medicine, and Psychiatry* 7:221–240.

Merleau-Ponty, Maurice. 1942. *La Structure du Comportement.* Presses Universitaires de France, Paris.

————. 1962. *Phenomenology of Perception.* Routledge and Kegan Paul, London.

————. 1964a. "Cezanne's Doubt." In *Sense and Non-sense.* Northwestern University Press, Evanston, Ill.

————. 1964b. *The Primacy of Perception.* Northwestern University Press, Evanston, IL.

————. 1968. "The Intertwining—The Chiasm." In *Maurice Merleau-Ponty, the Visible and the Invisible,* edited by Claude Lefort, pp. 130–155. Northwestern University Press, Evanston, Ill.

Messing, Simon. 1959. "Group Therapy and Social Status in the Zar Cult of Ethiopia." In *Culture and Mental Health: Cross-Cultural Studies,* edited by M. K. Opler. Macmillan, New York.

Metraux, Alfred. 1946. "The Concept of a Soul in Haitian Vodou." *Southwestern Journal of Anthropology* 2:83–92.

———. 1959. *Voodoo in Haiti*. Oxford University Press, New York.

Mills, C. Wright. 1940. "Situated Actions and Vocabularies of Motives." *American Sociological Review* 5:904–913.

Moerman, Daniel. 1979. "Anthropology of Symbolic Healing." *Current Anthropology* 20:59–80.

———. 1992. "Minding the Body: the Placebo Effect Unmasked." In *Giving the Body its Due*, edited by Maxine Sheets-Johnstone, pp. 69–88. SUNY Press, Albany.

———. 1997. "Physiology and Symbols: the Anthropological Implications of the Placebo Effect." In *The Anthropology of Medicine: From Culture to Method*, edited by Lola Romanucci-Ross, Daniel E. Moerman, and Laurence R. Tancredi, pp. 240–253. Bergin & Garvey, Westport, Conn.

Moerman, Michael, and Masaichi Nomura (editors). 1990. *Culture Embodied*. National Museum of Ethnology, Osaka, Japan.

Monfouga-Nicolas, Jacqueline. 1972. *Ambivalence et Culte de Possession*. Editions Anthropos, Paris.

Montagu, Ashley. 1978. *Touching*. Harper and Row, New York.

Morgan, Lynn. 1989. "When Does Life Begin: a Cross-Cultural Perspective on the Personhood of Fetuses and Young Children." In *Abortion Rights and Fetal "Personhood,"* edited by Ed Doerr and James Prescott. Centerline Press and Americans for Religious Liberty, New York.

Munn, Nancy. 1986. *The Fame of Gawa: a Symbolic Study of Value Transformation in a Massim (Papua New Guinea) Society*. Cambridge University Press, Cambridge.

Munroe, Robert L., Ruth H. Munroe, and John W. M. Whiting. 1973. "The Couvade: a Psychological Analysis." *Ethos* 1:30–74.

Murase, Takso. 1982. "Sunao: a Central Value in Japanese Psychotherapy." In *Cultural Conceptions of Mental Health and Therapy*, edited by A. J. Marsella and G. M. White. D. Reidel, Dordrecht.

Murphy, Jane. 1964. "Psychotherapeutic Aspects of Shamanism on St. Lawrence Island, Alaska." In *Magic, Faith, and Healing*, edited by Ari Kiev, pp. 53–83. The Free Press, New York.

Nagel, Thomas. 1979. "Subjective and Objective." In *Mortal Questions*. Cambridge University Press, Cambridge.

NCO Newsletter. 1980. "Gallup Poll Reveals 29 Million Charismatics in U.S.," p. 1.

———. 1981. "Diocesan Liaisons Surveyed about Renewal in U.S." 6:1.

Neher, A. 1962. "A Physiological Explanation of Unusual Behavior in Ceremonies Involving Drums." *Human Biology* 34:151–160.

Ness, Robert. 1980. "The Impact of Indigenous Healing Activity: an Empirical Study of Two Fundamentalist Churches." *Social Science and Medicine* 14B:167–180.

Ness, Robert, and Ronald Wintrob. 1980. "The Emotional Impact of Fundamentalist Religious Participation: an Empirical Study of Intragroup Variation." *American Journal of Orthopsychiatry* 50:302–315.

Neu, Jerome. 1977. *Emotion, Thought, and Therapy*. Routledge and Kegan Paul, London.

Noll, Richard. 1983. "Shamanism and Schizophrenia: a State-Specific Approach to the 'Schizophrenic Metaphor' of Shamanic States." *American Ethnologist* 10:443–459.

Obeyesekere, Gananath. 1981. *Medusa's Hair: an Essay on Personal Symbols and Religious Experience*. University of Chicago Press, Chicago.

O'Gorman, Frances. 1977. *Aluanda: a Look at Afro-Brazilian Cults*. Livraria Francisco Alves Editora S. A., Rio de Janeiro.

O'Neill, John. 1985. *Five Bodies: The Shape of Modern Society*. Cornell University Press, Ithaca.

Ong, Aihwa. 1996. "Cultural Citizenship as Subject-Making: Immigrants Negotiate Racial and Cultural Boundaries in the United States." *Current Anthropology* 37:737–762.

Ong, Walter. 1967. *The Presence of the Word*. Simon and Schuster, New York.

Onwuejeogwu, Michael. 1971. "The Cult of the *Bori* Spirits among the Hausa." In *Man in Africa*, edited by Mary Douglas and Phyllis Kaberry. Anchor, Garden City, Ga.

Opler, M. K. 1959. "Cultural Differences in Mental Disorders: an Italian and Irish Contrast in the Schizophrenias—U.S.A." In *Culture and Mental Health*, edited by M. K. Opler. Macmillan, New York.

Ortner, Sherry. 1974. "Is Female to Male as Nature is to Culture?" In *Woman, Culture, and Society*, edited by Michelle Zimbalist Rosaldo and Louise Lamphere. Stanford University Press, Stanford.

———. 1984. "Theory in Anthropology since the Sixties." *Comparative Studies in Society and History* 26:105–145.

Ots, Thomas. 1991. "Phenomenology of the Body: the Subject-Object Problem in Psychosomatic Medicine and the Role of Traditional Medical Systems." In *Anthropologies of Medicine: a Colloquium of West European and North American Perspectives,* edited by Beatrix Pflederer and Gilles Bibeau, pp. 43–58. Wieweg, Wiesbade.

———. 1994. "The Silenced Body—the Expressive *Lieb:* on the Dialectic of Mind and Life in Chinese Cathartic Healing." In *Embodiment and Experience,* edited by Thomas Csordas. Cambridge University Press, Cambridge.

Otto, Rudolph. 1958 [1917]. *The Idea of the Holy.* Oxford University Press, New York.

Pandolfi, Mariela. 1990. "Boundaries Inside the Body: Women's Suffering in Southern Peasant Italy." *Culture, Medicine, and Psychiatry* 14:255–274.

———. 1991. "Memory Within the Body: Women's Narrative and Identity in a Southern Italian Village." In *Anthropologies of Medicine: a Colloquium of West European and North American Perspectives,* edited by Beatrix Pflederer and Gilles Bibeau. Wieweg, Wiesbade.

Patterson, C. H. 1985. "What is the Placebo in Psychotherapy?" *Psychotherapy* 22:163–169.

Pattison, E. Mansell. 1974. "Ideological Support for the Marginal Middle Class: Faith Healing and Glossolalia." In *Religious Movements in Contemporary America,* edited by Irving Zaretsky and Mark Leone. Princeton University Press, Princeton.

Pattison, E. Mansell, N. Lapins, and Hans Doerr. 1973. "Faith Healing: a Study of Personality and Function." *Journal of Nervous and Mental Disorders* 157:397–409.

Persinger, Michael. 1983. "Religious and Mystical Experiences as Artifacts of Temporal Lobe Functions a General Hypothesis." *Perceptual and Motor Skills* 57:1255–1262.

———. 1984a. "People who Report Religious Experiences May Display Enhanced Temporal Lobe Signs." *Perceptual and Motor Skills* 58:963–975.

———. 1984b. "Propensity to Report Paranormal Experiences is Correlated with Temporal Lobe Signs." *Perceptual and Motor Skills* 59:583–586.

Persinger, Michael, and P. M. Vaillant. 1985. "Temporal Lobe Signs and Reports of Subjective Paranormal Experiences in a Normal Population: a Replication." *Perceptual and Motor Skill* 60:903–909.

Peters, Larry. 1981. *Ecstasy and Healing in Nepal: an Ethnopsychiatric Study of Tamang Shamanism.* Undena Publications, Malibu.

Peters, Larry, and Douglass Price-Williams. 1980. "Toward an Experiential Analysis of Shamanism." *American Ethnologist* 7:398–418.

Piaget, Jean. 1967. *Six Psychological Studies.* Vintage, New York.

Picone, Mary J. 1986. "Buddhist Popular Manuals and the Contemporary Commercialization o Religion in Japan." In *Interpreting Japanese Society: Anthropological Approaches,* edited by Joy Hendry and Jonathan Webber, pp. 157–165. Oxford University Press, Oxford.

Pressel, Esther. 1973. "Umbanda in Sao Paulo: Religious Innovation in a Developing Society." In *Religion, Altered States of Consciousness, and Social Change,* edited by Erika Bourguignon, pp. 264–320. Ohio State University Press, Columbus.

———. 1977. "Negative Spirit Possession in Experienced Umbanda Spirit Mediums." In *Case Studie in Spirit Possession,* edited by Vincent Crapanzano and Vivian Garrison. Wiley, New York.

Pressley, Sue Ann. 1993. "Navajos Fight Fear with Faith." In *Washington Post,* pp. A1, A12, Washington D.C.

Price-Williams, Douglass. 1987. "The Waking Dream in Ethnographic Perspective." In *Dreaming Anthropological and Psychological Perspectives,* edited by Barbara Tedlock. Cambridge University Press Cambridge.

Prince, Raymond. 1964. "Indigenous Yoruba Psychiatry." In *Magic, Faith, and Healing,* edited by Ari Kiev pp. 84–120. The Free Press, New York.

———, ed. 1966. *Trance and Possession States.* R. M. Bucke Memorial Society, Montreal.

———. 1974. "The Problem of Spirit Possession as a Treatment for Psychiatric Disorders." *Etho* 2:315–333.

———. 1976. "Psychotherapy as the Manipulation of Endogenous Healing Mechanisms: a Transcultura Survey." *Transcultural Psychiatric Research Review* 13:115–133.

———. 1980. "Variations in Psychotherapeutic Procedures." In *Handbook of Cross-Cultural Psycholog* edited by H. C. Triandis and J. C. Draguns. Allyn and Bacon, Boston.

————. 1982. "Shamans and Endorphins." *Ethos* 10:409–423.

Radin, Paul. 1927. *Primitive Man as Philosopher.* Dover Publications, New York.

————. 1937. *Primitive Religion.* Dover Publications, New York.

Ragland-Sullivan, Ellie. 1986. *Jacques Lacan and the Philosophy of Psychoanalysis.* University of Illinois Press, Urbana.

Ramirez de Jara, Maria Clemencia and Carlos Ernesto Pinzon Castano. 1992. "Sibundoy Shamanism and Popular Culture in Colombia." In *Portals of Power: Shamanism in South America,* edited by E. Jean Matteson Langdon and Gerhard Baer, pp. 287–304. University of New Mexico Press, Albuquerque.

Ramos, Arthur. 1939. *The Negro in Brazil.* Associated Publishers, Inc., Washington, D.C.

Rappaport, Roy A. 1999. *Ritual and Religion in the Making of Humanity.* Cambridge University Press, Cambridge.

Redliner, Irwin, and Clarissa Scott. 1979. "Incompatibilities of Professional and Religious Ideology: Problems of Medical Management and Outcome in a Case of Pediatric Meningitis." *Social Science and Medicine* 13(2B):89–93.

Reichard, Gladys. 1944. *Prayer: the Compulsive Word.* Princeton University Press, Princeton.

————. 1950. *Navajo Religion: A Study of Symbolism.* Princeton University Press, Princeton.

Rhodes, Lorna Amarasingham. 1980. "Movement among Healers in Sri Lanka: a Case Study of a Sinhalese Patient." *Culture, Medicine, and Psychiatry* 4:71–92.

Ribeiro, R. 1952. *Afrobrasileiros do Recife: um Estudo de Ajustamennto Social.* Instituto Joaquim Nabuco, Recife.

————. 1959. "Analises Socio-Psicologico de la Posesion en los Cultos Afro-Brasileiros." *Acta Neuropsiquiatrica Argentina* 5:249–262.

Ricoeur, Paul. 1979 [1971]. "The Model of the Text: Meaningful Action Considered as a Text." In *Interpretive Social Science,* edited by Paul Rabinow and William Sullivan, pp. 73–102. University of California Press, Berkeley.

Ritenbaugh, Cheryl. 1982. "Obesity as a Culture-Bound Syndrome." *Culture, Medicine, and Psychiatry* 6:347–362.

Rivers, W. H. R. 1901. "Introduction and Vision." In *Report of the Cambridge Expedition to the Torres Straits,* edited by A. C. Haddon. The University Press, Cambridge.

Roberts, J. K. A., M. Robertson, and M. R. Trimble. 1982. "The Lateralising Significance of Hypergraphia in Temporal Lobe Epilepsy." *Journal of Neurology, Neurosurgery, and Psychiatry* 454:131–138.

Roberts, Stephen. 1996. "From a Different Perspective: A Student's View." *Gratefully Yours* (National Library of Medicine on-line newsletter), 1–3.

Robertson, Roland, and JoAnn Chirico. 1985. "Humanity, Globalization, and Worldwide Religious Resurgence: A Theoretical Exploration." *Sociological Analysis* 46:219–242.

Rodriguez, Nina. 1935. *O Animisma Fetichista dos Negros Bahianos.* Civilizacao Brasileiro, Rio de Janero.

Rosaldo, Michelle. 1984. "Toward an Anthropology of Self and Feeling." In *Culture Theory,* edited by R. Shweder and R. Levine. Cambridge University Press, Cambridge.

Rose, Susan D. 1987. "Women Warriors: The Negotiation of Gender in a Charismatic Community." *Social Analysis* 48:245–258.

Roseman, Marina. 1991. *Healing Sounds from the Malaysian Rainforest: Temiar Music and Medicine.* University of California Press, Berkeley.

————. 1996. "Pure Products Go Crazy: Rainforest Healing in a Nation-State." In *The Performance of Healing,* edited by Christine Laderman and Marina Roseman, pp. 233–269. Routledge, New York.

Rouget, Gilbert. 1980. *La Musique et la Transe.* Gallimard, Paris.

Sacks, Oliver. 1985. *Migraine.* Simon and Schuster, New York.

Salisbury, C. G. 1937. "Disease Incidence Among the Navajos." *Southwestern Medicine* (July): 230–233.

Samarin, William. 1972. *Tongues of Men and Angels: The Religious Language of Pentecostalism.* Macmillan, New York.

————. 1979. "Making Sense of Glossolalic Nonsense. Beyond Charisma: Religious Movements as Discourse." In *Social Research,* edited by J. Fabian, 46:88–105.

Samuels, Andrew. 1985. "Counter Transference, the 'Mundus Imaginalis' and a Research Project." *Journal of Analytical Psychology* 30:47–71.

Sandner, D. 1979. *Navajo Symbols of Healing.* Harcourt, Brace, Jovanovich, New York.

Sandoval, M. 1979. "Santeria as a Mental Health Care System: An Historical Overview." *Social Science and Medicine* 13:137–151.

Sargant, William. 1973. *The Mind Possessed*. Lippincott, New York.

Satava, Richard. 1992. "Robotics, Telepresence, and Virtual Reality: A Critical Analysis of the Future o Surgery." *Minimally Invasive Therapy* 1:357–363.

———. 1994. "Human Interface Technology: An Essential Tool for the Modern Surgeon." *Surgica Endoscopy* 8:817–820.

———. n.d. "The Modern Medical Battlefield: Sequitur on Advanced Medical Technology." Unpublished ms., Advanced Research Projects Agency, Arlington, Va.

Saunders, L. 1977. "Variants in Zar Experience in an Egyptian Village." In *Case Studies in Spiri Possession*, edited by Vincent Crapanzano and Vivian Garrison. Wiley, New York.

Scarry, Elaine. 1885. *The Body in Pain: The Making and Unmaking of the World*. Oxford University Press New York.

Scheff, Thomas. 1979. *Catharsis in Healing, Ritual, and Drama*. University of California Press, Berkeley.

Scheper-Hughes, Nancy. 1990. "Mother Love and Child Death in Northeast Brazil." In *Cultura Psychology: Essays on Comparative Human Development*, edited by J. Stigler, R. Shweder, and C. Herdt pp. 542–565. Cambridge University Press, Cambridge.

Scheper-Hughes, Nancy, and Margaret Lock. 1987. "The Mindful Body: A Prolegomenon to Future Work in Medical Anthropology." *Medical Anthropology Quarterly* 1:6–41.

Schieffelin, Edward. 1985. "Performance and the Cultural Construction of Reality." *American Ethnologis* 12:707–724.

———. 1996. "Evil Spirit Sickness, the Christian Disease: The Innovation of a New Syndrome o Mental Derangement and Redemption in Papua New Guinea." *Culture, Medicine, and Psychiatr* 20:1–39.

Schilder, Paul. 1950. *The Image and Appearance of the Human Body*. International Universities Press New York.

Schodt, Carolyn. 1989. "Parental-Fetal Attachment and Couvade: A Study of Patterns of Human Environment Integrality." *Nursing Science Quarterly* 2:88–97.

Schoffeleers, M. 1991. "Ritual Healing and Political Acquiescence: The Case of the Zionist Churche in Southern Africa." *Africa* 6:1–25.

Schrader, Ann. 1994. "Human Anatomy to be On-Line Soon: C.U. Work Creating Database fo Project." *Denver Post*, June 6: B1

Schumaker, John F. 1992. *Religion and Mental Health*. Oxford University Press, New York.

Schutz, Alfred. 1967. *The Phenomenology of the Social World*. Northwestern University Press, Evanston.

———. 1970. *On Phenomenology and Social Relations; Selected Writings*. University of Chicago Press Chicago.

Schwartz, Maureen Trudelle. 1995. "The Explanatory and Predictive Power of History: Coping witl the 'Mystery Illness.'" *Ethnohistory* 42:375–401.

Schwartz-Salent, Nathan. 1987. "The Dead Self in Borderline Personality Disorders." In *Pathologies o the Modern Self*, edited by David M Levin, pp. 114–162. New York University Press, New York.

Selye, Hans. 1956. *The Stress of Life*. McGraw-Hill, New York.

Shapiro, Kennah Joel. 1985. *Bodily Reflective Modes: A Phenomenological Method for Psychology*. Duke University Press, Durham, N.C.

Sharp, Lesley. 1990. "Possessed and Dispossessed Youth: Spirit Possession of School Children ir Northwest Madagascar." *Culture, Medicine, and Psychiatry* 14:330–364.

———. 1993. *The Possessed and Dispossessed: Spirits, Identity, and Power in a Madagascar Migrant Town* University of California Press, Berkeley.

Shweder, Richard. 1986. "Divergent Rationalities." In *Metatheory in Social Science: Pluralisms an Subjectivities*, edited by D. Fiske and R. Shweder. University of Chicago Press, Chicago.

Shweder, Richard, and Edward Bourne. 1982. "Does the Concept of the Person Vary Cross-Culturally?' In *Cultural Conceptions of Mental Health and Therapy*, edited by A. J Marsella and G. M White. D Reidel, Dordrecht.

———. 1984. "Does the Concept of the Person Vary Cross-Culturally?" In *Culture Theory: Essays ol Mind, Self, and Emotion*, edited by R. Shweder and R. LeVine. Cambridge University Press Cambridge.

Singer, Milton. 1958. "From the Guest Editor." *Journal of American Folklore* 71:191–204.

Slater, E., W. Beard, and E. Glithero. 1963. "The Schizophrenia-Like Psychoses of Epilepsy." *British Journal of Psychiatry* 109:95–150.

Solomon, James. 1984. "Getting Angry: The Jamesian Theory of Emotion in Anthropology." In *Culture Theory: Essays on Mind, Self, and Emotion*, edited by R. Shweder and R. LeVine. Cambridge University Press, Cambridge.

Sombrero, Tweedy. 1996. "Two Paths." In *Native and Christian: Indigenous Voices on Religious Identity in the United States and Canada*, edited by James Treat, pp. 232–235. Routledge, New York.

Sontag, Susan. 1978. *Illness as a Metaphor.* Farrar, Straus, and Giroux, New York.

Stephen, Michele. 1989. "Self, the Sacred Other, and Autonomous Imagination." In *The Religious Imagination in New Guinea*, edited by Gilbert Herdt and Michele Stephen. Rutgers University Press, New Brunswick.

Stephen, Michele, and Luh Ketut Suryani. 2000. "Shamanism, Psychosis and Autonomous Imagination." *Culture, Medicine, and Psychiatry* 24:5–40.

Stewart, Omer C., and David F. Aberle. 1984. *Peyotism in the West.* University of Utah Press, Salt Lake City.

Stoller, Paul. 1989. *The Taste of Ethnographic Things: The Senses in Anthropology.* University of Pennsylvania Press, Philadelphia.

Stone, Allucquere Roseanne. 1995. *The War of Desire and Technology at the Close of the Mechanical Age.* MIT Press, Cambridge.

Strathern, Andrew. 1996. *Body Thoughts.* University of Michigan Press, Ann Arbor.

Strathern, Andrew, and Marilyn Strathern. 1971. *Self-Decoration in Mount Hagen.* Backworth, London.

Strathern, Marilyn. 1988. *The Gender of the Gift: Problems with Women and Problems with Society in Melanesia.* University of California Press, Berkeley.

———. 1992. *Reproducing the Future: Essays on Anthropology, Kinship, and the New Reproductive Technologies.* Routledge, New York.

Straus, Anne S. 1977. "Northern Cheyenne Ethnopsychology." *Ethos* 5(3):326–357.

Straus, Erwin. 1963. *The Primary World of Senses.* Free Press, New York.

———. 1966. "The Upright Posture." In *Phenomenological Psychology.* Basic Books, New York.

Suleiman, Susan (editor). 1986. *The Female Body in Western Culture.* Harvard University Press, Cambridge.

Tambiah, Stanley. 1985 [1981]. "A Performative Approach to Ritual." In *Culture, Thought, and Social Action*, pp. 123–166. Harvard University Press, Cambridge.

Tambiah, S. J. 1977. "The Cosmological and Performative Significance of a Thai Cult of Healing Through Meditation." *Culture, Medicine, and Psychiatry* 1:97–132.

Taussig, Michael. 1980a. *The Devil and Commodity Fetishism in South America.* University of North Carolina Press, Chapel Hill.

———. 1980b. "Folk Healing and the Structure of Conquest in Southwest Colombia." *Journal of Latin American Lore* 6:217–278.

———. 1987. *Shamanism, Colonialism, and the Wild Man: A Study in Terror and Healing.* University of Chicago Press, Chicago.

———. 1992. "Tactility and Distraction." In Michael Taussig, *The Nervous System.* Routledge, pp. 144–148. New York.

Torrey, E. Fuller. 1972. *The Mind Game: Witchdoctors and Psychiatrists.* Emerson Hall, New York.

Treicher, Paula A. 1998. *The Visible Woman: Imaging Technologies, Gender, and Science.* New York University Press, New York.

Tsiaras, Alexander. 1997. *Body Voyage: A Three-Dimensional Tour of a Real Human Body.* Warner, New York.

Tucker, Gary, Trevor Price, Virginia Johnson, and T. McAllister. 1986. "Phenomenology of Temporal Lobe Dysfunction: A Link to Atypical Psychosis—A Series of Cases." *Journal of Nervous and Mental Disease* 174:348–356.

Turkle, Sherry. 1995. *Life on the Screen: Identity in the Age of the Internet.* Simon and Schuster, New York.

Turner, Bryan S. 1984. "The Body and Religion." In *The Body and Society.* Basil Blackwell, Oxford.

———. 1992. *Regulating Bodies: Essays in Medical Sociology.* Routledge, London.

Turner, Judith, and Joan Romano. 1984. "Review of Prevalence of Coexisting Chronic Pain and Depression." *Advances in Pain Research and Therapy* 7:123–130.

Turner,Terence. 1980. "The Social Skin." In *Not Work Alone: A Cross-Cultural View of Activities Superfluous to Survival,* edited by Jeremy Cherfas and Roger Lewin. Sage Publications, Beverly Hills.

Turner,Victor. 1964. "An Ndembu Doctor in Practice." In *Magic, Faith, and Healing,* edited by Ari Kiev, pp. 230–263. The Free Press, New York.

———. 1966. "Colour Classification in Ndembu Ritual." In *Anthropological Approaches to the Study of Religion,* edited by M. Banton, pp. 47–84. Praeger, New York.

———. 1969. *The Ritual Process: Structure and Anti-Structure.* Aldine, Chicago.

———. 1974. *Dramas, Fields, and Metaphors.* Cornell University Press, Ithaca.

Turner,Victor, and Edward Bruner (editors). 1986. *The Anthropology of Experience.* University of Illinois Press, Urbana.

Tyler, Steven. 1988. *The Unspeakable.* University of Wisconsin Press, Madison.

Tyler, Stephen A. 1987. *The Unspeakable: Discourse, Dialogue, and Rhetoric in the Postmodern World.* University of Wisconsin Press, Madison.

Tymieniecka, Anna-Teresa (editor). 1988. *Maurice Merleau-Ponty, le Psychique et le Corporel.* Aubier, Paris.

Van der Leeuw, Gerardus. 1938. *Religion in Essence and Manifestation.* Allen and Unwin, London.

Van Dijk, Jose. n.d. "The Visible Human Project as Anatomical Theater." Unpublished manuscript.

Varella, J. 1972. *Cozhinha de Santo (Culinaria de Umbanda e Candomble).* Editora Espiritualista, Rio de Janeiro.

Vernant, Jean-Pierre. 1989. "Dim Body, Dazzling Body." In *Fragments for a History of the Human Body, Part One,* edited by Michel Feher, pp. 18–47. Zone, New York.

Von der Heydt,Vera. 1970. "The Treatment of Catholic Patients." *Journal of Analytic Psychology* 15:72–80.

Waldby, Catherine. 1997a. "Revenants: The Visible Human Project and the Digital Uncanny." *Body and Society* 3:1–16.

———. 1997b. "The Body and the Digital Archive: The Visible Human Project and the Computerization of Medicine." *Health* 1:227–243.

———. 1999. "IatroGenesis: The Visible Human Project and the Reproduction of Life." *Australian Feminist Studies* 14:77–90.

Waldram, J. 1993. "Aboriginal Spirituality: Symbolic Healing in Canadian Prisons." *Culture, Medicine, and Psychiatry* 17:345–362.

———. 2000. "The Efficacy of Traditional Medicine: Current Theoretical and Methodological Issues." *Medical Anthropology Quarterly.* 14:603–625.

Walker, Sheila S. 1972. *Ceremonial Spirit Possession in Africa and Afro-America.* Brill, Leiden.

Wallace, Anthony. 1956. "Revitalization Movements." *American Anthropologist* 59:264–281.

Ward, C. 1980. "Spirit Possession and Mental Health: A Psych-Anthropological Approach." *Human Relations* 33:149–163.

Waxman, Stephen, and Norman Geschwind. 1974. "Hypergraphia in Temporal Lobe Epilepsy." *Neurology* 24:629–636.

———. 1975. "The Interictal Behavior Syndrome of Temporal Lobe Epilepsy." *Archives of General Psychiatry* 32:1580–1586.

Wayman, Alex. 1966. "The Religious Meaning of Possession States (With Indo-Tibetan Emphasis)." In *Trance and Possession States,* edited by R. Prince. Bucke Memorial Society, Montreal.

Weber, Max. 1963. *The Sociology of Religion.* Beacon Press, Boston.

Werblowsky, R. J. Zwi. 1991. "Mizuko Kuyo: Notulae on the Most Important 'New Religion' of Japan." *Japanese Journal of Religious Studies* 18:295–354.

Werner, Oswald. 1965. "Semantics of Navajo Medical Terms." *International Journal of American Linguistics* 31:1–17.

Wikan, Unni. 1991. "Toward an Experience-Near Anthropology." *Cultural Anthropology* 6:285–305.

Williams, Paul V. 1979. *Primitive Religion and Healing: A Study of Folk Medicine in North-East Brazil.* Rowman and Littlefield, Totowa, New Jersey.

Witherspoon, Gary. 1977. *Language and Art in the Navajo Universe.* University of Michigan Press, Ann Arbor.

Wyman, Leland C. 1970. *Blessingway.* University of Arizona Press, Tucson.

Wyman, Leland C., and Clyde Kluckhohn. 1938. *Navaho Classification of Their Song Ceremonials,* Menasha, Wisconsin.

Young, Allan. 1975. "Why Amhara get Kureynya: Sickness and Possession in an Ethiopian Zar Cult." *American Ethnologist* 2:567–584.

————. 1976. "Some Implications of Medical Beliefs and Practices for Social Anthropology." *American Anthropologist* 78: 5–24.

————. 1982. "The Anthropology of Sickness." In *Annual Review of Anthropology*, edited by B. Siegel. Annual Review, Inc., Palo Alto.

Young, Robert, and William Morgan. 1987. *The Navajo Language: A Grammar and Colloquial Dictionary.* University of New Mexico Press, Albuquerque.

Zempleni, Andras. 1966. "La Dimension Therapeutique du Cultes des Rab, Kdop, Tuuru, et Samp: Rites de Possession Chez les Lebou et Wolof." *Psychiat. Afr.* 11:295–439.

————. 1985. "La 'Maladie' et ses 'Causes.' " *L'Ethnographie* 35:13–44.

————. 1988. "How to Say Things with Assertive Acts?" Paper presented at Twelfth International Congress for Anthropological and Ethnological Sciences, Zagreb.

# NAME INDEX

Aberle, David 148
Ackerman, Michael 274, 279–80
Agamben, Giorgio 282, 284
Alchon, S. A. 161

Badler, Norman 265
Bakker, James and Tammy 64
Bartes, Roland 241
Basso, Keith 293n.3
Beard, A. W. 235
Becker, Anne 252
Benedict, Ruth 216
Benjamin, Walter 73
Benthall, Jonathan 241
Blacking, John 241, 253
Blake, William 283
Boorstin, Daniel 278–79
Bourdieu, Pierre 60, 62–63, 81, 86,
    241–42, 256, 257, 258, 258–59,
    290n.20
Bourguignon, Erika 285n.5
Bruner, Edward 242

Campbell, E. J. M. 215
Castano, Carlos Ernesto Pinzon 161
Charon, Rita 253
Comaroff, Jean 16, 242, 288n.33
Corbin, Henry 254
Corin, Ellen 242
Crandon, Libbet 161
Crapanzano, Vincent 51
Crawford, Robert 67
Csordas, Thomas J. 242, 243

Daniel, E. Valentine 249–50
Derrida, 242
Descartes, René 59, 289n. 4
Devisch, Rene 242
Dewhurst, Kenneth 235

Douglas, Mary 241, 242

Eliade, Mircea 70, 82
Ellis, Albert 285n. 7
Estroff, Sue 292n.11

Fabricant, Shiela 89
Fernandez, James 35, 291n.25
Field, Karen 75
Finkler, Kaja 51
Foucault, Michel 241, 282, 286n.9
Frank, Gelya 242
Friedman, Jonathan 163

Gailly, Antoine 242
Gaines, Atwood 285n.1
Garrison, V. 249
Geertz, Clifford 82, 242
Geller, Jay 286n.15
Gellner, David 161
Geschwind, Norman 229
Ginsburg, Faye 94
Good, Byron 57, 243, 275, 288n.34
Good, Mary-Jo DelVecchio 57, 288n. 34

Hallowell, Irving 58, 59, 66, 289n.6
Hanks, William 243
Haraway, Donna J. 236–37
Harwood, Alan 249
Heidegger, Martin 219, 228, 236, 237,
    284
Heim, Michael, 260
Helsel, Sandra Kay 265
Hillerman, Tony 144
Hoffman, Helene 274

Jackson, Hughlings 235
Jackson, Michael 242
James, William 191

# SUBJECT INDEX

Lightning Source UK Ltd.
Milton Keynes UK
06 October 2009

144640UK00001B/41/A